T0327331

INTERFERENCE ANALYSIS

INTERFERENCE ANALYSIS

MODELLING RADIO SYSTEMS FOR SPECTRUM MANAGEMENT

John Pahl
Transfinite Systems Ltd, UK

Library of Congress Cataloging-in-Publication Data

Names: Pahl, John, author.
Title: Interference analysis : modelling radio systems for spectrum
    management / John Pahl.
Description: Chichester, UK ; Hoboken, NJ : John Wiley & Sons, 2016. |
    Includes bibliographical references and index.
Identifiers: LCCN 2015044553 (print) | LCCN 2015049974 (ebook) |
    ISBN 9781119065289 (cloth) | ISBN 9781119065319 (ePDF) | ISBN 9781119065326 (ePub) |
    ISBN 9781119065296 (online) | ISBN 9781119065319 (Adobe PDF)
Subjects: LCSH: Radio–Interference–Mathematical models. |
    Radio–Transmitters and transmission–Mathematical models. |
    Electromagnetic waves–Transmission–Mathematical models. | Radio frequency
    allocation–Management.
Classification: LCC TK6552 .P24 2016 (print) | LCC TK6552 (ebook) |
    DDC 621.382/24–dc23
LC record available at http://lccn.loc.gov/2015044553

A catalogue record for this book is available from the British Library.

Cover image: fotographic1980/Getty

Set in 10/12pt Times by SPi Global, Pondicherry, India

1   2016

To my family

Acknowledgements

The radio spectrum management industry is full of bright and friendly people, and the best bit of advice I can give is to be prepared to ask questions and then listen to the suggestions from others. There are so many people I would like thank and these are just a few in alphabetical order: Tony Azzarelli, David Bacon, Malcolm Barbour, Joe Butler, Ken Craig, Ian Flood, Paul Hansell, Dominic Hayes, Chris Haslett, Philip Hodson, Whitney Lohmeyer, Karl Löw, Bill McDonald, Steve Munday, John Parker, Tony Reed, John Rogers, François Rancy, Kumar Singarajah and Alastair Taylor – in addition the team at Wiley in particular Victoria Taylor, Tiina Wigley, Sandra Grayson, Purushothaman Saravanan and Nivedhitha Elavarasan.

My thanks also go to:

- ITU for permission to quote documents such as Radio Regulations and Recommendations
- Ofcom for permission to quote documents such as Consultation Documents and Technical Frequency Assignment Criteria
- Ofcom and Ordnance Survey for permission to use the Ofcom 50m terrain and land use database in the production of coverage prediction figures
- Transfinite Systems for use of their simulation tools and time to write this book
- General Dynamics for provision of the measured gain pattern data

Every attempt has been made to avoid typographic errors: if any should be spotted, the author would greatly appreciate notice via email at johnpahl@transfinite.com.

Additional Credits

Front cover screenshot credit: *Visualyse Professional*, overlay credit: *NASA: Visible Earth*
Screenshots from *Visualyse Professional, Visualyse GSO, Visualyse Coordinate, Visualyse PMR, Visualyse Spectrum Manager* and *Visualyse EPFD* are credit: Transfinite Systems Ltd www.transfinite.com

Overlay from NASA: *Visible Earth* full credit: NASA Goddard Space Flight Center Image by Reto Stöckli (land surface, shallow water, clouds). Enhancements by Robert Simmon (ocean color, compositing, 3D globes, animation). Data and technical support: MODIS Land Group; MODIS Science Data Support Team; MODIS Atmosphere Group; MODIS Ocean Group Additional data: USGS EROS Data Center (topography); USGS Terrestrial Remote Sensing Flagstaff Field Center (Antarctica); Defense Meteorological Satellite Program (city lights).

Contents

Foreword

Radiocommunications is the generic term to describe the various uses of the radio-frequency spectrum that have gradually become an integral part of our daily life in the last 30 years.

Television and sound broadcasting, satellite communications, radionavigation systems (such as GPS), mobile telephones or smartphones, Wi-Fi or Bluetooth systems or garage door openers, radars, emergency or defence communications, aircraft or maritime communications, radio relays, meteorological radiosondes or satellites, scientific or Earth exploration satellites, radio astronomy and deep space missions are only a few examples of the ever-increasing number of systems and applications that rely on spectrum to exist.

The associated investments represent trillions of dollars and are increasing every day as gathering and exchange of data also increase. The task of ensuring a viable ecosystem for the coexistence of these investments in the short, medium and long terms is entrusted to the International Telecommunication Union (ITU), which celebrated its 150th anniversary on 17 May 2015.

The objective of the ITU in this regard is to ensure the rational, efficient, equitable and economical use of the natural resources of the radio-frequency spectrum and satellite orbits. This is done by the application and regular updating of the ITU Radio Regulations, the international treaty that regulates the use of these resources by all countries in the world. The overriding objective of these regulations is to ensure operation of the radiocommunication systems of all countries free of harmful interference, thereby protecting these systems and the associated investments and providing the assurance that existing and future investments will be protected in the future.

To this end, any change foreseen in spectrum use is duly scrutinized by a population of experts coming from all parts of the world to attend frequent and multiple meetings of the Study Groups and Working Parties of the ITU Radiocommunication Sector. These experts literally dedicate their lives to building the future of radiocommunication systems and applications. Mr John Pahl is one of them. Over the last 20 years, he has played a key role in pushing the state of the art in analysis of interference between complex systems and developing appropriate regulations and best practices for their use of spectrum.

His book benefits from his long experience in world-level discussions within the technical, operational and regulatory decision-making process of the ITU Radiocommunication Sector. It covers the various aspects that need to receive careful consideration in assessing the interference that may occur among radiocommunication systems, at the design or coordination stage of these systems.

With the increased use of spectrum required to satisfy the growing demand for orbit and spectrum resources, more efficient use of these resources will come with increased complexity in system design, regulations, frequency assignment and coordination. Mr John Pahl's book will certainly be a gold mine for the current and future generations of spectrum managers, communication system designers and regulators in their day-to-day work to continue to deliver viable radiocommunication services and meet the growing expectations of the world's population in this regard.

François Rancy
Director, ITU Radio Sector

Preface

Photo credit: the author

We were on the lookout for ice.

I was in a 32 foot sailing yacht with writer and explorer Tristan Gooley, undertaking a double-handed sail from Scotland through the Faroes up to 66° 33′ 45.7″ N and the midnight sun. Now sailing out of the Arctic Circle we were approaching Iceland from the north, heading for the Denmark Straits, where ice flowed south. The Admiralty Pilot warned of bergs but the ice charts we had sailed with were over a week old. We needed an update.

So I reached for the Iridium satellite phone and rang a number in Greenland. A polite voice reassured us that as long as we kept within 50 nm of Iceland we should be okay.

Though I'd never had need of a satellite phone before that call, it was a technology I'd been involved in for nearly 20 years. It was by working on one of the other non-GSO mobile-satellite systems that I learnt the techniques and engineering principles of interference analysis. It turned out that there was a lot to cover: dynamics, antennas, link budgets, service objectives, thresholds, methodologies, modulations, coverage and much, much more.

On that voyage we didn't get to see any icebergs but the following year would make up for it when I sailed from Iceland over to Greenland and then down its east coast to Tasiilaq, passing close to the one in the photo on the previous page.

A berg like this drifted into the Atlantic in 1912 to sink the *RMS Titanic*, which radioed in distress for a rescue that was to come too late. Just a few months after this disaster, the International Radiotelegraph Conference in London was spurred to agree on common frequencies, and this led to what we now call the Radio Regulations.

I would have to learn about those too, first studying the ITU-R Regulations, Recommendations and Reports, then writing some of my own, getting them approved within the ITU-R, understanding the processes and, where necessary, chairing meetings.

Interference analysis involves engineering and regulation, and this book will by its nature cover both.

My hope is that it will assist those who want to learn about these topics and help others to avoid some of the potential icebergs.

John Pahl

1

Introduction

All radio systems share the same electromagnetic spectrum. This means that each radio receiver is detecting not just its wanted signal but all other signals transmitted at the same time anywhere – not just on this planet, but anywhere, even in space. If there are aliens out there using radio technology, their signals will also be added to the mix.

But it is rare for us to experience interference into our communication devices – such as radios, televisions and mobile phones – on a day-to-day basis. This is not an accident but the result of years of hard work by radio engineers and regulators to ensure that the signals from one user of the radio spectrum do not degrade significantly another user or, as is more commonly described, cause interference into the receiver.

Interference analysis is the study of how one or more radio systems can degrade the operation of other users of the radio system. It includes techniques to predict the level of interference and whether that could be tolerated or would represent a serious degradation, otherwise known as harmful interference.

This subject builds upon other specialist topics, such as antenna design and propagation, and often involves analysis of scenarios that includes different types of radio system. Therefore when undertaking interference analysis, it is necessary to become familiar with a wide range of other topics plus mathematical modelling techniques, statistics and geometry.

The objective of this book is to be useful to anyone involved in interference analysis – to help understand the various techniques and methodologies that could be used in studies. The approach in this book is to give an integrated view of interference analysis, describing all the key issues necessary to generate results. It will consider different types of interference and metrics to determine whether the interference could be accepted or would be considered harmful.

Interference Analysis: Modelling Radio Systems for Spectrum Management, First Edition. John Pahl.
© 2016 John Wiley & Sons, Ltd. Published 2016 by John Wiley & Sons, Ltd.
Companion website: www.wiley.com/go/pahl1015

1.1 Motivations and Target Audience

There are different motivations for undertaking interference analysis, in particular:

- System design: to optimise a radio system to allow the maximum service while reducing interference, whether within the system, to other systems or from other systems
- Regulatory: to identify what radio services can share with other radio services and hence allow them to be included in the tables of allocation used by spectrum managers
- Frequency assignment: to determine if a regulator can issue a new licence (e.g. to a taxi company) without it causing or suffering harmful interference
- Coordination: during discussions between two radio system operators (or countries) to identify ways to protect each other's receivers from the other's transmissions.

This book is aimed at anyone with an interest in interference between radio systems, in particular those operating with frequency above about 30 MHz. This could include:

- A member of a national spectrum management regulator is meeting with representatives from a neighbouring country to make a bilateral agreement that can be used to protect each country from interference. What would be a suitable interference threshold level to agree and how would it be checked/calculated?
- A mobile operator's spectrum management team wants to open up new bands for next generation broadband. They need to argue at an international level, in particular at the International Telecommunication Union, that such an introduction would be the most effective use of the scarce radio spectrum.
- A satellite operator is meeting with another satellite operator to ensure that neither causes harmful interference into the other. At the coordination meeting they will agree the signal levels that can be transmitted on each satellite beam: what should they propose and how can they check the suggestions from the other company?
- A consultant is working for the aeronautical community to help identify bands that can be used to provide broadband services to commercial aircraft. They must undertake sharing scenarios and convince regional bodies, such as Europe and North America, that these will be safe.
- The national spectrum management regulator receives a request for a new licence to provide a service (e.g. land mobile or fixed link). How can they check it would not cause or suffer interference to/from existing licensees?
- Students studying electrical engineering, communication systems or simulation techniques, who wish to learn more about interference analysis and modelling radio systems, together with academics undertaking research on these topics.

1.2 Book Structure

The book is structured as follows:

Chapter 2. Motivations: considering the reasons why people undertake interference analysis, including the regulatory framework, international organisations and working methods.

Chapter 3. Fundamental Concepts: covers the basics of radio engineering, including modulations, access methods, antennas, noise calculations, the underlying geometry and dynamics, link budgets and their attributes.

Chapter 4. Propagation Models: the model of how radio waves propagate between transmitter and receiver (whether wanted or interfering) can make a huge difference to the results, so it is important to understand the various propagation models that are available and when they should be used.

Chapter 5. The Interference Calculation: how to use the concepts above to calculate interference, including aggregation effects, polarisation adjustment, co-frequency vs. non-co-frequency, thresholds, interference apportionment and possible mitigation techniques.

Chapter 6. Interference Analysis Methodologies: the complexity of the analysis can vary from static analysis, where the answer is a single number, to dynamic, Monte Carlo and area analysis, each with strengths and weaknesses. To help explain each of the different approaches, worked examples will show how they can be used to analyse sharing between:
- A deployment of base stations of a Long Term Evolution (LTE) network into satellite Earth stations in parts of C-band
- A non-GSO mobile-satellite service (MSS) system and point-to-point fixed links.

Chapter 7. Specific Algorithms and Services: some services and sharing scenarios have well-defined publically available algorithms to calculate performance including interference analysis. Examples would be broadcasting, private mobile radio, white space and satellite coordination.

1.3 Chapter Structure and Additional Resources

Each chapter is structured starting with a summary of its contents and ending with pointers for further reading. Within each chapter are examples of the calculations involved to, as far as possible, allow the reader to reproduce the analysis undertaken. There are also available additional resources to increase understanding of the topics being analysed. These include:

- Spreadsheets to support standard calculations, such as link budgets and geometry conversions
- Example simulation files configured for the scenario under discussion.

These can be found by following the link on the Wiley web site page for this book. Wherever possible the international system of units are used for all calculations.

1.4 Case Study: How to Observe Interference

It is generally hard to accidentally create interference between different types of consumer radio systems. It should be noted that deliberately causing interference into another's radio system is considered a criminal offence in most legal systems and so should not be attempted. But you could try observe the impact of interference into one of your own radio receivers in a licence exempt band and see if you can detect any change in behaviour.

A good frequency band to experiment with is the one used by Wi-Fi at around 2.4 GHz as this has a range of different uses including microwave ovens. These are shielded to reduce

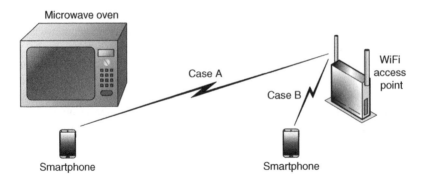

Figure 1.1 Domestic test set-up for detecting interference

Table 1.1 Results of tests for case A: Smartphone by microwave oven

Microwave oven	Off	On
Test 1	9.38 Mbps	0.72 Mbps
Test 2	9.40 Mbps	—[a]

[a] The smartphone did not complete the speed test reporting 'Network Communication Issues'.

Table 1.2 Results of tests for case B: Smartphone by Wi-Fi access point

Microwave oven	Off	On
Test 1	9.40 Mbps	7.24 Mbps
Test 2	9.38 Mbps	6.43 Mbps

emissions outside the device but there is usually some leakage that can be used as a source of interference into communications equipment. In particular, this can lead to issues for sensitive services such as radio astronomy. In 2015 the Parkes Radio Telescope in Australia investigated unusual signals it classed as 'perytons' and the source was discovered to be an on-site microwave oven emitting pulses at 1.4 and 2.4 GHz (Petroff et al., 2015).

For example, consider the two set-ups in Figure 1.1 where a smartphone was configured with an application to test the speed of the broadband link accessed via a Wi-Fi connection. Initially in case A the smartphone was positioned 0.5 m from a microwave oven and about 5 m from the access point. The Wi-Fi access point was configured to use the 2.4 GHz band rather than other frequencies (e.g. those around 5 GHz).

The throughput was tested twice for each of the cases when the oven is on or off with results as in Table 1.1.

In case B the smartphone was moved to be located 0.5 m from the access point and about 5 m from the microwave oven. The corresponding results are shown in Table 1.2.

From these tests the following results can be deduced:

- The microwave oven can degrade the performance of the smartphone's communication link
- The degree of degradation varies depending upon the distances from the smartphone to the Wi-Fi access point and microwave oven.

You could try this yourself at home, though you are likely to get different numbers depending upon equipment types and broadband link. Note that no electronic device should be placed inside a microwave oven.

Note that while there was a degradation of the communication service from around 9.4 to 6–7 Mbps, this data rate could still be considered usable. One of the key questions about interference analysis is what counts as an acceptable level and what would be 'harmful interference'.

This is an example of using measurement to detect how interference can degrade a communication service: the objective of this book is to describe tools and methodologies that can predict whether interference would or would not occur.

Important Note

All the systems and their parameters used in this book are for illustration purposes only. Any study based upon ideas in this book should check references, in particular for more recent developments.

2

Motivations

This chapter considers the question of the motivations for interference analysis. It puts the subject in its context and describes the framework within which interference analysis is often undertaken, in particular the work of the International Telecommunication Union (ITU).

It begins by considering why we analyse interference, the drivers that lead to requirements for interference analysis and the different types of interference analysis and then looks at the international and regional regulatory organisations. Given the importance of these institutions, there is a description of the working methods involved and how the results of studies documented in input papers are handled.

2.1 Why Undertake Interference Analysis?

In the 1920s, in the United States, there was a boom in commercial radio stations so that by 1926 there were 536 stations transmitting but only 89 channels available. With such congestion, each would turn their power up to maximum to drown out their competitors. The result was chaos, with radio becoming 'a tower of Babel', and according to the New York Times, all you could hear was something like 'the whistle of a peanut stand' (Goodman, n.d.).

It was a classic example of the tragedy of the commons (Wikipedia, 2014c), a concept developed by the economist Garrett Hardin in which if a resource that has value is freely available, it will be over-utilised to the point at which it becomes unusable. In this case, the radio spectrum had value, as it permitted the radio broadcasters to operate: indeed their business model would fail without it. The problem was that uncontrolled use led to interference between radio systems, which meant that the operators were unable to achieve their required quality of service (QoS).

Interference Analysis: Modelling Radio Systems for Spectrum Management, First Edition. John Pahl.
© 2016 John Wiley & Sons, Ltd. Published 2016 by John Wiley & Sons, Ltd.
Companion website: www.wiley.com/go/pahl1015

The solution was to develop a regulatory framework to control access to the radio spectrum. In the United States, this led to the Radio Act of 1927, which created a government agency, the Federal Radio Commission (FRC), with the authority to manage the radio spectrum by ensuring that transmissions are licensed, and these licences are issued in a way that allows each operator to meet their requested QoS, in particular by limiting interference. Similar legal structures and spectrum management organisations were created around the world integrated into the international framework of the ITU's **Radio Regulations** (RR).

A founding principle of these regulatory structures is the need for spectrum efficiency, i.e. to use the limited natural resource of the radio spectrum as efficiently as possible. Often the limiting factor on its utilisation is interference, and so the need to understand, predict and manage interference is central to spectrum management for organisations, nations and the global community.

There is an alternative to interference analysis – measurement. Rather than predicting interference levels using complex calculations and algorithms, it is possible to make field measurements of the actual levels of interference and use these instead.

This is indeed possible in some cases and is used to support interference analysis – in particular by verifying the predictions – but is not generally used, because:

1. Measurement is much more expensive than prediction. In a few minutes using standard software tools on a PC, it is possible to calculate interference levels over a large area, including possibly thousands of test locations. To do this by deploying equipment, transmitters and receivers would be prohibitively expensive, particularly for satellite systems
2. Measurement can be very time-consuming. For example, terrestrial point-to-point links, used for backhaul, often require high availability, with an average annual availability of 99.999% not uncommon. This clearly would require at least a year's worth of measurements with as many as 2,000,000 measurement samples to ensure statistical significance, and it is unlikely the operator wanting the link would be prepared to wait that long.

For this reason, in most cases, it is much more efficient in both time and money to undertake interference analysis rather than make measurements.

2.2 Drivers of Change

The main reasons to undertake interference analysis tend to relate to the introduction of new radio systems, and it is worth noting the different types of forces of change behind this. It is well known that the telecommunications industry is in a period of unprecedented rapid development and that it is continually finding new and innovative ways of using the radio frequency spectrum. Change can result from many external pressures including:

- **Economic**: as mass production reduces the price of equipment, it becomes possible to introduce new services or, as countries develop and their GDP increases, they can afford new advanced systems
- **Technological**: the movement towards higher frequencies, the use of constellations of non-GSO satellites and the development of the web or machine to machine (M2M) communications

- **Market**: as users request new services, higher data rates or increased mobility or need new types of scientific measurements
- **Data**: as the behaviour of radio waves and equipment are better understood, how they are characterised in propagation models can be improved.

While these are the 'big picture' pressures driving the industry, there can be more personal motivations. The mobile satellite system Iridium was in part inspired due to a question asked by the wife of Motorola executive Bary Bertiger. While on holiday in the Bahamas, Karen Bertiger asked whether he could devise a way for her to phone home wherever she was – even on holiday on a remote island. It was a good time to ask such a question as there were rapid technological changes that would soon allow systems to be designed to provide such a service.

While usually there are one or more organisations that instigate change, there may be many more who are affected by the changes proposed or have alternative, conflicting, suggestions.

Changes can be minor – for example, a parameter of a propagation model is updated to reflect new measurements. Or a change can dominate the ITU for years, such as the development of IMT-2000 standards or the proposals for non-GSO fixed-satellite service (FSS) such as Teledesic and SkyBridge.

Whether the instigator or respondent, either way, it is essential to be sure that proposed changes do not harm your organisation. With billion-dollar industries dependent upon assured access to the radio spectrum, the cost of *not* analysing interference can be very high indeed.

2.3   The Regulatory Framework

To manage the many types of changes, a complex regulatory framework has been developed that comprises both national and international components. To get an idea of the scale of the framework, one of the key documents is the ITU's RR, which comprises four volumes:

1. Articles (436 pages)
2. Appendices (826 pages)
3. Resolutions and Recommendations (524 pages)
4. Recommendations incorporated by reference (546 pages).

There are in total 2,332 pages in the 2012 English version of the RR, and this is only one of the regulatory texts that must be considered – though admittedly it is one of the most important documents.

One way to give an initial feel of the main concepts is by considering an example of a private mobile radio (PMR), sometimes called business radio (BR), such as those that might be used by a transportation company to keep in touch with its drivers. It has an existing network of base stations operating at around 420–430 MHz, and it wishes to install another to extend its coverage, as shown in Figure 2.1.

What is the process by which the PMR operator gains authorisation to transmit at the new base station?

The first steps are taken by the operator of the planned new base station. They must determine the characteristics of the system, such as its location and power and whether it could

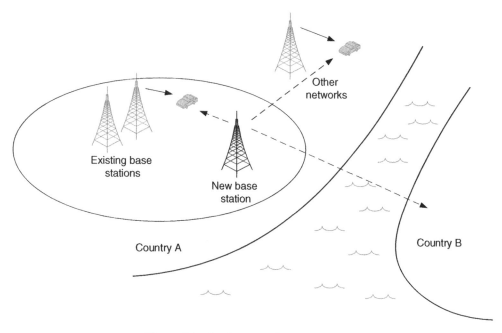

Figure 2.1 Deployment of new base station

operate on the same frequency as existing base stations or require a new channel. This would require interference analysis to be undertaken as part of the system design. When the PMR operator is happy with a set of parameters, they can be submitted to the national regulator of the relevant country, in this case A, as a licence application.

Each country is responsible for the radio systems operating in its territory and must check these systems would meet its own national rules and also those agreed internationally. National rules are likely to include constraints on what frequency bands can be used by PMR systems and to ensure that the new licence would not cause or suffer unacceptable degradation into or from existing licences. The selection of the frequency to be assigned to this licence is therefore likely to include interference analysis.

In addition the new system could cause or suffer interference with radio systems in neighbouring Country B. There is likely to be a process called international **coordination** by which such new assignments can be assessed as to whether there could be a problem – this too could include interference analysis.

If all these steps are passed successfully, then the operator will be issued a licence for the new base station on a specific frequency (either proposed by the operator or selected during the frequency assignment process, as in Figure 2.12).

As can be seen, this process involved a whole series of checks:

• That the new base station would not cause harmful interference into another of the operator's radio systems
• That the frequencies involved had been allocated to the land mobile service (LMS)

- That the frequencies proposed or selected would not cause interference that would unacceptably degrade this or other assignments
- That the new base station would meet agreed coordination agreements with neighbouring countries.

But how are frequency bands allocated to specific services and how do countries define coordination agreements? These two regulatory instruments will be the result of years of study, which is likely to involve extensive interference analysis. Furthermore the assignment methodology used by the national regulator, which involved interference analysis, will have had to be developed, again a process that could have taken many years.

The principle regulatory instruments in this case were:

- The definition of services (e.g. LMS)
- The national table of allocations (e.g. in the United Kingdom, the 420–430 MHz band is available to fixed and mobile including some programme making and special events)
- The international table of allocations (e.g. 420–430 MHz to fixed and mobile except aeronautical mobile)
- Frequency assignment methodology (e.g. in the United Kingdom, the MASTS algorithm described in Section 7.2)
- The coordination method to be used with neighbouring countries, which could involve bilateral agreement(s), a regional process (such as the Harmonised Common Methodology (HCM) Agreement) or the use of the process described in Article 9 of the RR.

These various instruments are described in more detail in the following sections.

2.4   International Regulations

2.4.1   History and Structure

The ITU is the oldest of the UN's specialist agencies. It was founded on 17 May 1865 as the International Telegraph Union, set up to standardise rules on handling telegraphy between countries (ITU, n.d.). Initially the ITU's bureau operated from Bern in Switzerland and incidentally was there during the time Einstein worked at that city's patent office, until it was moved to Geneva in 1948 where it remains to this day.

Early in the twentieth century, a series of events pointed to the need for the regulation of radio waves and telegraphic systems. Firstly when Prince Henry of Prussia was returning from a visit to the United States, his courtesy radio message was rejected as his ship's radio equipment was of a different type and nationality from that used on shore. Another more serious incident was the sinking of the *RMS Titanic*, where the nearest ship, the *SS Californian*, was not keeping radio watch, leading to heavy loss of life.

In response, the International Radiotelegraph Conference of 1912 (and other such meetings) created the framework that is still used to this day, namely, the RR and the concept of agreed wavelengths for particular services. In 1932 it was agreed that the organisation's name should be changed on 1 January 1934 to the ITU.

An overview of the current structure of the ITU is shown in Figure 2.2, focussing on the parts relevant to radiocommunications.

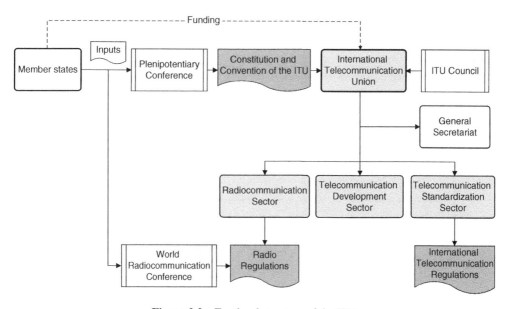

Figure 2.2   Top level structure of the ITU

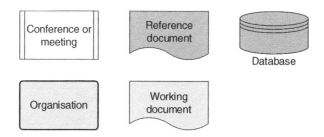

Figure 2.3   Key to ITU organisation figures

The key used for Figures 2.2 and 2.4 is given in Figure 2.3.

The most important document is the Constitution and Convention of the ITU (2011), which defines its objectives and working methods. This document is agreed at one of the ITU's Plenipotentiary Conferences, the highest level meeting of the ITU by the member states, that is, UN recognised countries.

It is not usually necessary for those undertaking interference analysis to get involved at this level, though it is useful to be aware of where the ITU's structure and procedures are defined and agreed. For example, the purposes of the Union is to:

1a) to maintain and extend international cooperation among all its Member States for the improvement and rational use of telecommunications of all kinds;

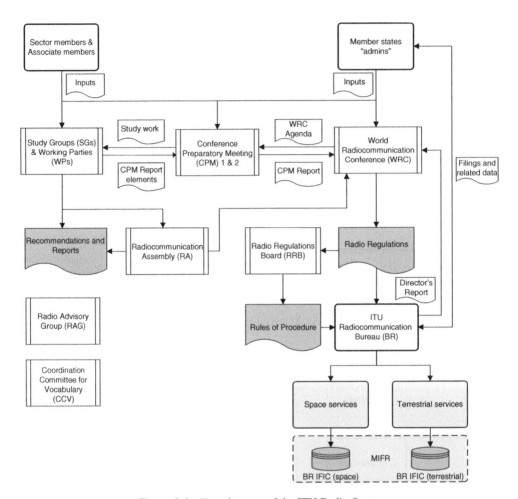

Figure 2.4   Key elements of the ITU Radio Sector

In particular, the Union shall:

2a) *effect allocation of bands of the radio-frequency spectrum, the allotment of radio frequen-*
      cies and the registration of radio-frequency assignments and, for space services, of any
      associated orbital position in the geostationary-satellite orbit or of any associated char-
      acteristics of satellites in other orbits, in order to avoid harmful interference between
      radio stations of different countries.

The objectives in this key sentence – concepts such as allocation of bands and registration
of assignments as ways to avoid harmful interference – are critical to the use of interference
analysis in regulatory studies and are described in more detail in the sections in the succeeding text.

The work of the ITU is split into three sectors:

1. 'R Sector': the radio sector coordinates work relating to radiocommunication services including international management of the radio-frequency spectrum and satellite orbits.
2. 'D Sector': the telecommunication development sector works to promote best practices in emerging market's telecommunications activities, focussing on issues such as the digital divide and best corporate practices.
3. 'T Sector': the telecommunication standardisation sector agrees the recommendations that define how services work, ranging from the Internet to voice communications.

To support the work of the ITU, there is a general secretariat. In addition, the ITU Council '*acts as the Union's governing body in the interval between Plenipotentiary Conferences. ITU Council also prepares a report on the policy and strategic planning of the ITU*'. The ITU also organises the ITU Telecom World convention and conference plus regional variants.

2.4.2   The Radiocommunication Sector

The part of the ITU that is most relevant to interference analysis is the radiocommunication sector, with key elements shown in Figure 2.4: if this looks complicated, then the reality is even more so!

The best way to follow the various interactions is to consider the main documents and data sets that must be managed within the ITU system:

* The RR
* The Recommendations and Reports
* Databases for terrestrial and space services
* The Rules of Procedure.

Each of these are considered in the following subsections together with other groups and committees.

2.4.2.1   The Radio Regulations

The RR are described in more detail in Section 2.4.3 and are the main regulatory document within the international process. They hold the status of a treaty document and are binding on each of the member states of the ITU.

The RR are agreed at a **World Radiocommunication Conference** (WRC), often simply called the Conference, which is described as sovereign in the sense that it can make decisions without being constrained by other ITU-R bodies.

A key document in the work of the ITU-R is the agenda for each Conference, which is agreed at the end of the previous one. It is not unknown for Conferences to discuss issues not on its agenda if there is support from member states, but it is usually felt that the Conference already has more than enough work and hence should focus on the previously agreed agenda. In addition the agenda item process allows due time for studies to be completed within the ITU-R.

The agenda for the next Conference, which itself can require extensive discussions, is an output of each WRC and then discussed at the first **Conference Preparatory Meeting** (CPM). This identifies the studies required to help answer the agenda items and allocates work within the ITU-R.

The **Study Groups** (SGs) and their **Working Parties** (WPs) then analyse the issues involved in each agenda item, and this is where there can be considerable work including interference analysis. These studies are usually provided as input papers as the result of work by:

- Member States, typically the organisation responsible for spectrum management in that country, hence are described as **administrations** or admins
- Sector and Associate Members: the ITU provides these two routes for non-state organisations including companies and trade associations to get involved in the work of the Union. They can contribute to studies and attend WRCs but not vote on decisions.

The results of the studies by the WPs are integrated into a single document, the **CPM Report**, at the second CPM, which is then submitted to the WRC. The period between WRCs is called a **cycle**, and it is usually dominated by the contents of the agenda.

2.4.2.2   The Recommendations and Reports

As well as the RR, additional support is available in the form of the **Recommendations** (Recs.) and **Reports**. These are advisory; resources that can be used by admins and those involved in the WPs and SGs, describing best practices, useful algorithms and methodologies, ways to describe radio systems, interference thresholds, terminology, etc.

Maintenance of the Recommendations and Reports is one of the prime tasks of the SGs and WPs along with providing input text into the CPM Report.

Some Recommendations are required to support the work of the Conference (and could be incorporated by reference into the RR), some are used to support equipment type approval, while, as will be seen, many others are helpful in interference analysis.

New Recommendations and Reports or their revisions are the result of input contributions and are approved either by correspondence or by the **Radiocommunication Assembly** (often just called the Radio Assembly or RA).

2.4.2.3   The Terrestrial and Space IFICs

One of the key tasks for the ITU BR is to maintain a register of radio assignments for both terrestrial and space systems. An assignment recorded with a favourable finding in its Master International Frequency Register (MIFR) gives it a degree of regulatory protection from harmful interference from any future assignments. The principle is that the first to file is protected from future assignments from other countries, and each new application submitted to the BR is circulated to all admins to allow them the opportunity to identify any potential problems. This coordination process is described in more detail in Section 2.10.

The data is distributed to admins in two International Frequency Information Circulars (IFICs):

- **The Terrestrial IFIC**
- **The Space IFIC.**

These databases are updated on a regular basis as new filings are submitted to the BR from administrations.

The BR will follow the process in the **Rules of Procedure** (described in the following subsection), which will include checks that the assignment is in conformity with the RR.

2.4.2.4   The Rules of Procedure

Whereas the RR contains regulations, these must be converted into day-to-day procedures to handle filings of radio assignments submitted from administrations to the BR. This task is the responsibility of the **Radio Regulations Board** (RRB), as identified in Article 14 of the Constitution of the ITU:

2. The duties of the Radio Regulations Board shall consist of:
a) the approval of Rules of Procedure, which include technical criteria, in accordance with the Radio Regulations and with any decision which may be taken by competent radiocommunication conferences. These Rules of Procedure shall be used by the Director and the Bureau in the application of the Radio Regulations to register frequency assignments made by Member States. These Rules shall be developed in a transparent manner and shall be open to comment by administrations and, in case of continuing disagreement, the matter shall be submitted to the next world radiocommunication conference.

2.4.2.5   Other Groups and Committees

In addition to the bodies and documents described earlier, other groups and committees include:

- **Radiocommunication Advisory Group** (RAG) among other things provides guidance for the work of the SGs and recommends measures to foster cooperation and coordination with other organisations and with the other ITU sectors
- **Coordination Committee for Vocabulary** (CCV) ensures there is consistency with vocabulary, including abbreviations and initials and related subjects (e.g. units).

Note that Figure 2.4 does not include all the interactions between these and other groups and committees and that there can be changes to this structure. For example, prior to 2015 the RA had established a Special Committee (SC) on Regulatory/Procedural Matters, as described in Resolution ITU-R 38 (suppressed at RA-15).

2.4.3   Radio Regulations

The RR is not a boxed set you can read from cover to cover. However they are crucial so it is worth getting familiar with as much of the content as possible, or at least, the way it is structured. This section covers the key topics, but nothing beats flipping through reading regulations, which relate to bands where your organisation has an interest.

The RR are updated soon after each WRC, so the three most recent are:

1. RR 2004 (ITU, 2004)
2. RR 2008 (ITU, 2008)
3. RR 2012 (ITU, 2012a).

Most of this book is based upon the RR of 2012 with, where appropriate, comments regarding changes at WRC-15.

2.4.3.1 Principles, Terminology and Services

The RR starts with some principles, in particular:

• Article 0.2: to ensure efficient use of spectrum by limiting '*the number of frequencies and the spectrum used to the minimum essential to provide in a satisfactory manner the necessary services*'.
• Article 0.3: This is because '*radio frequencies and the geostationary-satellite orbit are limited natural resources*'.
• Article 0.4: Furthermore all '*stations, whatever their purpose, must be established and operated in such a manner as not to cause harmful interference*'.

Then there is a definition of the key terms and services, in particular the following three terms highlighted in bold:

1.16 **allocation** *(of a frequency band): Entry in the Table of Frequency Allocations of a given frequency band for the purpose of its use by one or more terrestrial or space radiocommunication services or the radio astronomy service under specified conditions. This term shall also be applied to the frequency band concerned.*
1.17 **allotment** *(of a radio frequency or radio frequency channel): Entry of a designated frequency channel in an agreed plan, adopted by a competent conference, for use by one or more administrations for a terrestrial or space radiocommunication service in one or more identified countries or geographical areas and under specified conditions.*
1.18 **assignment** *(of a radio frequency or radio frequency channel): Authorization given by an administration for a radio station to use a radio frequency or radio frequency channel under specified conditions.*

The concepts of allocation, allotment and assignment are fundamental to the process of spectrum management.

The table of allocations (discussed in the next section) allocates frequency bands to specific services. These services have been selected because studies, in particular interference analysis, showed that they are compatible, that is, that actual radio systems (e.g. assignments) could be introduced in a way that avoids harmful interference, most likely subject to a defined process or set of constraints. An assignment process could involve interference analysis of the specific scenarios involved, such as for the land mobile example in Section 2.3. As the scenario has been defined in advance (via the table of allocations), a procedure or algorithm can be defined to determine if specific assignments are compatible.

Allotments are used to reserve channels and/or geostationary orbit (GSO) positions (slots) in advance of their actual implementation. This gives the administration involved flexibility and avoids the issue that spectrum resources could be fully utilised by early adopters leaving insufficient access to the radio spectrum for other administrations, in particular those in emerging markets. As well as reserving GSO frequency and orbital slots, the allotment process is also used for the management of terrestrial broadcasting, such as in the GE06 regional conference.

The table of allocations identifies where various services can operate, with the services defined at the start of the RR. These include terrestrial services such as:

- Amateur service
- Broadcasting service (BS)
- Fixed service (FS)
- Mobile service (MS), plus variations such as LMS, aeronautical mobile service (AMS), maritime mobile service (MMS)
- Radiodetermination service
- Radiolocation service
- Radionavigation service plus variations such as maritime radionavigation service and aeronautical radionavigation service
- Standard frequency and time signal service.

There are also space services, often with terrestrial equivalents, such as:

- Amateur-satellite service
- Broadcasting-satellite service (BSS)
- Earth exploration-satellite service (EESS)
- Fixed-satellite service (FSS)
- Inter-satellite service
- Meteorological-satellite service
- Mobile-satellite service (MSS) plus variations such as aeronautical mobile-satellite service (AMSS)
- Radiodetermination-satellite service
- Radiolocation-satellite service
- Radionavigation-satellite service (RNSS) plus variations such as maritime radionavigation-satellite service and aeronautical radionavigation-satellite service
- Space operations service
- Space research service
- Standard frequency and time signal-satellite service.

Finally there is a service that could be either terrestrial or space based:

- Radio astronomy service.

There are variations and sub-variations of some of these services. For example, aeronautical services are often subdivided into route (R) and off-route (OR), where the former are civilian airliners flying on predetermined routes. Satellite services can be subdivided into Earth–space and space–Earth directions.

These services are closely linked to the type of station used for that service, so that a mobile station is one operating in the mobile service, a fixed station in the FS and so on.

Bands can also be **identified** for use by a particular technology, such as the International Mobile Telecommunications (IMT). This has no specific regulatory implication but is a marker that regulators view the band as suitable for that technology, in particular to encourage equipment manufacturers as part of harmonisation (regional or global) processes.

It should be noted that there has been some debate about whether this classification is the most effective and where the boundary between fixed and mobile should be. For example, recently there has been significant analysis of sharing scenarios involving Earth stations on mobile platforms (ESOMPs) – in particular Earth stations on ships and aircraft. These are mobile but communicate with space stations operating within the fixed-satellite service and are also known as Earth stations in motion (ESIMs).

One argument in favour of such blending of fixed and mobile is that a key factor in any sharing scenario is the gain patterns used, and that has been the basis of proposals for alternative approaches. Rather than distinguishing between fixed and mobile, the proposals discussed the use of directive or non-directive antennas, with a minimal gain value separating the two. There was not sufficient support for these ideas to be adopted, but it is worth noting the discussion as the ideas involved impact on many interference analysis studies.

Of particular relevance to this book are the definitions of interference, namely:

1.166 *interference: The effect of unwanted energy due to one or a combination of emissions, radiations, or inductions upon reception in a radiocommunication system, manifested by any performance degradation, misinterpretation, or loss of information which could be extracted in the absence of such unwanted energy.*

1.167 *permissible interference[1]: Observed or predicted interference which complies with quantitative interference and sharing criteria contained in these Regulations or in ITU-R Recommendations or in special agreements as provided for in these Regulations.*

1.168 *accepted interference: Interference at a higher level than that defined as permissible interference and which has been agreed upon between two or more administrations without prejudice to other administrations.*

1.169 *harmful interference: Interference which endangers the functioning of a radionavigation service or of other safety services or seriously degrades, obstructs, or repeatedly interrupts a radiocommunication service operating in accordance with Radio Regulations.*

In this book, the most commonly used of these terms will be interference and harmful interference.

---

[1] **1.167.1** and **1.168.1:** The terms 'permissible interference' and 'accepted interference' are used in the coordination of frequency assignments between *administrations*.

2.4.3.2 Table of Allocations

The cornerstone of the RR is the **table of allocations** to be found in Article 5. This allocates frequency bands to services, often with allocations varying in the three **ITU regions** of:

1. Region 1: Europe and Africa including Russia
2. Region 2: Americas
3. Region 3: Asia including Iran and China.

There can also be country-specific allocations defined via footnotes.
Services are **allocated** with one of at least two different priorities:

1. **PRIMARY**: services identified in capital letters have higher priority
2. **Secondary**: services identified in lowercase have lower priority and hence should not cause harmful interference into primary services and would have to accept interference from them.

There can be finer resolutions: for example, there are the so-called super-primary services where there are footnotes or other regulatory text identifying that other co-primary services must provide protection against harmful interference.

The concept of primary or secondary service will have a significant (and sometimes explicit) impact on the threshold used to identify harmful interference.

An example of the table of allocations (ITU, 2012a) is given in Table 2.1, and it is worth noting the following:

- When the same allocations and footnotes apply to all three regions, the table is formatted with the frequency range in the left column and allocations in a combination of the central and right columns
- When there are different allocations and/or footnotes in each region, the table is formatted with the frequency at the top of the column for that region
- The units of the frequency range is given in the title, in this case MHz
- There can be significant variations in the allocations between regions in both the boundaries and services – for example, look at the mobile allocations in this extract
- In general services, prefer global to regional allocations as that assists in the harmonisation of equipment, but it is not always possible to achieve
- It is worth checking all the footnotes, just in case.

The subcategorisations of services assist in interference mitigation. For example, it is harder to share with aeronautical services as their emissions can cover a larger area. Hence it is not unusual to see 'mobile excluding aeronautical' in the table of allocations. Similarly defining a satellite service as 'space-to-Earth' means that the Earth stations will be receiving but not transmitting, defining what type of interference scenario has to be analysed.

The IMT sharing with satellite Earth station example in Chapter 6 considers services in this band, while Footnote 5.430A is described in Example 5.48. Note that this part of the table of allocations was significantly modified at WRC-15 (ITU, 2015) with additional mobile allocations and identification for IMT, mostly via footnote.

Table 2.1   Extract from RR 2012 table of allocations 2 700–4 800 MHz

Allocation to services		
Region 1	Region 2	Region 3
3 100–3 300	RADIOLOCATION Earth exploration-satellite (active) Space research (active) 5.149 5.428	
3 300–3 400 RADIOLOCATION 5.149 5.429 5.430	**3 300–3 400** RADIOLOCATION Amateur Fixed Mobile 5.149	**3 300–3 400** RADIOLOCATION Amateur 5.149 5.429
3 400–3 600 FIXED FIXED-SATELLITE (space-to-Earth) Mobile 5.430A Radiolocation	**3 400–3 500** FIXED FIXED-SATELLITE (space-to-Earth) Amateur Mobile 5.431A Radiolocation 5.433 5.282	**3 400–3 500** FIXED FIXED-SATELLITE (space-to-Earth) Amateur Mobile 5.432B Radiolocation 5.433 5.282 5.432 5.432A
	3 500–3 700 FIXED FIXED-SATELLITE (space-to-Earth) MOBILE except aeronautical mobile Radiolocation 5.433	**3 500–3 600** FIXED FIXED- SATELLITE (space-to-Earth) MOBILE except aeronautical mobile 5.433A Radiolocation 5.433
3 600–4 200 FIXED FIXED-SATELLITE (space-to-Earth) Mobile		**3 600–3 700** FIXED FIXED- SATELLITE (space-to-Earth) MOBILE except aeronautical mobile Radiolocation 5.435
	3 700–4 200 FIXED FIXED-SATELLITE (space-to-Earth) MOBILE except aeronautical mobile	

2.4.3.3 Articles

The table of allocation is just one **article** of many (though an important one). Some of the most important ones for those involved in interference are mentioned here.

The non-interference clause of 4.4 (ITU, 2012a) is always worth remembering:

4.4 Administrations of the Member States shall not assign to a station any frequency in derogation of either the Table of Frequency Allocations in this Chapter or the other provisions of these Regulations, except on the express condition that such a station, when using such a frequency assignment, shall not cause harmful interference to, and shall not claim protection from harmful interference caused by, a station operating in accordance with the provisions of the Constitution, the Convention and these Regulations.

In other words, you can operate a service not consistent with the table of allocations but should there be a problem, you would have no protection and would have to cease transmissions.

Some of the other key articles can be found in the following:

- Article 9: Procedure for effecting coordination with or obtaining agreement of other administrations. In particular it is worth noting Subsection IIA on requirement and request for coordination, Articles 9.6–9.21 and also Appendix 5 discussed further in the next section
- Article 11: Notification and recording of frequency assignments identifies the procedures by which assignments can be added to the BR's Master International Frequency Register (MIFR). This could involve coordination as described by Article 9
- Article 21: Terrestrial and space services sharing frequency bands above 1 GHz. These contain power flux density (PFD) limits that must be met by space systems and therefore involve two sets of interference analysis:
  1. Studies to identify what would be a suitable PFD level to protect terrestrial services from harmful interference
  2. Studies to determine whether a specific space system would meet or exceed the PFD levels in this article
- Article 22: Space services. There are a range of measures in this article to protect space systems, most particularly the equivalent power flux density (EPFD) metric to control interference into GSO satellite systems from non-GSO constellations (as described in Section 7.6). Again, there could be two types of interference analysis:
  1. Studies to identify what would be suitable EPFD levels to protect GSO satellite systems from harmful interference
  2. Studies to determine whether a specific non-GSO constellation would meet or exceed the EPFD levels in this article.

Other articles could be of interest and use depending upon the service – for example, the structure of a Mayday distress signal used by maritime services as part of the Global Maritime Distress and Safety System (GMDSS) is given in Article 32.13C.

2.4.3.4 Appendices

The **Appendices** build on the articles and give regulations for specific scenarios. Some worth considering include:

- Appendix 1: Classification of emissions and necessary bandwidths. This is a format to describe carriers and is discussed in Section 5.1
- Appendix 3: Maximum permitted power levels for unwanted emissions in the spurious domain, as discussed in Section 5.3.6. These are not very stringent and most systems should operate significantly better than this, so it is usually better to look for emission masks in other documents (e.g. ETSI or 3GPP standards)
- Appendix 4: Consolidated list and tables of characteristics for use in the application of the procedures. This defines the data that must be provided to the ITU when filing an assignment, whether terrestrial or space system. It can be a good source of parameters of actual systems when undertaking interference analysis
- Appendix 5: Identification of administrations with which coordination is to be effected or agreement sought under the provisions of Article 9. This section defines when coordination is required and under what circumstances so it is well worth being familiar with its contents
- Appendix 7: Methods for the determination of the coordination area around an Earth station in frequency bands between 100 MHz and 105 GHz. This is effectively an interference analysis methodology, as described in Section 7.4. As it is a coordination trigger, it is designed to be conservative so that it is more likely to show that there is a problem when there is not rather than to overlook scenarios that require further analysis
- Appendix 8: Method of calculation for determining if coordination is required between geostationary-satellite networks sharing the same frequency bands. This is a GSO to GSO satellite coordination trigger algorithm as discussed in Section 7.5
- Appendices 30, 30A and 30B: these are various plans for the BSS and FSS including feeder links, that is, frequencies and GSO slots are reserved for administrations and defined the process by which allotments can be brought into use.

2.4.3.5 Resolutions

WRCs pass **resolutions** to identify that work should be undertaken by the ITU. For example, Resolution 233 (WRC-12) *Studies on frequency-related matters on International Mobile Telecommunications and other terrestrial mobile broadband applications* initiated work with the Joint Task Group 4-5-6-7 to analyse sharing between IMT and other services to identify potential frequency bands that could be used by these mobile services. The JTG received 715 input contributions containing the results of interference analysis between many different types of service.

A resolution is structured with sections similar to this:

- Considerings
- Emphasizings
- Notings
- Recognizings
- Resolves
- Invites ITU-R
- Encourages/invites administrations.

The critical parts are typically contained in the last three, which request the actual work to be done by the ITU-R and administrations.

A particularly important resolution is the agenda for the next conference. For example, WRC-12-approved Resolution 807 (WRC-12) Agenda for the 2015 *World Radiocommunication*

Conference. Typically this cross-references the resolution for specific work items, so that, for example, Agenda Item 1.1 was:

1.1 to consider additional spectrum allocations to the mobile service on a primary basis and identification of additional frequency bands for International Mobile Telecommunications (IMT) and related regulatory provisions, to facilitate the development of terrestrial mobile broadband applications, in accordance with Resolution 233 (WRC-12).

2.4.4 World Radiocommunication Conference

The study cycle starts and ends with a WRC, a 4-week conference typically held in Geneva (though not always, WRC 2000 was held in Istanbul). The purpose of the WRC is to update the RR and the output of the WRC, called the Final Acts, is the delta between the old version of the RR and the new as in Figure 2.5.

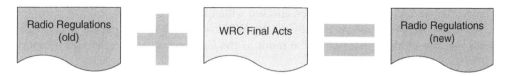

Figure 2.5   Update of the Radio Regulations via the WRC Final Acts

The WRC has a very wide remit, as under the terms of the ITU Constitution, it can not only revise the RR (including the table of allocations) but also address *any* radiocommunication matter of worldwide character.

Just prior to the WRC is the Radiocommunication Assembly (usually lasting a week), and it is followed by the first CPM (which is also a week), meaning that delegates who go to all three meetings can be in Geneva for 6 weeks. The second CPM is held sometime between 6 months and a year before the WRC. This gap is often criticised as it could be used to undertake further studies, but it can help in offline discussions to iron out agreements on some issues before the WRC itself.

Typically the period between WRCs is around 3–4 years and is the single opportunity for making changes to the RR. All work in the cycle between WRCs is building up to this moment, and so there is an intensity about WRCs that other meetings cannot capture – a combination of high politics and radio engineering.

The Conference is an inter-government treaty negotiation meeting so issues like who can speak and delegate credentials become important. Delegates have been known to be told they have 'diplomatic flu' and hence sent home if they speak out of line. A lot of the real work of the Conference is undertaken in the corridors or over coffees where compromises are reached and text drafted.

The top level of meetings and documentation is translated into all of the ITU's official languages of Arabic, Chinese, English, French, Russian and Spanish. However, most of the detailed work in the drafting groups is undertaken in English. Note that if there is a difference between the texts, then the French version is taken as being definitive, so it is worth double-checking what it says.

The first week is the most relaxed, and often when those with specific interests to promote host events to entertain and inform delegates. During this period, most of the work is handled at top level, known as the plenary, involving setup of the Conference structures, main committees and allocating documents to groups. There might also be proposals to add to the agenda provided by the previous Conference, an increase in workload that is likely to meet resistance.

In the later weeks, work will intensify, going on long into the evening and weekend, possibly overnight.

The inputs include:

- Contributions from administrations, sometimes grouped by regional body (see Section 2.8)
- The CPM Report
- Inputs from the BR, such as the Director's Report.

The CPM Report is the main reference for technical issues and summarises the work of the various SGs and WPs. The WRC, as a sovereign body, is not obliged to use the work in the CPM Report, but it does have more weight than (say) additional technical work provided by administrations, input that has not been checked within the SGs and WPs.

The CPM Report is meant to provide the WRC with technical input on the agenda items, but the ultimate decision is the remit of the Conference. Hence the CPM Report will provide options, such as:

- Option A: add a new primary allocation for service ABC in band XYZ
- Option B: add a secondary allocation for service ABC in band PQR
- Option C: no change.

Where appropriate, inputs will contain draft regulatory text for inclusion in the WRC Final Acts identifying changes to the RR using:

- ADD: add text to the RR
- SUP: suppress, that is, delete text from the RR
- MOD: modify text in the RR.

As noted previously, the WRC also decides the agenda for the next Conference, which starts the cycle all over again.

Note that in addition to WRCs, there can also be Regional Radiocommunication Conferences (RRCs), which are usually more focussed on specific spectrum planning aspects, for example, the GE06 Agreement for Digital Broadcasting. Also, prior to WRC 93, they were called World Administrative Radio Conferences (WARCs).

2.4.5 Study Groups and Working Parties

The SGs and WPs are responsible for:

- Work identified by WRCs (e.g. by agenda item) to be completed by the next Conference, with results documented in the CPM Report
- Management of the Recommendations and Reports of the ITU-R, as described in the following section.

This work is assigned to one of the WPs grouped by topic as SGs, which, for the cycle from WRC 2012 to WRC 2015, used the structure shown in Figure 2.6. Note that in the cycle starting after WRC-15, there was no JTG 4-5-6-7 as its work was complete, but there was an additional task group, namely TG 5/1. For issues that cross SG remits, there can be Joint Task Groups (JTGs), Joint Expert Groups (JEGs) or Joint Rapporteur Groups (JRGs). Occasionally this SG/WP structure is rearranged, for example, SG 6 used to be SG 10 and SG 11.

The structure is defined in the RA Resolution ITU-R 4, Structure of Radiocommunication Study Groups (ITU, 2012b).

2.4.6 Recommendations and Reports

One of the key resources for those undertaking interference analysis are the Recommendations and Reports of the ITU-R. These documents can be anything from a couple of pages to several hundred and are created within the SG and WP process.

A proposal for a new Recommendation typically goes through a series of stages in the WP as it becomes refined, and assessed by technical experts (which could include those of other WPs informed via liaison statements):

- Working document (WD) towards a preliminary draft new recommendation (PDNR)
- PDNR
- Draft New Recommendation (DNR).

The DNR is forwarded by the WP to its head SG for approval either by correspondence or to the Radiocommunication Assembly. A similar process is used for revisions to an existing approved Recommendation. The cycle for a Report is similar though these documents have a lesser status so in general the approval process is not as vigorous. In addition, Reports do not need to be translated into all of the ITU's official languages.

They can be identified by a letter and a number, such as Recommendation (Rec.) ITU-R P.452. Recommendations are often referenced in short form with just a letter and number, as in P.452, though it is better to be precise and prefix with 'Rec. ITU-R'. The letter identifies the series, which is helpful for tracking down its background, and can be one of the values in Table 2.2. Each number is unique, so if there is a P.452, there would not be (say) a S.452.

As Recommendations can be revised, there can also be a revision number. The initial release is either labelled '-0' or, for older Recommendations, has no version number. This version number is incremented by one for each revision. In some cases, the identification of minor editorials can result in small changes in released documents without a change in reference, though this is very rare (e.g. there were two variants of Rec. ITU-R P.2001-0).

Recommendations and Reports are good sources of:

- Propagation models
- Gain patterns
- Methodologies and algorithms
- Interference thresholds
- System parameters
- Terminology.

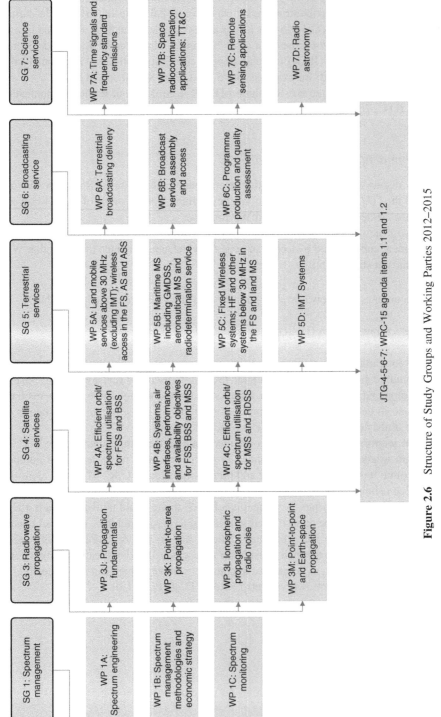

Figure 2.6 Structure of Study Groups and Working Parties 2012–2015

Table 2.2 ITU-R Recommendations letter identifiers

BO	Satellite delivery
BR	Recording for production, archival and play-out; film for television
BS	Broadcasting service (sound)
BT	Broadcasting service (television)
F	Fixed service
M	Mobile, radiodetermination, amateur and related satellite services
P	Radiowave propagation
RA	Radio astronomy
RS	Remote sensing systems
S	Fixed-satellite service
SA	Space applications and meteorology
SF	Frequency sharing and coordination between fixed-satellite and fixed service systems
SM	Spectrum management
SNG	Satellite news gathering
TF	Time signals and frequency standard emissions
V	Vocabulary and related subjects

2.5   Updating the Radio Regulations and Recommendations

The RR, Recommendations, Reports and Rules of Procedure together represent the key reference documents of the ITU-R and define the process to manage the introduction of a wide range of different types of radio system. However, these are not static but continually evolving regulatory instruments, updated in response to changes within the industry, as described in Section 2.2.

A major change can require modifications to all these documents. An overview of the process is shown in Figure 2.7, with key steps:

1. The starting point is the idea in industry or another organisation for a new service or revision to an existing service that is not covered by the existing regulations: this could include the requirement for more spectrum be allocated to an existing service.
2. The proposing organisation is likely to undertake their own studies to determine options to proceed, looking at possible changes, in particular selection of frequency bands.
3. When a combination is found that has a good chance of being accepted by the ITU-R and also provides a positive business case, the regulatory process can begin. This is aided if there is a WRC agenda item covering this topic under which studies and CPM text can be developed: if not, one is likely to be required.
4. Sharing studies are then undertaken at the ITU-R SGs and WPs to identify if this new service could share spectrum with existing services and if so under what conditions. It could be that sharing constraints would be so severe to make it infeasible for the new service to operate and so there could be several iterations, potentially taking years, until a suitable solution is identified (if at all).
5. This is likely to require interference analysis to analyse compatibility with existing services undertaken both by the organisation proposing the change (and a supporting national

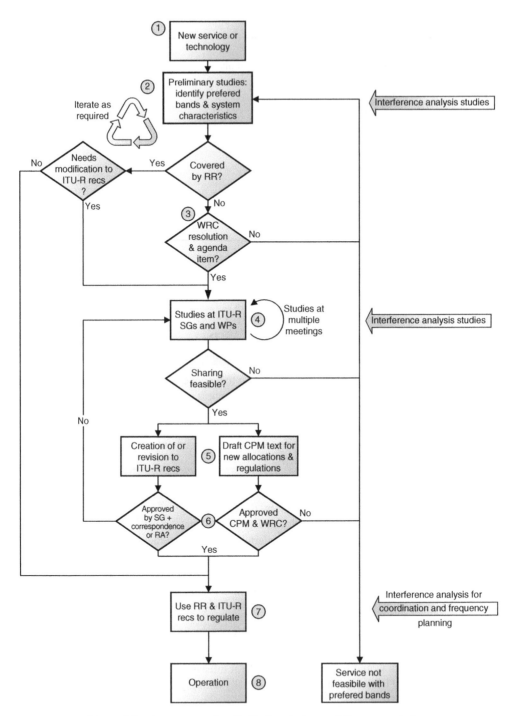

Figure 2.7   Process to update Radio Regulations and Recommendations

administration) and by incumbents who want to ensure their systems can continue to operate without harmful interference.

6. If a suitable solution is found, then the SGs and WPs will develop the necessary regulatory tools, which are likely to involve the development or revision of Recommendations and Reports, plus writing draft CPM text to forward to the WRC.
7. These documents would then have to go through the approval process, either by correspondence or at the RA for Recommendations and Reports, or at the WRC for changes to the RR.
8. Finally, the new regulatory instruments would have to be utilised by the new service to bring their system into use, a step that is likely to involve additional analysis to verify that they meet the constraints developed within the WPs and SGs.

To be successful, there is likely to be significant ground work undertaken at the national and most likely also the regional levels as described in Sections 2.7 and 2.8.

2.6 Meetings and Presenting Results

Given the importance of the ITU-R to interference analysis studies, it is worth considering its working methods, in particular at the Working Party (WP) level. The agreed working methods and structures are defined and documented in Radiocommunication Assembly Resolutions (ITU, 2012b), in particular:

- RA Resolution ITU-R 1: Working methods for the Radiocommunication Assembly, the Radiocommunication Study Groups and the Radiocommunication Advisory Group
- RA Resolution ITU-R 2: CPM.

The ITU-R has also prepared a useful document that gives guidelines to working practices (ITU-R, 2013a).

The majority of the time is spent within WPs or JTGs, which typically take between 1 and 2 weeks. The working practices can vary, but this section gives a flavour of a typical meeting. The work can be thought of as splitting into three stages:

1. Distribution of input documents from the top level (plenary) to subgroups and then drafting groups
2. Work within the drafting groups to prepare output documents, initially in the form of temporary WP documents called TEMPs
3. Polishing and agreement of output documents via the subgroups to plenary and hence, where applicable, on to other groups.

Figure 2.8 shows the first step, the distribution of input documents to the drafting groups where the output documents are created.

In this figure, note that:

- Inputs can come from member states or sector members (most commonly from administrations, i.e. countries) but also documents from other WPs called liaison statements

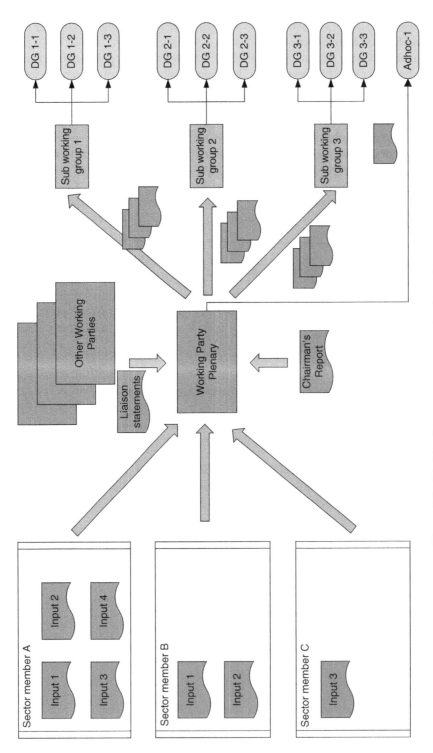

Figure 2.8 Working Party structure: processing input documents

- The Chairman's Report is the repository of work from previous meetings that is not ready for distribution to other groups and so is 'kept within the WP' for the time being. This allows iterations and improvements until the work is either discarded or sent on as (say) a revision to a Recommendation
- The work is split into subgroups and then drafting groups as not everyone at the meeting will be interested in every document. Hence it is more efficient to present each document once at a lower level rather than at plenary
- However, having too many groups operating in parallel can be difficult if delegates want to follow multiple topics. The subgroup is a balance between plenary level and drafting group
- Plenary is responsible for setting up the subgroups (including selecting their chairs) and allocating documents to them
- Sometimes the input documents do not fit a simple structure and so there can be 'ad hoc' drafting groups on specific topics that report direct to plenary.

It is important that input contributions are presented in suitable detail. At the first presentation (either in plenary or subgroup), the presentation can be brief, just to give a heads-up to other delegates as to what the document relates to (band, service, etc.) and also what is the intention (working document, liaison statement, DNR, etc.). When presented in the drafting group, it can be described in more detail, going through (say) the study's assumptions, methodology, findings, implications, next steps and so on.

A key part of the meeting will be the discussions in the drafting group between those that have papers that have an overlapping interest or those that have an interest in one of the documents. Generally there needs to be agreement for Recommendations to be approved to go forward to the SG, but CPM text allows for alternative views to be presented as it is for the Conference to make a final decision.

The meeting will have a time table that will depend upon the number of subgroups, drafting groups and length of the meeting, but an example is shown in Figure 2.9. In this case, the meeting is just over a week long and so includes a weekend. The working day is split into sessions, two in the morning, two in the afternoon and possibly later meetings if required. Often there can be multiple drafting groups being held at the same time in different rooms, though this figure has been simplified to only show two.

It is typical for some meetings to be required during the weekend in order to complete the work in time, but schedules will vary. The figure shows, as hashed blocks, examples of the time slots being reserved to allow for delegate's religious activities during:

- Friday lunchtime
- Saturday morning
- Sunday morning.

This procedure is consistent with Resolution 111 of the Plenipotentiary Conference (Busan, 2014; ITU, 2014).

The key concern for the chair of the meeting and the subgroups is to complete the work in the time provided. There are likely to be progress meetings, here shown on Monday morning, to ensure there are no unforeseen issues that could delay completion, and if any outputs are available their processing could be expedited.

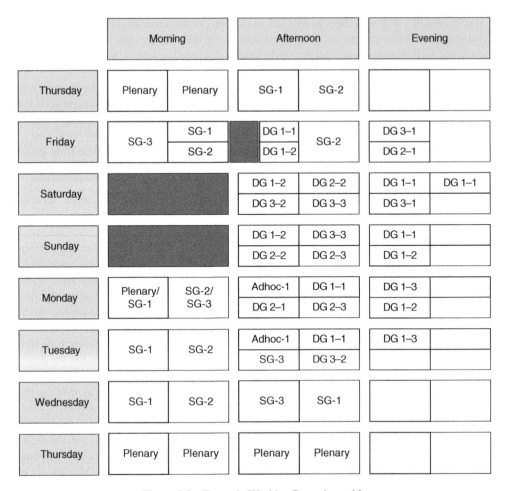

Figure 2.9   Example Working Party time table

The output documents, called TEMPs, are generated by the drafting groups and then processed up to plenary via the subgroups. Outputs can be:

- Working documents to be attached to the chairman's report for future meetings
- Liaison statements to other WPs
- Draft new Recommendations and Reports or draft revisions to existing Recommendations and Reports to be sent to the relevant SG
- Draft CPM text to go forward to that meeting.

Typically there needs to be at least two drafting group meetings: one to agree what the outputs would be and another to approve their format, with time spent between them used to draft their contents.

The output document process, shown in Figure 2.10, is almost a mirror image of the workflow in Figure 2.8.

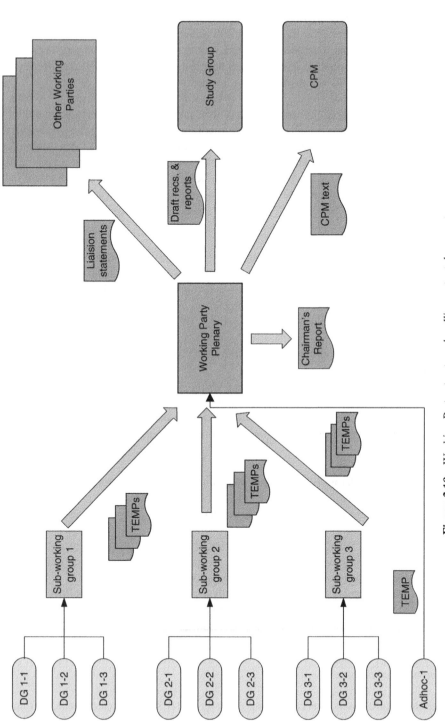

Figure 2.10 Working Party structure: handling output documents

To complete this work, it will be necessary to agree a chair for each sub-group or drafting group. It can be useful to chair a drafting group or take responsibility for the creation of the output documents so that it can be crafted in a way that takes account of your organisation's interests. But it is also important to be aware of the needs of other organisations: meetings work most effectively when all voices are given a fair hearing.

As with all ITU groups, a lot of the real work happens between formal meetings, where delegates can describe in depth the assumptions in their papers or give more background on why they hold the position they do. These corridor or coffee break discussions are invaluable when attending ITU-R WPs, and it is worth taking the time to put names and faces to input documents. Chairs are in general happy if they hear that an agreement has been reached and text prepared since their last meeting, as long as all those with an interest (which could include the chair) have been consulted and their views incorporated in the draft output document(s).

2.7 National Regulators

Each nation has sovereignty over the radio spectrum in its own country, subject to the constraints in the RR. Within each country, an organisation is given responsibility for managing that spectrum including issuing licences to transmit, undertaking coordinations with neighbouring countries and negotiations at ITU meetings.

Examples of these national regulators include:

- Australia: Australian Communications and Media Authority (ACMA)
- Canada: Industry Canada
- France: Agence Nationale des Fréquences (ANFR)
- The United Kingdom: Office of Communications (Ofcom)
- The United States: Federal Communications Commission (FCC), with the National Telecommunications and Information Administration (NTIA) providing support for government uses of spectrum.

Each national regulator will have a set of spectrum management tools such as:

- National frequency plan or frequency allocation table (FAT), a domestic version of the ITU's table of allocations
- Legal framework by which they can issue licences and prosecute those transmitting illegally
- Licensing procedures, such as the technical frequency assignment criteria (TFACs) in the United Kingdom or the FCC Rules and Regulations
- Database of assignments, sometimes publically available (the FCC and ACMA provide web interfaces), though more typically access is restricted to government-related organisations
- Monitoring stations capable of checking spectrum usage.

The licensing procedure, as used in the example in Section 2.3, can include multiple types of interference analysis, such as:

- Checking that the new licence will not cause or suffer harmful interference to/from other licences
- Checking whether coordination could be required with adjacent countries.

Each regulator will therefore also have to handle requests for coordination from other countries. Inter-country arrangements are sometimes formalised into bilateral agreements, for example, identifying PFD limits on each other's borders. Calculating what PFD levels to use and what propagation models and associated percentages of time involves interference analysis techniques.

2.8 Regional and Industry Organisations

One of the reasons for an international component to management of the radio spectrum is that radio waves do not stop at national borders but can propagate into other countries, particularly if the transmit station is on an aircraft or even in space. Another reason is that the use of the radio spectrum is often driven by the availability and cost of equipment, and these costs can be reduced if the potential market is large due to measures such as harmonisation.

Getting agreement for harmonisation at a global level via the ITU-R can be a time-consuming business given the 200+ countries each with their own agendas and priorities. An intermediate solution is to work at a regional basis, focussing on markets and political blocks, work managed via a regional organisation.

Some examples of these regional organisations are:

- African Telecommunications Union (ATU)
- Asia Pacific Telecommunity (APT)
- Arab Spectrum Management Group (ASMG)
- European Conference of Postal and Telecommunications Administrations (CEPT) via the Electronic Communications Committee (ECC)
- Inter-American Telecommunication Commission (CITEL)
- Regional Commonwealth in the field of Communications (RCC).

These groups are welcomed by the ITU-R for, in general, they assist WRC in coming to a successful conclusion.

There are also other UN specialised agencies with an interest in radiocommunications such as the following:

- International Civil Aviation Organization (ICAO), created in 1944
- International Maritime Organization (IMO), created in 1948.

The degree of cohesion and work level varies between regional organisations. In particular the ECC is very active and has achieved considerable levels of agreement as to common positions to be held by European nations prior to the WRC. Just as the ITU-R SGs and WGs have a dual role of providing support to the WRC and development of Recommendations and Reports, the ECC groups:

1. Work to develop European Common Positions (ECPs), namely, agreed European positions at the next WRC
2. Work to develop ECC Recommendations, Reports and Decisions.

The working groups within the ECC are more fluid than ITU-R SGs and WPs, often changing depending upon the current topics. Note also that some older documents will refer to the European Radiocommunications Committee (ERC) rather than ECC.

A key difference between works at the ITU-R and the ECC level is that the latter can be heavily influenced by the primary political structure within Europe, namely, the European Union (EU). The EU's executive arm, the European Commission (EC), can mandate the ECC to undertake work it considers important and implement Decisions agreed within the ECC at the EU level. The EC has two groups on spectrum issues:

- The Radio Spectrum Policy Group (RSPG)
- The Radio Spectrum Committee (RSC).

In order to support a common European market, there is a standardised equipment approval process and regulatory body, the European Telecommunications Standards Institute (ETSI). As with the ECC, the EC can mandate ETSI to develop a standard to support a harmonised European market under EU legislation such as the Radio Equipment Directive (RED).

Conceptually, the top level interactions between the EC, ECC and ETSI are shown in Figure 2.11.

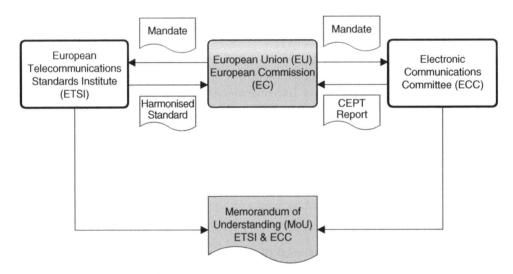

Figure 2.11   Top level interactions between the EC, ECC and ETSI

Note that the EU, CEPT and ETSI memberships overlap but are not the same: at the time this book was written the EU was composed of 28 countries, but CEPT includes all the EU countries plus 21 more, with 48 national regulatory administrations (ECC and ETSI, 2011). ETSI, however, has as many as 700 members including a wide range of organisation types.

There are also a number industry bodies that develop specific standards, such as the 3rd Generation Partnership Project (3GPP) responsible for the mobile broadband standards.

In terms of interference analysis, there are two points to note:

1. A significant part of the work of the ECC includes studies that involve interference analysis
2. The Decisions, Recommendations and Reports produced by the ECC are good sources of the types of information that can be found in the ITU-R Recommendations and Reports (e.g. gain patterns, propagation models, interference methodologies and thresholds).

So, for example, the extended Hata/COST 231 propagation model (described in Section 4.3.8) can be found in CEPT ERC Report 68: *Monte Carlo Radio Simulation Methodology* (CEPT ERC, 2002).

The ETSI standards are a useful source of antenna gain masks and transmit spectrum masks, and, in a limited number of cases, receive filter masks. These are invaluable for adjacent and out-of-band studies but tend to be conservative as they were developed by the manufacturers who have an interest in ensuring their equipment can readily be approved. There can also be a wide range of possible masks to choose from (in particular for fixed links and IMT base stations), increasing the degree to which the interference study that uses them is specific rather than generic.

2.9   Frequency Assignment and Planning

A key task for each national regulator is to ensure that transmissions in its territory are licensed in a way that is consistent with the RR. The complexity of this task will depend upon the demand for spectrum, how many licences are required and their characteristics. In an island country with low demand, it is possible to use simplistic methods to assign a frequency to each link – maybe just ensuring there is no frequency overlap between assignments.

However, most countries will have an excess of demand for spectrum over its availability. The radio spectrum, as is often said, is a limited natural resource, and it is important to ensure it is used in the most efficient way possible. A study commissioned by Ofcom (PA Knowledge Limited, 2009) identified strong demand for spectrum by almost every sector, resulting in difficulties for the regulator when supply is fixed.

The concept of efficient spectrum usage raises the question of how one can measure spectrum utilisation and hence measure efficiency, as is discussed further in Section 3.12.

Access to the radio spectrum can be managed using:

• Frequency and/or geographic block licences, such as those auctioned off for mobile broadband, whereby the whole of a geographic territory was available for (say) $2 \times 10\,MHz$ of spectrum, subject to flexible limits such as maximum equivalent isotropically radiated power (EIRP) density and block edge masks (as described in Section 5.3.8)
• Individual or site licence, such as the PMR example in Section 2.3. These are used for a range of services including fixed, amateur and satellite Earth stations
• Licence exempt, where operation is permitted without a licence as long as equipment meets specific requirements such as frequencies used and the maximum EIRP. An example would be Wi-Fi at 2.4 GHz.

Determining the conditions for a block licence or licence exempt classes is likely to require interference analysis, and further studies could be required during implementation and deployment of systems.

For site-specific licensing, there are a range of tests that can include interference analysis. An example site licensing process is shown in Figure 2.12.

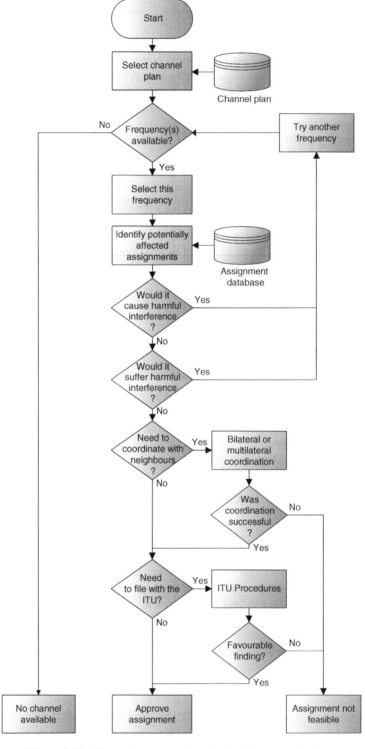

Figure 2.12 Example process to handle site licence applications

In some cases, the licence application will include a request for a specific frequency, while in others, a frequency band will have been specified based upon the equipment chosen or to be purchased. It is then the responsibility of the spectrum management organisation (typically the regulator but it could also be a private organisation) to select appropriate frequencies. This process involves iterating over a set frequencies (the channel plan) and for each selection analysing whether there could be harmful interference:

- From the new licence into existing licences
- Into the new licence from existing licences.

When (in some congested bands, this could be an if) frequencies are found that are acceptable, then there could be further checks for international coordination as shown in Figure 2.12 and discussed further in the following section. The licence could involve one or more transmitters, with each transmitter potentially requiring a different frequency and being defined as a separate assignment in terms of the ITU-R process.

The process by which a new site licence is planned and a frequency selected is usually well documented and can include calculations involving both the wanted (to ensure a link achieves its target receiver sensitivity level or coverage requirements) and interfering signals.

In the United States and Australia, some of these calculations are assumed to be undertaken by experts external to the regulator, and this can give greater flexibility in the process, though the regulator has less direct control.

Not all regulators use the same algorithms in the technical analysis part of their licensing processes. It is useful to compare two such examples, the planning of fixed links and PMR in the United Kingdom and Australia. The relevant documents are:

The United Kingdom:

- Ofcom's OfW 164: Business Radio Technical Frequency Assignment Criteria (Ofcom, 2008b)
- Ofcom's OfW 446: Technical Frequency Assignment Criteria for Fixed Point-to-Point Radio Services with Digital Modulation (Ofcom, 2013).

Australia:

- ACMA's LM 8: Frequency Assignment Requirements for the Land Mobile Service (ACMA, 2000)
- ACMA's FX 3: Microwave Fixed Services Frequency Coordination (ACMA, 1998).

While both have similar concepts in terms of propagation models, key parameters, calculation of wanted and interfering signals, there are differences as the United Kingdom has a higher average population density than Australia. This drives the demand for spectrum per square kilometre, and hence, the planning algorithms in the United Kingdom try to pack a greater number of licences per square kilometre. In these cases, it means that:

- The UK PMR planning algorithm is designed to handle overlapping coverage areas while the algorithm in LM3 is based upon geographic separation of co-frequency assignments

- In the UK, the transmit power used by point-to-point fixed links is selected by the regulator to be the minimum necessary to close the link to the requested availability, while in Australia, the operator can select higher EIRPs.

These algorithms are discussed in more detail in Sections 7.1 and 7.2.

2.10   Coordination

A key task in spectrum management is coordination, which can be required for both satellite and terrestrial systems at multiple levels:

- Nationally, between two or more organisations within a country
- Bilaterally or multilaterally, between two or more countries using mutually agreed procedures that avoid the need to involve the ITU-R
- At the ITU-R level, using the procedures in the RR.

The objective of the coordination is to find a way for a new radio system to operate so that it protects an existing radio system from harmful interference. There is usually explicitly or implicitly a first-come, first-served (FCFS) priority scheme in which the first to be registered has priority over later systems.

In each case, there is typically a trigger identifying that coordination is required, and examples of these would be:

- For co-primary GSO satellite networks, coordination is triggered when the interfering system would cause a noise rise of more than 6% of the thermal noise in the victim receiver or be within a certain angular separation of each other on the GSO arc. The methodology is described further in Section 7.5
- For terrestrial systems, the trigger could be a PFD threshold on a country's border. An example of this can be found in footnote 5.430A of the RR shown in Table 2.1, as described in Example 5.48
- Another approach for terrestrial systems is the generation of a coordination contour around the proposed new radio system within which there could be an interference issue. If this contour intersects another country or contains other co-frequency radio systems, then coordination would be required. The methodology is described further in Section 7.4.

Examples of the latter two scenarios are shown in Figure 2.13.

A key task is to ensure that the threshold is appropriate, balancing the two factors that:

1. The coordination trigger should be sufficiently low that it will not miss cases that should be subject to coordination as they could cause harmful interference
2. The coordination trigger should be sufficiently high that cases which are very unlikely to cause harmful interference are excluded.

The reason for the second point is that the coordination process can be time-consuming, particularly if it involves multiple parties. The ITU-R process in Article 9 of the RR discusses time periods of the order of months, and for this reason, it can be convenient for countries to come to their own arrangements for coordination via bilateral agreements.

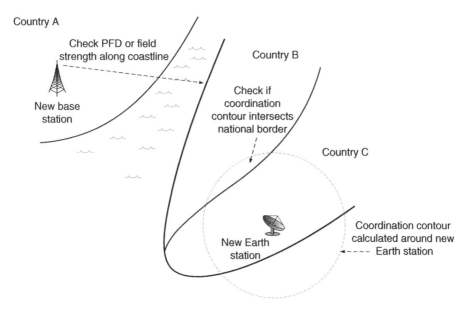

Figure 2.13   Example of scenarios involving coordination

These are typically agreed via memorandum of understandings (MoUs), and example of which is the publically available one between the United Kingdom and France:

• Memorandum of understanding concluded between the administrations of France and the United Kingdom on coordination in the 47–68 MHz frequency band (Ofcom and ANFR, 2004).

This document gives some examples of regulatory tools that can be contained in MoUs including:

• The use of ITU-R standard propagation model in Rec. ITU-R P.1546 described in Section 4.3.5
• The use of CEPT ECC standards via Recommendation T/R 25-08
• Concept of preferred frequencies, where each country has channels on which it has greater flexibility and can operate at higher powers
• The use of field strengths (similar to PFDs) as coordination triggers – or, more importantly, if the field strength is not exceeded, then there is no need to enter coordination
• Exchange of data in an agreed format (in this case, defined in the HCM Agreement)
• Constraints on the effective radiated power (ERP)
• Methods and time limits for communications including requests for coordination.

Within the coordination process itself, there are no definitive rules but the parties are encouraged to use the resources available in the community, such as ITU-R Recommendations and Reports or (within Europe) ECC Reports and Recommendations.

Groups of countries can develop their own coordination procedures and technical algorithms, an example of which is the HCM Agreement (HCM Administrations, 2013).

Whereas these examples have been between different countries, there can also be coordinations between organisations. These could be because the coordination between two nations that have filed satellite services is more easily managed via direct discussions involving the two companies. It could also be due to the coordination being required within a country between two operators (e.g. between a satellite Earth station owner and a fixed link organisation).

For example, in the United Kingdom, parts of the 28 GHz band were sold as regional spectrum blocks, and it could therefore be necessary for operators of different blocks to coordinate the installation of fixed links to prevent harmful interference. A coordination trigger of −102.5 dBW/MHz/m² was specified together with the requirement to use the propagation model in Rec. ITU-R P.452 (see Section 4.3.4). Spectrum block edge masks (described further in Section 5.3.8) in the licence terms and conditions were assumed to protect against adjacent band interference (Ofcom, 2007).

2.11 Types of Interference Analysis

It can be seen from the discussion throughout this section that there are a range of types of interference analysis, including:

- **System interference analysis**: ensuring a proposed system is designed in a way to minimise interference into itself or other services, which could be to assist a licence application (as in Section 2.3) or propose changes to the regulations (stage 1 of the flow chart in Figure 2.7)
- **Regulatory interference analysis**: to undertake studies to change the RR or develop new Recommendations and Reports, as described in Section 2.5
- **Frequency assignment interference analysis**: during the frequency assignment process, often as part of licensing procedures, as described in Section 2.9
- **Coordination interference analysis**: to work with other organisations to agree methods to ensure radio systems from each organisation can operate without causing harmful interference into the other, as described in Section 2.10.

Note that interference within a system is called **intra-system interference**, while between systems it is called **inter-system interference**.

2.12 Further Reading and Next Steps

This chapter has touched on a number of key regulatory documents that could be examined in greater detail, in particular:

- ITU RR
- ITU-R Recommendations and Reports
- CPM Report to a WRC
- CEPT ECC Decisions, Recommendations and Reports.

The next step is to look at the fundamental concepts behind interference analysis, the logical building blocks of all methodologies and algorithms.

3

Fundamental Concepts

Interference analysis involves a lot of adding up: the skill involves working out which numbers to add and why. This chapter gives an overview of the key summation, the link budget, describing its components, in particular the antenna gain, path loss, noise, modulations and associated link metrics, together with the underlying statistics, dynamics and geometry.

At the end you should understand the concept of the link budget and be able to calculate one from scratch.

Note that some of the key calculations in this chapter are provided as a spreadsheet available for download.

3.1 Radiocommunication Systems

Radiocommunication systems have come a long way from the first experiments of Hertz and Marconi to today's smartphone. The complexity of some of today's radiocommunication systems makes the task of analysing interference more complicated, but fundamentally it remains to be about the same concept: power.

Interference analysis is principally involved in issues relating to the strength of a radio signal, in particular the power received. Units are therefore watts or related terms, such as milliwatts or dBW (as described in the next section).

In order to analyse complex radiocommunication systems, it is usually necessary to simplify: in particular, to focus on those key aspects that influence the signal strength. If necessary, a model can be made more detailed, but it is good practice to start with something understood to build upon.

A generic simplified radiocommunication system is shown in Figure 3.1.

This block diagram could apply to a mobile phone signal, a Wi-Fi connection, a taxi company's despatch service or even a satellite link.

Interference Analysis: Modelling Radio Systems for Spectrum Management, First Edition. John Pahl.
© 2016 John Wiley & Sons, Ltd. Published 2016 by John Wiley & Sons, Ltd.
Companion website: www.wiley.com/go/pahl1015

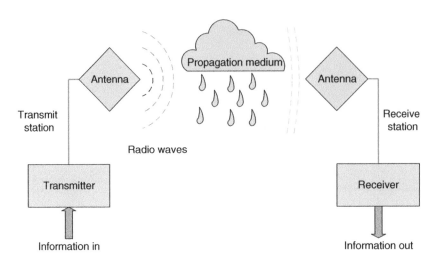

Figure 3.1   Generic radiocommunication system block diagram

The basic concepts of a radiocommunication system are:

- A transmitter uses power to create radio waves. The radio waves will have characteristics that can vary by frequency and time and that carry information
- These radio waves are transmitted via an antenna that sends different amounts of energy in different directions according to its gain pattern
- The radio waves radiate out from the transmit antenna, propagating through a medium that can include atmospheric effects such as rain and dust
- At the far end another antenna, with its own gain pattern, receives these radio waves
- The receiver detects the radio waves delivered by the antenna and attempts to decode the information stream
- The station is the combination of antenna + transmitter/receiver at either end of the link that could include additional elements such as feed loss.

Some radio systems will have basic block diagrams slightly different to this: those for radars and radio astronomy observatories are shown in Figures 3.2 and 3.3.

We need to start from these block diagrams and build a mathematical model that can be used to generate information about levels of interference between radio systems.

The key concept is that we need to be able to calculate the power of the radio waves transmitted by wanted and interfering systems, usually measured at a receiver. This means we need to calculate the key numbers shown in Figure 3.4.

In this simplified model, the power at the receiver, P_{rx}, will depend on:

- P_{tx} = the transmit power into the antenna, which itself will depend upon the transmitter power and the line feed to the transmit antenna
- $G_{tx}(rx)$ = the gain of the transmit antenna towards the receive antenna
- P_L = the propagation loss for the path between the transmit antenna and the receive antenna

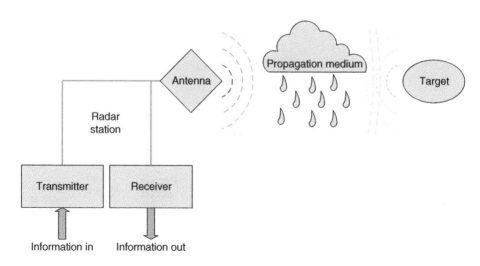

Figure 3.2   Generic radar system block diagram

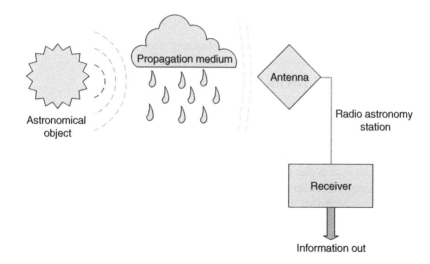

Figure 3.3   Radio astronomy observatory block diagram

- $G_{rx}(tx)$ = the gain at the receive antenna towards the transmit antenna
- $L_f(rx)$ = the loss on the line feed between the receive antenna and the receiver.

To be able to calculate these various terms, it will be necessary to build mathematical models of antennas and radio wave propagation. This in turn will require a geometric framework so that angles and distances can be calculated. These frameworks will have to be comprehensive enough to model the positions and motions of radio stations on the ground, in the air, on ships or in space.

These topics will be covered in this section, starting from basic concepts of radio waves and path loss and continuing to antennas, geometry and using these in the key concept of the link budget.

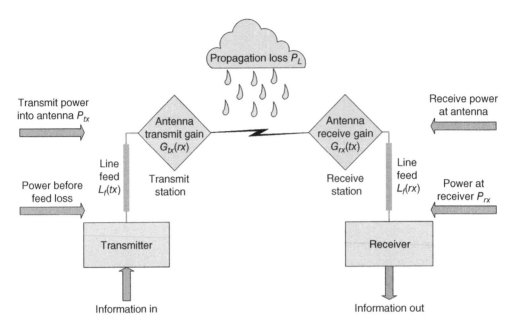

Figure 3.4 Key values in calculating received power

3.2   Radio Waves and Decibels

Radio waves are part of the electromagnetic spectrum that includes gamma rays, x-rays, ultra-violet, visible light and infrared. Most of the work of interference analysis involves electromagnetic radiation with frequencies no higher than those that are described as extremely high frequency (EHF) using the notation in Table 3.1.

The frequency unit is in cycles per seconds or hertz after the scientist Heinrich Rudolf Hertz who proved the existence of radio waves. At the time he considered his work purely theoretical, stating that he did not think that 'the wireless waves I have discovered will have any practical application'.

As can be seen in the nomenclature table, rather than describing all frequencies in hertz or Hz, units of kHz, MHz and GHz are used, where

1 kHz = 1 000 Hz
1 MHz = 1 000 000 Hz
1 GHz = 1 000 000 000 Hz.

Other nomenclatures are often used as some of the categories in Table 3.1 are large and cover many services, in particular the ones in UHF and SHF. For this reason alternatives are in regular use, such as those given in Table 3.2.

It should be noted that these are less well defined and can have alternative frequency ranges between the radar and space communities. However they are commonly used and this table is based upon information in Rec. ITU-R V.431 (ITU-R, 2000c). This recommendation also defines VHF/UHF Bands I, II, III, IV and V used by broadcasting services and others including private mobile radio (PMR).

Table 3.1   Nomenclature of frequency bands

Designator	Definition	Frequency Range
ULF	Ultra low frequency	300–3 000 Hz
VLF	Very low frequency	3–30 kHz
LF	Low frequency	30–300 kHz
MF	Medium frequency	300–3 000 kHz
HF	High frequency	3–30 MHz
VHF	Very high frequency	30–300 MHz
UHF	Ultra high frequency	300–3 000 MHz
SHF	Super high frequency	3–30 GHz
EHF	Extremely high frequency	30–300 GHz

Table 3.2   Alternative nomenclature of frequency bands

Designator	IEEE radar bands	Space radiocommunications
L	1–2 GHz	About 1–2 GHz
S	2–4 GHz	About 2–3 GHz
C	4–8 GHz	About 3–7 GHz
X	8–12 GHz	About 7–11 GHz
Ku	12–18 GHz	About 11–[a] GHz
K	18–27 GHz	—
Ka	27–40 GHz	About [a]–30 GHz
V	40–75 GHz	About 40[b]–50 GHz
W	75–100 GHz	—

[a] For space radiocommunications, the boundary between Ku and Ka is not well defined, but it is generally accepted that 14.5 GHz is within Ku band and 17.7 GHz is within Ka band.
[b] Sometimes the waveguide bands are used, namely Q band for 33–50 GHz, V band for 50–75 GHz and E band for 60–90 GHz.

In general it is better to be explicit about which frequencies are involved to avoid ambiguity but these terms are often used as a short hand.

The **frequency** and **wavelength** of a radio wave are connected via the speed of light as in Equation 3.1:

$$c = f\lambda \tag{3.1}$$

where
c = speed of light in m/s as per Table 3.5
f = frequency in Hz, where 1 Hz = 1/s
λ = wavelength in metres.

Example 3.1
For a frequency of $f = 3.6\,\text{GHz} = 3.6 \times 10^9$ Hz, the wavelength is

$$\lambda = \frac{3e^8}{3.6e^9} = 0.0833\,\text{m} = 8.33\,\text{cm}$$

Often interference analysis calculations can involve multiplication and/or division of very big or small numbers. It is generally easier to follow and understand numbers that are within everyday experience, for example, numbers between zero and (say) a few hundred, and add or subtract rather than multiply numbers so large or small they require scientific notation to capture them. For this reason most calculations are undertaken in log format, in particular using the **decibel** conversions between absolute form (abs), sometimes called linear, and decibel equivalent (dB):

$$X = 10\log_{10}(x) \qquad (3.2)$$

$$x = 10^{X/10} \qquad (3.3)$$

As far as possible, the notation that will be used in this book is that the lower case form of the variable is in absolute and the upper case in decibels.

It's also worth getting familiar with the most common examples of conversions between absolute and decibels, as in Table 3.3.

There are specific notation schemes that are used for decibel versions of standard units as shown in Table 3.4.

Table 3.3 Example conversions between absolute and decibel

Absolute	Decibel
2	3
10	10
$4 = 2 \times 2$	$3 + 3 = 6$
$5 = 10/2$	$10 - 3 = 7$
$20 = 2 \times 10$	$3 + 10 = 13$
$100 = 10 \times 10$	$10 + 10 = 20$

Table 3.4 Example unit conversions between absolute and decibel

Absolute units	Decibel units
Watts	dBW
Watts per square meter	dBW/m^2
Milliwatts	dBm
Gain relative to isotropic antenna	dBi
Gain relative to dipole antenna	dBd

The Key Numbers to Remember

The numbers in Table 3.5 are well worth remembering as they come up so often. In most cases link budgets are calculated to one decimal place, that is, 0.1 dB, and hence it is not usually necessary to remember these numbers to higher accuracy (or at least to within 0.05 dB). One reason for this is that in practice it is hard to measure signal strengths to an accuracy better than 0.1 dB.

Table 3.5 Key numbers to remember

Description	Number
Speed of light in a vacuum is officially $c = 299{,}792{,}458$ m/s but it's easier to remember it to one decimal place (Wikipedia, 2014b)	$3.0e8$ m/s
Boltzmann's constant is officially $1.3806488 \times 10^{-23}$ J/ K but is easier to remember in decibels (Wikipedia, 2014a)	-228.6 dBJ/K
$\times 2$ in decibels $= +3.010299957$ but just remember	$+3$ dB

Often bandwidths are given in MHz, and it is useful to be able quickly convert them into Hz in decibels:
    To convert from dB(MHz) to dB(Hz) add:
    To convert from dB(Hz) to dB(MHz) subtract:
	
	60 dB

To convert from dBW to dBm add: To convert from dBm to dBW subtract:	30 dB
The free space path loss equation is $L = 32.45 + 20 \log 10(d_{km}) + 20 \log 10(f_{MHz})$. This is derived in Section 3.3 and the constant is actually something like 32.447783 but it is sufficient to remember:	32.45 dB
Radius of the Earth. The Earth model used in interference analysis is typically spherical rather than an oblate spheroid (ellipsoid) and hence a single value is usually sufficient. The WGS 84 model (Wikipedia, 2014d) is the basis and the mean equatorial radius of 6378.137 km is simplified to a single figure of:	6378.1 km
To convert between an antenna gain in dBd (relative to a dipole) to dBi (relative to an isotropic antenna) the calculation is dBi = dBd + 2.15 dB (Kraus and Marhefka, 2003)	2.15 dB

3.3 The Power Calculation

Interference calculations are about the power of radio waves, and a starting point is to consider how the signal strength changes with distance from a transmitter. The simplest starting point is a transmitter with power p_{tx} in watts, radiating energy equally in all directions in a perfect vacuum as in Figure 3.5.

Antennas that radiate equally in all directions are called **isotropic**. From basic geometry the area of a sphere of radius r is

$$a_s = 4\pi r^2 \tag{3.4}$$

The power going through a unit of area at a distance of r from the transmitter, described as the power flux density or *PFD*, can therefore be calculated as

$$pfd = \frac{p_{tx}}{u_s} = \frac{p_{tx}}{4\pi r^2} \tag{3.5}$$

As interference analysis is primarily concerned with signal strength at a receiver, we need to convert this *PFD* by identifying the effective area of the receive antenna. For an isotropic antenna receiving at a wavelength of λ, it can be shown (Kraus and Marhefka, 2003) that this is

$$a_{e,i} = \frac{\lambda^2}{4\pi} \tag{3.6}$$

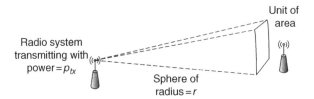

Figure 3.5   Isotropic radiator in a vacuum

The received signal at an isotropic antenna is therefore

$$s = pfd.a_{e,i} = \frac{p_{tx}}{4\pi r^2} \cdot \frac{\lambda^2}{4\pi} = \frac{p_{tx}}{\left(4\pi r / \lambda\right)^2} \tag{3.7}$$

It is usually easier to operate in decibels rather than absolute so these two equations become

$$PFD = P_{tx} - L_s \tag{3.8}$$

$$S = P_{tx} - L_{fs} \tag{3.9}$$

where the **spreading loss** L_s is

$$L_s = 10\log_{10}\left(4\pi r^2\right) \tag{3.10}$$

and the **free space path loss** L_{fs} is

$$L_{fs} = 10\log_{10}\left(\left(\frac{4\pi r}{\lambda}\right)^2\right) \tag{3.11}$$

Note how the power p_{tx} in watts has been converted into decibels, that is,

$$P_{tx} = 10\log_{10}(p_{tx}) \tag{3.12}$$

The free space path loss equation is fundamental and is often rearranged into a more memorable form in which the loss in decibels between two locations separated by distance d_{km} in km and at frequency f_{MHz} in MHz is as follows:

$$L_{fs} = 10\log_{10}\left(\left(\frac{4\pi r}{\lambda}\right)^2\right)$$

$$= 20\log_{10}\left(\frac{4\pi 1000 \cdot d_{km} f_{MHz} \cdot 10^6}{c}\right)$$

$$= 20\log_{10}\left(\frac{4\pi 10^9}{c}\right) + 20\log_{10} \cdot d_{km} + 20\log_{10} f_{MHz}$$

Hence

$$L_{fs} = 32.45 + 20\log_{10}d_{km} + 20\log_{10}f_{MHz} \tag{3.13}$$

Example 3.2
For a distance of $d = 2$ km and a frequency of $f = 3\,600$ MHz, the free space path loss would be

$$L_{fs} = 32.45 + 20\log_{10}2 + 20\log_{10}3600 = 109.6\,\text{dB}$$

Equation 3.13 gives the path loss in a pure vacuum, what is called free space path loss. This would be appropriate for communication between two satellites operating in space, but terrestrial communications, plus those between Earth and space, must consider the atmosphere and possibly terrain effects. This introduces a whole range of additional factors, and there are a number of propagation models that can be used to predict loss under a range of conditions. These are described in more detail in Chapter 4.

Note on Terminology

It is often useful for interference analysis to be able to compare a link's propagation loss against that it would have experienced in a vacuum (i.e. whether reduced due to attenuations or enhanced). It is also useful to have a different term for the loss in the *PFD* calculation to that in the signal strength calculation. However other terminology could be encountered: for example, sometimes the term 'spreading loss' is used to describe the path loss in Equation 3.13, while Rec. ITU-R P.341 (ITU-R, 1999a) uses 'free space basic transmission loss' and 'basic transmission loss'.
   This book will use the following terminology:

- **Spreading loss**: path loss in Equation 3.10 used to calculate the *PFD* received at a reference area across unobstructed vacuum
- **Free space path loss**: path loss in Equation 3.13 used to calculate signal strength received at an antenna across unobstructed vacuum
- **Propagation loss**: path loss used to calculate signal strength received at an antenna including the effects of the atmosphere and any obstructions.

Equations 3.8 and 3.9 for *PFD* and signal strength, respectively, did not include any gain at the transmitter or the receiver, as isotropic antennas were assumed. The gain is the ratio of the power at the input to an antenna to the power in the direction of the target. In absolute this is then a ratio

$$g = \frac{p_{out}}{p_{in}} \tag{3.14}$$

In decibels this term can therefore be added and so the two equations become

$$PFD = P_{tx} + G_{tx} - L_s \tag{3.15}$$

$$S = P_{tx} + G_{tx} - L_{fs} + G_{rx} \tag{3.16}$$

Along with Equation 3.13 these are some of the most important equations in interference analysis, and the concept of antenna gain is discussed in more detail in the following section. Note that the gain is implicitly in dBi as it compares the gain against that of an isotropic antenna.

The transmit power and the transmit gain are often included together as the **equivalent isotropically radiated power** or *EIRP*:

$$EIRP = P_{tx} + G_{tx} \tag{3.17}$$

Hence another form of these two equations are

$$PFD = EIRP - L_s \tag{3.18}$$

$$S = EIRP - L_{fs} + G_{rx} \tag{3.19}$$

In this equation S could represent either C = wanted or I = interfering signal, and there could be additional terms to take account of other losses such as those due to the line feed.

Note that most of the units of power in interference analysis are based on watts, whether dBW, milliwatts or dBm. A watt is defined as a joule per second; hence it is the rate of energy transfer. When it is necessary to measure this energy, an integration time is required, which usually is a shorter timescale than other variations in the scenario under consideration. However in some cases it can be necessary to consider the definitions of power in more detail, for example, when undertaking Monte Carlo simulations involving radar systems as described in Section 6.9.2.

Related to the *EIRP* is the **effective radiated power** (*ERP*), which is the radiated power compared to a half-wave dipole, and hence

$$EIRP = ERP + 2.14 \text{dB} \tag{3.20}$$

This book will standardise on *EIRP*s rather than *ERP*s and generally specify gains compared to an isotropic antenna (dBi) rather than gain compared to a dipole (dBd).

Example 3.3
The received signal at a distance of $d = 2$ km of a link operating at $f = 3\,600$ MHz using a transmit power of 1 mW (i.e. −30 dBW) with a 36.9 dBi transmit antenna gain and a receiver with antenna gain of 36.9 dBi would be

$$S = P_{tx} + G_{tx} - L_{fs} + G_{rx}$$

$$= -30 + 36.9 - 109.6 + 36.9 = -65.8 \ dBW$$

3.4   Carrier Types and Modulation

3.4.1   Overview

A key characteristic of a radiocommunication system is how it encodes the information to be transmitted to the receive station using the radio wave or carrier. The process of encoding is called modulation and so at the transmit end there will be a modulator and at the receive end a demodulator. Equipment that can both modulate and demodulate a signal are called modems.

There are a large number of methods to modulate information onto a radio wave carrier including analogue and digital techniques. As with antennas, it is important to focus on those aspects of carrier types and modulations that are relevant to interference analysis, in particular the carrier's data rate, bandwidth and variation of the power density (shape) in the frequency domain. The modulation will also determine the thresholds the system must meet in order to provide the required quality of service and avoid harmful interference.

It is useful to get an idea of what is involved in the modulation process so as to have a mental picture and grasp of the key concepts when carriers are discussed. Therefore:

- Section 3.4.2 describes analogue modulations such as amplitude modulation (AM)
- Section 3.4.3 describes digital modulations, in particular binary phase sequence keying (BPSK) and quadrature phase shift keying (QPSK)
- Section 3.4.4 looks at frequency hopping and orthogonal frequency division multiplexing (OFDM) techniques
- Given the range of possible digital modulation methods, Section 3.4.5 discusses factors to consider when selecting one
- There is a brief description of pulse modulation in Section 3.4.6
- Finally, filtering is described in Section 3.4.7.

3.4.2   Analogue Modulation

The earliest method to communicate via radio waves was simply to switch the carrier on and off in order to create Morse code characters. Transmitting data in this way is inefficient and in 1901 a major advance was the development, by Reginald Fessenden, of a way to combine signals of two frequencies using a multiplier, a technique that he called heterodyning. This approach is the basis of almost all radiocommunication systems, whereby a lower intermediate frequency (IF) carrier is created and modified (or modulated) and then raised up to a higher transmitting frequency. At the receiver there is the reverse path by which the frequency is lowered and the signal demodulated.

The communication system transmits radio waves even for null data (e.g. zeros for digital systems) and therefore is described as continuous wave (CW), with the information coded onto the radio wave form. This coding was originally analogue, with simple wave forms such as amplitude modulation (AM) where the strength of the underlying carrier is changed in a way that represents the message, as in Figures 3.6, 3.7 and 3.8.

The carrier signal at time $= t$ can be described mathematically (Rappaport, 1996) using

$$S_{AM}(t) = A_c[1 + m(t)]\cos(2\pi f_c t) \tag{3.21}$$

where
$m(t)$ is the message at time t
A_c is the amplitude of the AM signal
f_c is the frequency of the carrier.

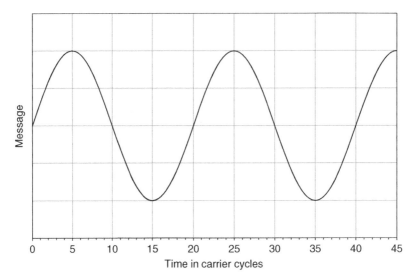

Figure 3.6   Amplitude modulated message

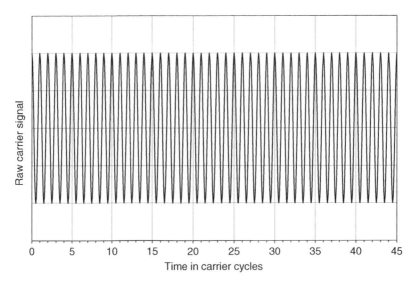

Figure 3.7   Unmodified carrier signal

Typically the bandwidth of an AM signal is twice the frequency of the modulating message, though filtering can be used to reduce this, for example, by removing one of the sidebands (described as single sideband, or SSB).

Other analogue modulations have been used, in particular frequency modulation (FM), whereby the frequency of the carrier signal is altered rather than the amplitude. The shape of the resulting wave form often depended upon the message type, so that, for example, an FM voice carrier would have a different shape to that of a television carrier. This would have

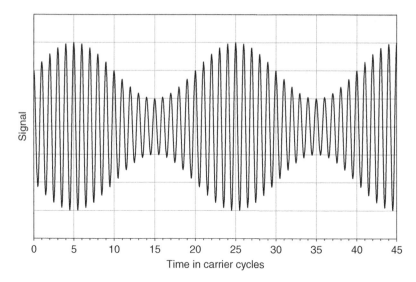

Figure 3.8   Resulting amplitude modulated signal

a significant impact on interference analysis in both the shape of the interfering signal and the thresholds for interference at the receiver.

Most modern systems tend to use digital modulations as described in the following section. However legacy communication systems using analogue modulations continue to operate, in particular PMR voice communications as described in Section 7.2.

3.4.3   Digital Modulation

Digital modulation provides communication systems with a wide range of advantages over analogue, including the ability to:

- Operate with lower received signal-to-noise (S/N) ratios
- Include coding to reduce transmission errors
- Include encryption
- Multiplex multiple types of communication
- Multiplex multiple users' communications
- Handle fading better
- Be managed within software.

One benefit for interference analysis is that the resulting carrier shape in the frequency domain of a digitally modulated signal is nearly square, which makes modelling bandwidth factors simpler in many scenarios (see Section 5.2). Also the performance of the radio system can be calculated from the modulation involved, which tends to come from a limited well-documented set.

The simplest form of digital modulation is probably binary phase shift keying (BPSK) in which the phase of the carrier is modified between two states. Consider the unmodified carrier in Figure 3.9 and then how it has been modulated in Figure 3.10.

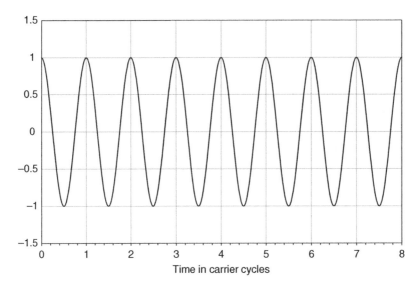

Figure 3.9   Unmodulated carrier

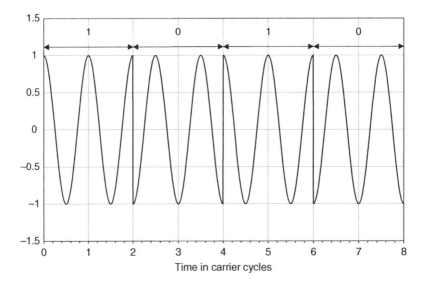

Figure 3.10   BPSK modulated carrier

The modulated carrier shape is either unchanged or the inversion of the unmodified carrier. In mathematical terms, it has either one of $\{0, \pi\}$ added to the wave function, as follows:

$$\text{Binary } 0: \quad S_{BPSK}(t) = A_c \ \cos(2\pi f_c t + \pi) \quad 0 \le t < T_b \tag{3.22}$$

$$\text{Binary } 1: \quad S_{BPSK}(t) = A_c \ \cos(2\pi f_c t) \quad 0 \le t < T_b \tag{3.23}$$

where the amplitude A_c is a function of the energy per bit E_b and bit duration T_b:

$$A_c = \sqrt{\frac{2E_b}{T_b}} \tag{3.24}$$

Note that the carrier wave frequency is larger (sometimes considerably greater) than the message frequency.

The phase modulations are often shown graphically via the **constellation diagram**, such as those for BPSK and QPSK in Figure 3.11.

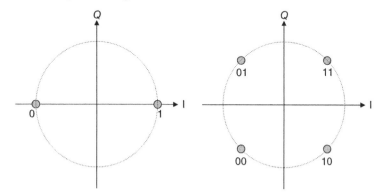

Figure 3.11   Constellation diagrams for BPSK (left) and QPSK (right)

BPSK can be seen to have two states and hence each symbol codes one bit of information, either $\{0, 1\}$. Higher-order modulations can have more bits per symbol and result in more complex wave forms. For example, the constellation diagram for quadrature phase shift keying (QPSK) can be seen to have four states, resulting in two bits per symbol $\{00, 01, 10, 11\}$ for each of the phases.

There are multiple mappings of phases to symbols: this is an example of Gray coding where adjacent values differ by just one bit, and the resulting wave form is

$$S_{QPSK}(t) = A_c \ \cos\left(2\pi f_c t + \frac{(i-1)\pi}{2}\right) \quad i = \{1,2,3,4\} \quad 0 \le t < T_s \tag{3.25}$$

This is shown graphically as in Figure 3.12.

BPSK and QPSK are closely related and have the same shape in the frequency domain, that is, the same power spectral density (PSD), which is given by

$$P_{BPSK}(t) = \frac{E_b}{2}\left[\left(\frac{\sin\pi(f - f_c)T_b}{\pi(f - f_c)T_b}\right)^2 + \left(\frac{\sin\pi(-f - f_c)T_b}{\pi(-f - f_c)T_b}\right)^2\right] \tag{3.26}$$

Example 3.4

The power density (as the PSD is also called) of a BPSK carrier with data rate = symbol rate = 30 Mbps is shown in Figure 3.13.

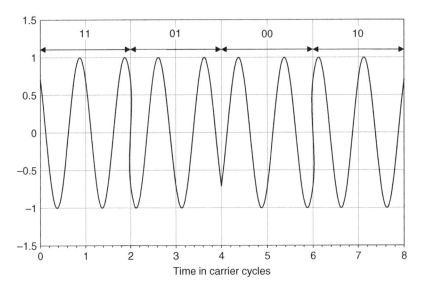

Figure 3.12   QPSK modulated carrier

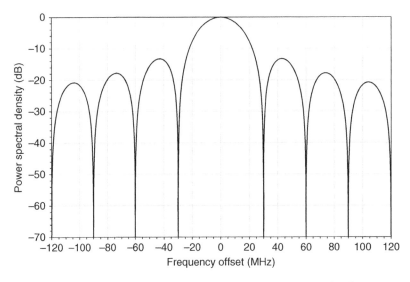

Figure 3.13   Power spectral density of 30 Mbps BPSK signal

In practice the signal actually transmitted when using a BPSK or QPSK carrier will have a very different shape than this theoretical prediction. In order to manage interference, most radio systems employ filters (as in Section 3.4.7) to reduce emissions outside of an operating bandwidth, typically described as the necessary bandwidth. The degree of filtering will depend upon factors such as:

- Costs: higher-quality filters tend to be more expensive
- Size: in some devices (e.g. handsets) there is limited space available
- Requirements: if there is a sensitive service in an adjacent frequency band, then there will be a greater need for filtering.

These will vary from service to service and will be factors that must be considered when determining the spectrum mask that a transmitter has to meet, as discussed in Section 5.3.1.

A key attribute of a carrier is the relationship between the energy per bit (E_b) and the bit error rate (*BER*). The higher the energy per bit compared to the noise per Hz (N_0), the more likely it is that the demodulator correctly identifies the phase of the modulation and hence selects the 'right' bit. The concept of noise is discussed further in Section 3.6 and E_b/N_0 in Section 3.11.

For both BPSK and QPSK, the *BER* can be calculated (Bousquet and Maral, 1986) using

$$BER = \frac{1}{2} erfc \sqrt{\frac{E_b}{N_0}} \qquad (3.27)$$

This curve is shown in Figure 3.14.

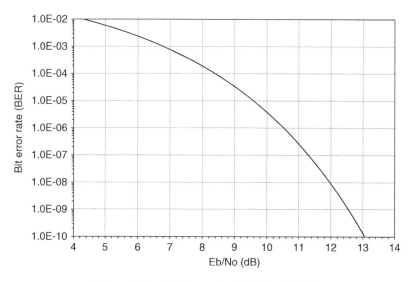

Figure 3.14   BER curve for BPSK and QPSK

Example 3.5
A DPSK carrier with $E_b/N_0 = 10.5$ dB would have a *BER* ~ 1e − 6.

There are a number of variants on BPSK and QPSK, including:

- Additional phases, such as 8PSK, which has eight phases
- Differential phase shift keying or DPSK in which bit differences are coded rather than the bits themselves.

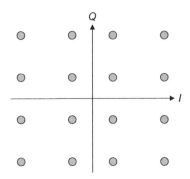

Figure 3.15 Constellation diagram for $M = 16$ QAM

A significant variation is to modify both the amplitude and phase, such as for quadrature amplitude modulation (QAM), which for $M = 16$ has a constellation diagram as shown in Figure 3.15. The constellation diagram could also be rotated to improve resilience, as in the DVB-T2 standard.

The impact on these different types of digital modulation is to change:

- The carrier shape, that is, how the power spectral density varies in frequency. As noted, the actual transmitted PSD will also depend upon filtering, but the necessary bandwidth will depend upon the modulation and data rate
- The *BER* curve, that is, what performance can be achieved for a given E_b/N_0. A factor to consider will be the code scheme that can correct for errors but at the cost of reduced payload.

Part of the system design process is to select which would be the most appropriate modulation scheme as discussed in Section 3.4.5. Additional information on digital modulations can be found in Annex 6: Digital phase modulation of Rec. ITU-R SM.328 (ITU-R, 2006i).

3.4.4 Frequency Hopping and OFDM

The aforementioned examples in Sections 3.4.2 and 3.4.3 of analogue and digital modulations assumed a single frequency was used to transmit the carrier. There are occasions when multiple frequencies are used, in particular for frequency hopping and OFDM transmissions.

The motivation for frequency hopping systems is to reduce the likelihood of frequency overlap, which can be used:

- To reduce interference or, for military systems, the impact of jamming
- To reduce the likelihood of intercept
- To handle fading better, as only a subset of bits are affected, so that with coding the payload signal can be recovered.

Frequency hopping was invented by film star Hedy Lamarr in Hollywood, where she arrived in 1938 after escaping from Europe. She was working with composer George Antheil who had

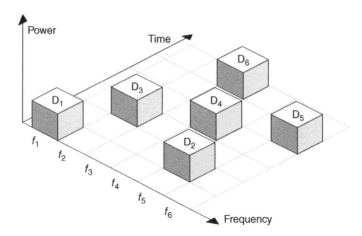

Figure 3.16   Frequency hopping system

developed pianos that could play automatically, synchronised in pairs. Lamarr realised that the same concept could be used for radio systems whereby the transmit and receive stations switch operating frequency in sequence, as in Figure 3.16.

The frequencies are selected using a pseudorandom sequence with a key that is known to both the transmit and receive stations. Without this key it becomes nearly impossible to follow the hops between frequencies and hence intercept or jam the signal.

A variation of this method is direct sequence, whereby the whole bandwidth is used but the signal modified using a key, again known to both the transmit and receive stations. This is discussed further in Section 3.5.5 on the code division multiple access (CDMA) method.

One advantage of these techniques is in the management of multipath fading, which can be very sensitive to frequency, so two nearby channels can have very different fade depths. The propagation effects of multipath on multiple frequency carriers can be to reduce the received signal in just a few channels while adjacent frequencies are almost unaffected.

This behaviour was one of the motivations for the development of orthogonal frequency division multiplexing (OFDM). This is used for a wide range of radiocommunication applications including terrestrial digital TV (described in Section 7.3) and 4G mobile phone systems using the LTE standard.

Rather than having a single carrier carrying all the data, the information is split across multiple subcarriers. The frequencies for these subcarriers are selected to be orthogonal in the sense that they can transmit without impacting the reception of other subcarriers and in general are separated in frequency by

$$\Delta f = \frac{k}{T_U} \tag{3.28}$$

where T_U = is the symbol duration in seconds and k is typically set to 1. This means the total bandwidth of N subcarriers is

$$B \cong N\Delta f \tag{3.29}$$

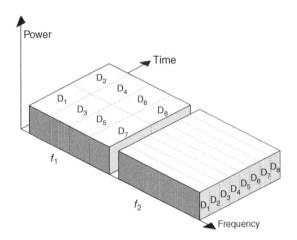

Figure 3.17   OFDM structure

An example of OFDM is shown in Figure 3.17 where:

- f_1 is an OFDM carrier with four subcarriers each with two data sets per frame
- f_2 is a single wideband carrier with eight data sets in the frame.

The *BER* can be monitored for each subcarrier and used by the error correcting algorithms to identify which are providing the 'best' bits against those which are suspect so should be rejected. In addition, the symbol duration time is longer, reducing multipath interference and facilitating single frequency networks.

Another benefit of this approach is that it facilitates multiple access, as discussed in Section 3.5.6.

3.4.5   Digital Modulation Selection

Given the wide range of possible digital modulation types, how should one be selected? A summary of the theoretical performance of the frequency shift keying (FSK), phase shift keying (PSK) and quadrature amplitude modulation (QAM) techniques is given in Table 3.6 taken from Rec. ITU-R F.1101 (ITU-R, 1994a) excluding coding in all cases.

Note how a single threshold is given for each modulation and variant: the value of a $BER = 10^{-6}$ given in the table is a reference value but that required for a system's operation could be either lower or higher. Accepting a higher *BER* can be one method to facilitate sharing between radio systems. The total noise is assumed to include interference and so the S/N value could be used as the *C/N* or *C/(N + I)* threshold.

In many discussions it is assumed that the best performance, as in highest bandwidth efficiency, is the one that transmits the greatest data for a given bandwidth. Hence QPSK (i.e. 4-state PSK) would be better than BPSK (i.e. 2-state PSK) as it has half the bandwidth, and 64-QAM would be even better, as it only requires $R_d/6$ compared to $R_d/2$ for QPSK.

However it is also necessary to consider the additional requirement for transmit power needed to close the link, which will depend upon the threshold *S/N* required to achieve a

Table 3.6 Theoretical performance of selected modulations

System	Variant	Signal to total noise (S/N) in dB required for $BER = 10^{-6}$	Bandwidth (Hz) given data rate R_d (bps)
FSK	2-State FSK	13.4	R_d
	3-State FSK	15.9	R_d
	4-State FSK	23.1	$R_d/2$
PSK	2-State PSK	10.5	R_d
	4-State PSK	13.5	$R_d/2$
	8-State PSK	18.8	$R_d/3$
	16-State PSK	24.4	$R_d/4$
QAM	16-QAM	20.5	$R_d/4$
	32-QAM	23.5	$R_d/5$
	64-QAM	26.5	$R_d/6$
	128-QAM	29.5	$R_d/7$
	256-QAM	32.6	$R_d/8$
	512-QAM	35.5	$R_d/9$

specific *BER*. It can be seen that QPSK requires an extra 3 dB over BPSK, while 64-QAM would need as much as 16 dB more power.

This can have a significant impact during interference analysis as it means that, for example, interference from a 64-QAM link would be 16 dB greater than that for a BPSK link, all other things being equal. It can also have an impact on the range covered by a given transmitter. Mobile base stations often have maximum power limits for environmental considerations (including to limit human exposure to RF radiation as in Table 5.23), and so the coverage will be smaller using higher modulations rather than something like BPSK.

This is one reason why some systems employ adaptive modulations, in that they can automatically modify the modulation used to adjust for their environment. An example would be Wi-Fi that can up the modulation for users close to the access point and then reduce it for those that are shielded or further away. This can have an impact on the carrier shape and hence interference into adjacent channels, as discussed in Section 5.5.3.

The upper bound or channel capacity C_a for a specified bandwidth B and ratio of the signal to noise s/n (S/N in dB) can calculated using Shannon's channel coding theorem as (Rappaport, 1996)

$$C_a = B \cdot \log_2\left(1 + \frac{s}{n}\right) \qquad (3.30)$$

Example 3.6
A link has an $S/N = 20$ dB and bandwidth of 200 kHz (as used for GSM): the maximum theoretical channel capacity is therefore 1.33 Mbps – much greater than given by that standard.

Another factor to consider is resilience, both to interference and fading, and the degree of forward error correction (FEC) coding employed. This involves the transmission of additional redundant bits that can be used to correct errors but at the cost of reducing the payload data rate. But by combining FEC with a lowering of *BER* requirements, the threshold *S/N* can be reduced significantly.

The selection of the preferred modulation and degree of coding can require extensive study during system design that is likely to include both the wanted service and interference impact, both into and from other systems.

3.4.6 Pulse Modulation and UWB

A variation on the continuous wave modulations described in the previous sections is pulse modulation, where information is sent in a series of short bursts or pulses. The result can be communication systems that operate with low power densities over extremely large band-widths, what is called ultra wideband (UWB). Multiple access can be provided using random-ization of the pulses creating a noise-like signal.

UWB is defined as being when the bandwidth exceeds the lesser of 500 MHz or 20% of the centre frequency, and this can reach several GHz. The advantage of such systems is that when the transmit power is spread over a wide frequency range, then the power density can become significantly lower than the receiver noise density. This facilitates operating without causing harmful interference into licensed services, though there can be issues if there are large numbers of UWB devices (particularly for aggregate interference scenarios into satellite uplinks) or if the device is very close to a sensitive receiver. Applications of UWB include short-range commu-nications and also radar, for example, short-range radars (SRR) on vehicles.

If UWB devices transmit at a sufficiently low power, then they can be permitted to operate on a licence-exempt basis. Within the European Union a range of maximum mean power densities were given in a commission decision (Commission of the European Communities, 2007), while in the United States the powers are defined by FCC Part 15 Rules (Federal Communications Commission, 2002). In both cases the highest value in selected bands is −41.3 dBm/MHz, while the limits in other bands are more stringent. Pulses of UWB systems can be generated in a way that ensures equipment meets emission levels that vary by frequency band.

3.4.7 Filtering

Figure 3.13 showed the theoretical power spectral density of a 30 MHz BPSK signal without filtering. However most radio systems employ filters to reduce interference to and from other systems. A number of filtering processes are available, including Gaussian, Butterworth and Nyquist. Gaussian filtering can be modelled by the normalised function (Rappaport, 1996):

$$A_{rx}(\Delta f) = 10\log_{10}\left[e^{-\frac{1}{2}\left(\frac{x}{\sigma}\right)^2}\right] \tag{3.31}$$

The standard deviation term σ in this equation relates to the full bandwidth B (i.e. twice half the bandwidth from the central frequency) at which the attenuation has reached a specified level X_{dB} down from peak, using

$$\sigma = \frac{B}{\sqrt{0.8X_{dB}\ln(10)}} \tag{3.32}$$

Typically the filter bandwidth is defined for $X_{dB} = 3$ dB down, though other values could be used (e.g. -30 dB down at offset B from the centre frequency).

The attenuation due to an nth order Butterworth filter with 3 dB bandwidth B_{3dB} can be calculated using

$$A_{rx}(\Delta f) = 10\log_{10}\left[1 + \left(\frac{\Delta f}{B_{3dB}}\right)^{2n}\right]$$
(3.33)

Example 3.7

Figure 3.18 shows the Gaussian and Butterworth $n = 4$ filters that have a 3 dB bandwidth of 10 MHz.

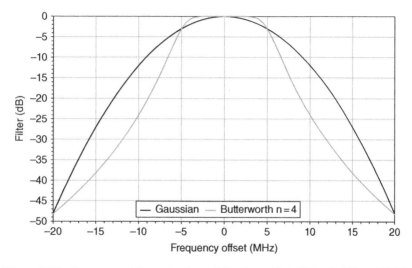

Figure 3.18   Gaussian and Butterworth $n = 4$ filtering with 3 dB bandwidth $= 10$ MHz

Another common filter is Nyquist defined by frequency f_n and roll-off factor r_{of} with function in absolute:

$$
\begin{aligned}
a_{rx}(\Delta f) &= 1 & f &< f_n(1 - r_{of}) \\
a_{rx}(\Delta f) &= 0.5\left(1 + \cos\left[\frac{\pi}{2r_{ro}}\left(\frac{f}{f_n} - 1 + r_{of}\right)\right]\right) & f_n(1 - r_{of}) &< f < f_n(1 + r_{of}) \\
a_{rx}(\Delta f) &= 0 & f &> f_n(1 + r_{of})
\end{aligned}
$$
(3.34)

For example, ETSI TR 101 854 (ETSI, 2005) uses the Nyquist filter together with other assumptions to derive receive spectrum masks.

3.5   Multiple Access Methods

3.5.1   Overview

In many situations, multiple users located within a geographic area need to share access to the same part of the radio spectrum. This is particularly the case with organisations providing a commercial radiocommunication service who would like to have as many users as possible and provide them with the best possible quality of service. A number of methods are used to enable the same block of radio spectrum to be accessed by multiple users, as will be discussed in this section.

A key issue to consider is the impact of these different access methods on interference analysis. The main factors to assess are how the access method impacts the:

* Number of stations that need to be modelled
* Number of simultaneously transmitting stations
* How the power increases with number of users, including the ratio between peak power and averaged over time
* How the bandwidth increases with number of users.

Note that a separate issue is how busy each user is: for the purposes of this section, it is assumed that all nodes are fully loaded. More information on traffic modelling is contained in Section 5.7.

A typical scenario would be a mobile communications network, with a base station providing bidirectional access to multiple users as in Figure 3.19.

Two factors need to be considered:

1. Is each sector independent of the others, so spectrum reuse is limited to those users within each sector? For example, each sector could have different frequencies assigned to it.
2. Are the communications from the base station to the handsets (downlink) using the same frequency as from the handsets to the base station (uplink)?

For the second question, one solution is to use different blocks of spectrum for the downlink and uplink directions, so-called paired blocks, as in Figure 3.20. These allow **duplex**

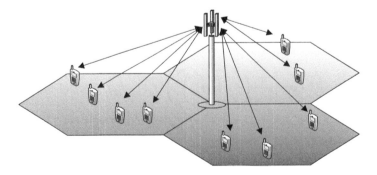

Figure 3.19   Base station supporting nine users in three sectors

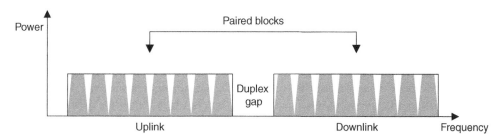

Figure 3.20   Paired blocks for duplex operation

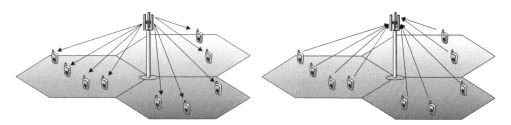

Figure 3.21   Separate downlink and uplink scenarios

operations, that is, in both directions simultaneously, rather than **simplex** where a single unpaired block is used for both directions. For the former case, there can be a **duplex gap** between the blocks of frequencies used for the uplink and downlink directions to give enough frequency separation to protect against non-co-frequency intra-system interference. In this example, frequency separation would be used at the base station to ensure that transmissions to the mobile do not cause harmful interference into the weaker signal received from the mobile.

If a paired block arrangement is used, then there are, for interference analysis purposes, two separate scenarios to model, one involving the base station transmitting and the other the mobiles transmitting, as in Figure 3.21. The choice of whether to assign the downlink to the upper or lower block can depend upon the sharing environment as it is typically easier to share with low power low height mobiles than base stations.

The main multiple access methods are:

- Collision sensing multiple access (CSMA) in Section 3.5.2
- Frequency division multiple access (FDMA) in Section 3.5.3
- Time division multiple access (TDMA) in Section 3.5.4
- Code division multiple access (CDMA) in Section 3.5.5
- Orthogonal frequency division multiple access (OFDMA) in Section 3.5.6.

There can also be combinations of these, where multiple access is provided by (say) multiple frequencies and time slots.

3.5.2   Collision Sensing Multiple Access

Collision sensing multiple access (CSMA) is an approach that uses a single channel for all user communications with no control or synchronisation. It can operate in simplex mode where all communications (uplink and downlink) use the same channel, as in Wi-Fi. However it could be used in duplex mode, such as all satellite return links using the same channel with the ALOHA protocol. Another example would be shared PMR channels, where congestion means that low-activity users are licensed to operate on the same frequency channel at nearby geographic locations, as described in Section 7.2.

There are a number of variations possible, such as the slotted ALOHA scheme in which communication attempts are only permitted at specific times.

Before transmitting, nodes sense the channel to detect whether any other user is operating, only accessing the channel when it appears to be clear. This can lead to collisions if two nodes attempt to access the channel simultaneously, though there are a number of methods that can reduce this including collision detection (CD) and collision avoidance (CA), where transmissions are terminated if collisions are detected and a random time waited until retry.

One potential problem with pure sensing is that nodes can be hidden from each other and hence transmit simultaneously leading to collisions despite initially checking the channel for activity as in Figure 3.22.

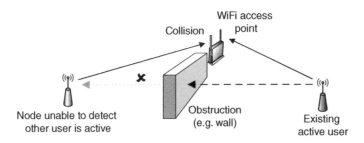

Figure 3.22   Hidden node problem – collision in Wi-Fi network

In practice this is often a secondary effect. The author was involved in a study commissioned by Ofcom (Aegis Systems Ltd and Transfinite Systems Ltd, 2004), which attempted to measure interference thresholds between Wi-Fi networks. However it was found that each network's activity was being detected by the others, resulting in time being shared efficiently between each, making it hard to detect interference effects.

The ability of multiple users to fully access a radio channel depends upon factors such as signalling overhead and time between sensing for activity. In another study for Ofcom (Ofcom, 2014), this one into PMR applications, it was noted that as occupancy of a channel shared between N_{sys} users reached 50%, overhead and wait time-related issues meant that delays to access the radio channel increased exponentially, limiting the maximum activity in these cases to

$$A_f \cong \frac{50}{N_{sys}} \tag{3.35}$$

For interference analysis purposes, therefore, it can be acceptable to consider at most a single station as being active simultaneously, with the activity of that station dependent upon the traffic model used.

In single frequency systems with different station types, such as an access point and user node, it might be necessary to model both station types, with activity split between the two depending upon the ratio of uplink/downlink traffic ratios. An alternative approach would be to select the worst case, having done an initial analysis as to which that would be, though that could lead to an overestimation of interference. In each direction and for each station type, power is likely to be determined by other factors such as range and service required.

If the stations are at significantly different locations, so that the interference they cause would vary depending upon which is active, it is necessary to model multiple stations with one being selected as active for each simulation sample (e.g. at random).

With a shared channel other metrics become important, such as:

- The probability of a request to access the channel being blocked by another's transmissions
- Average delay when blocked to gain access to the channel.

An example of when these metrics need to be assessed during interference analysis is when PMR systems with low activity have licences to operate on the same frequency at nearby locations, requiring them to share the same channel. An example of a methodology to assess the channel sharing ratio of PMR systems is described in Section 7.2.

3.5.3   *Frequency Division Multiple Access*

Frequency division multiple access (FDMA) is when a separate frequency is made available to each user as in Figure 3.23. Here there are six users, $\{U_1 \dots U_6\}$, each one operating on a different frequency $\{f_1 \dots f_6\}$.

In this case, as the number of users increases:

- The bandwidth of each user channel is fixed, but the total bandwidth increases linearly with the number of users

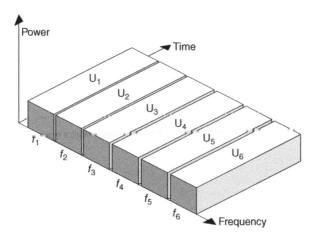

Figure 3.23   Frequency division multiple access

- The power per carrier is fixed, but the total power over all carriers increases by the number of users.

As each user has exclusive access to an individual radio channel, there is no danger of blocking due to other users. To provide simultaneous two-way communication, each user would need access to two channels, one to transmit and another to receive, that is, duplex operation.

Note that the transmit power of each user could depend upon location and service provided by the carrier: for example, higher powers would be required at the edge of coverage and for higher modulations.

This approach is often inefficient in its use of the radio spectrum, as when a user is not accessing its channel, it is not available to other users. In general with traffic that varies significantly (e.g. voice or web browsing), it is unlikely that all users require their maximum data rate simultaneously. Therefore, by sharing or multiplexing a channel, the overall capacity in terms of maximum users can be increased, the multiplex gain. For this reason TDMA, CDMA or OFDMA (described in the following sections) are more commonly used for scenarios that have variable traffic, such as mobile phone networks. Hence while the earliest mobile phone standards were based upon FDMA techniques, the later ones would use more efficient access methods.

FDMA concepts are used in the licensing of fixed links and PMR applications, in that a single frequency (simplex) or pair of frequencies (duplex) are provided by a regulator for use by a specific end user. Fixed links are used for applications such as backhaul and comprise a pair of transmit and receive stations as described in Section 7.1.

3.5.4   Time Division Multiple Access

In time division multiple access (TDMA), a single channel is shared between different users in the time domain as shown in Figure 3.24.

In this case the same channel is used by multiple users. Typically this involves defining a frame, that is, a fixed number of bytes transmitted over a defined time period. Within the frame are slots that can be assigned to users.

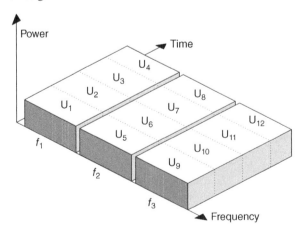

Figure 3.24   Time division multiple access

For example, the 2G mobile standard GSM has the following characteristics (Rappaport, 1996):

Bandwidth: 200 kHz
Frame period: 4.615 mS
User/frame: 8.

GSM can also combine slots together and frames into multiframes and then superframes.

For TDMA it is important to note that there should only ever be one user active at one time and place (e.g. in a sector) in both directions. Hence for modelling purposes it could be sufficient to model just the one station rather than one for each of the time slots.

If the cell is lightly loaded, then it could be that some of the frames are inactive, and hence there will be a difference between the average power over the frame duration and peak power when a frame is being used. In Monte Carlo analysis it is easier to model this as a user being selected at random to be either active or inactive unless there is a mechanism to average interference at the victim receiver.

The minimum number of TDMA channels required will increase in steps as the number of users increase as in

$$N_c = roundup \left[\frac{N_{users}}{N_{users/frame}} \right] \tag{3.36}$$

The total bandwidth then increases as multiples of the TDMA channel bandwidth as the number of users increases.

3.5.5   Code Division Multiple Access

In code division multiple access (CDMA), a single channel is shared between different users, with isolation provided via use of codes, as shown in Figure 3.25.

Here multiple users are allocated to the same frequency channel and could operate in the same geographic area at the same time. Each of the users is isolated from each other by

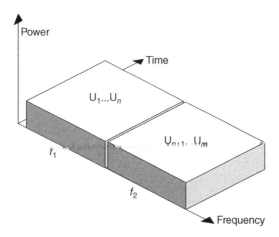

Figure 3.25   Code division multiple access

spreading the input data over multiple CDMA chip cycles using a unique code. At the receiver this code is used to integrate the inputs over multiple chip cycles. Signals from other users with different codes will almost cancel out, so that each one results in just a slight rise in the victim receiver's noise level. The higher rate of the chip cycles (compared to the data rate) results in a wider bandwidth, which is then shared between multiple users.

Example 3.8
Consider a simplified case with a 32-bit code and two users, and a single bit of data to be transmitted, in this case a '1'. The code comprises 32 bits {0, 1}, with a different set assigned to each of the two users. The code plus the data bit (here a '1') are transmitted and detected by both the wanted receiver RX-1 and the unwanted receiver RX-2. To decode the data stream, both receivers use their respective code, resulting in a set of 32 bits for each. These are then integrated so that the bits are added up over the 32 chips. For the wanted system the bits add up continuously, but if a different code is used, some will add and others subtract, resulting in a much lower resulting signal (though in most cases it won't be zero). This unwanted signal will result in an increase in the receiver noise, but there will be sufficient processing to ensure the wanted signal can be extracted, as in Figure 3.26.

The process of using codes to generate a wideband signal with higher chip rate is called spreading and can be a useful technique in its own right in addition to being used for multiple access.

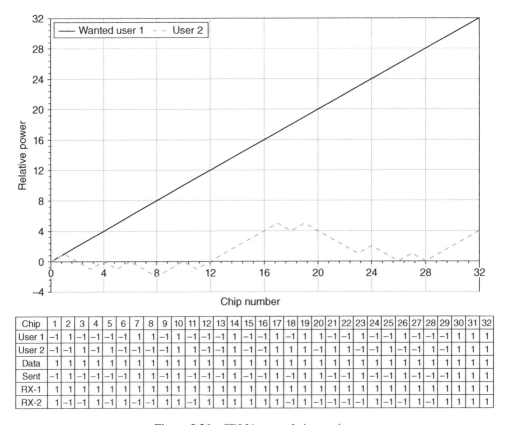

Chip	1	2	3	4	5	6	7	8	9	10	11	12	13	14	15	16	17	18	19	20	21	22	23	24	25	26	27	28	29	30	31	32
User 1	−1	1	−1	−1	−1	−1	1	1	−1	1	−1	−1	−1	1	−1	−1	1	−1	1	1	−1	−1	1	−1	−1	1	−1	−1	−1	1	1	1
User 2	−1	−1	1	−1	1	−1	−1	−1	−1	1	1	−1	1	1	1	−1	1	1	1	−1	1	1	−1	−1	1	−1	−1	1	−1	1	1	1
Data	1	1	1	1	1	1	1	1	1	1	1	1	1	1	1	1	1	1	1	1	1	1	1	1	1	1	1	1	1	1	1	1
Sent	−1	1	−1	−1	−1	−1	1	1	−1	1	−1	−1	−1	1	−1	−1	1	−1	1	1	−1	−1	1	−1	−1	1	−1	−1	−1	1	1	1
RX-1	1	1	1	1	1	1	1	1	1	1	1	1	1	1	1	1	1	1	1	1	1	1	1	1	1	1	1	1	1	1	1	1
RX-2	1	−1	−1	1	−1	1	−1	−1	1	1	−1	1	−1	1	−1	1	1	−1	1	−1	−1	−1	−1	1	−1	−1	1	−1	1	1	1	1

Figure 3.26   CDMA example integration

For example, the average power spectral density tends to be lower as the total power is spread over a wide bandwidth, which means it can be used as a way of reducing the interference caused. However that typically implies a reduction in capacity as a fully loaded CDMA system should have a similar power spectral density to a TDMA system with an identical number of users. CDMA systems can make planning easier, as frequencies can be reused in adjacent sectors with isolation provided via codes. Other advantages of CDMA include increased security, increased difficulty to detect and resilience against jamming.

The aggregation of the signal over the integration time gives a processing gain that can be calculated using

$$G_p = 10 \log_{10} \left(\frac{R_c}{R_d} \right) \tag{3.37}$$

where

R_c = chip rate
R_d = data rate.

This can be used to adjust the threshold when comparing against a wanted signal level, C.

Example 3.9
A wideband CDMA (WCDMA) data link operating at 144 kbps is being transmitted across a wideband carrier with chip rate = 3.84 Mcps using QPSK. The threshold for QPSK is $C/N = 13.5$ dB. The processing gain $G_p = 10 \log_{10}(3840/144) = 14.3$ dB. Therefore, theoretically, the link could operate with an actual $C/N = -0.8$ dB.

It is not unusual for signals that are widely spread to have negative thresholds, and it is typically the threshold that is adjusted for spreading rather than the C. This often includes other factors such as implementation losses and diversity gain.

Note that there are costs, such as in the processing overhead required at either end. There are also power implications from the need to continually transmit/receive a signal compared to a TDMA user that can switch off during parts of the time frame accessed by other users.

In addition the noise level increases with traffic levels, reducing the range of a cell, a process described as cell breathing. The noise rise on the uplink can be calculated using (Holma and Toskala, 2010).

$$N_r = 10 \log_{10} \left(\frac{1}{1-\eta} \right) \tag{3.38}$$

where

$$\eta = \frac{(e_b/n_0)}{(R_c/R_d)} N v (1+i) \tag{3.39}$$

Here all the calculations are in absolute with:

N = number of users
i = factor to take account of interference from other sectors (for three-sector macro cells, a suitable value could be $i = 0.65$)
v = traffic factor, which could be 1 for continuous data and 0.5 for voice traffic.

Table 3.7 Example voice and data service parameters from Rec. ITU-R M.1654

Service	Voice	Data
E_b/N_0 (dB)	5.0	1.5
i	0.55	0.55
W (Mcps)	3.84	3.84
Rate (Mbps)	0.0122	0.144
v	0.65	1.0

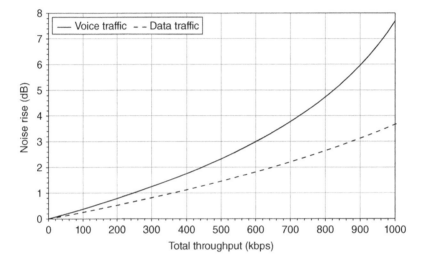

Figure 3.27 Increase in noise in WCDMA uplink

Example 3.10
An omni macro base station is providing services as per Table 3.7 from Rec. ITU-R M.1654 (ITU-R, 2003c) to the users in a rural cell. The noise rises on the uplink as the total throughput per service in the cells increases as per Figure 3.27. See Section 5.10.5 for use of the noise rise figure to derive a threshold for interference analysis.

The wideband CDMA (WCDMA) standard for 3G mobile services has attributes:

Channel bandwidth: 5 MHz
Chip rate: 3.84 Mcps
Frame length: 10 mS (38,400 chips).

While the number of users per channel can increase using different codes, these can be limited, and hence after a set number of users, it could be necessary to provide an additional channel. For example, with WCDMA there is a maximum of 512 codes per base station and frequency. In this case there are likely to be multiple handsets all operating on the same frequency at the same time and hence would need to be modelled as being simultaneously active.

Some CDMA systems spread their signal over such a wide bandwidth they can be classified as UWB as discussed in Section 3.4.6.

3.5.6 Orthogonal Frequency Division Multiple Access

In orthogonal frequency division multiple access (OFDMA), a single channel is split into several subcarriers. Time slots and subcarriers are considered resources that can be dynamically and flexibly allocated to users, with isolation provided via use of orthogonal frequencies (see Section 3.4.4), as shown in Figure 3.28.

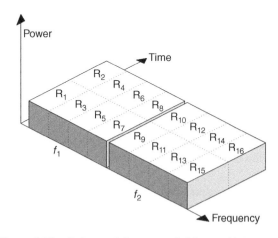

Figure 3.28   Orthogonal frequency division multiple access

The allocation of resources (time and subcarriers) will depend upon the number of users and the traffic required by each. This will vary significantly, with the characteristics of (say) voice being very different from high data rate services such as video.

There are in general two approaches to interference analysis involving OFDMA systems:

1. Model communications at the subcarrier levels, aggregating where necessary depending upon traffic requirements. In this case, each channel would only be used by a single user at a specific time
2. Model at the wider channel level, averaging the power and taking into account the fact that there could potentially be multiple users operating within the same OFDM channel.

The approach to use will depend upon the scenario involved, and a key factor would be the victim bandwidth. For narrowband receivers it could be more appropriate to model OFDMA via subcarriers, but for wideband receivers that would detect an entire (say) LTE 10 MHz channel and all the subcarriers in it, the second approach could be used.

3.6   Noise Temperature and Reference Points

If you were to switch on a receiver far off in space, so remote you couldn't detect anything created by humans, it would still pick up a very faint signal. This is the noise of the receiver and it is a limiting factor on the information that can be transmitted over a radiocommunication system.

All matter radiates energy according to its temperature. The remnants of the Big Bang give the cosmos a temperature of about 2.7 K, very close to absolute zero of 0 K. The microwave radiation corresponding to this temperature was detected in 1965 by Arno Penzias and Robert Wilson using the Holmdel Horn Antenna in New Jersey, United States, which was originally developed for NASA's Echo satellite communication system.

Noise can come from a wide variety of sources. Rec. ITU-R P.372 (ITU-R, 2013c) identifies the following:

- Radiation from lightning discharges (atmospheric noise due to lightning)
- Aggregated unintended radiation from electrical machinery, electrical and electronic equipment, power transmission lines or internal combustion engine ignition (man-made noise)
- Emissions from atmospheric gases and hydrometeors
- The ground or other obstructions within the antenna beam
- Radiation from celestial radio sources.

Many of these sources can be simplified into a generic model, though there are specific scenarios where changes in the background noise become significant, such as:

- Increase in noise of satellite systems with receive antennas that point towards the sun
- Increase in noise in satellite systems during a rainstorm where the wanted signal is absorbed and then reradiated as noise (see Section 4.4.2)
- Increase in noise for urban areas where there are a very large number of radio systems producing unintended radiation across the spectrum, in particular for bands below about 300 MHz.

This is separate from the noise rise due to intra-system interference that occurs within WCMDA networks due to other users, which is classified as interference with noise-like characteristics, rather than inherent noise.

Noise is often characterised as having a uniform (flat) power spectral density (often just called power density) and the noise per hertz n_0 relates to the temperature T via Boltzmann's constant, as in

$$n_0 = kT \qquad (3.40)$$

k is usually given in dB as -228.6 dBW/K/Hz, and so in dB this is

$$N_0 = 10\log_{10}T - 228.6 \qquad (3.41)$$

The noise N over a receiver bandwidth B in Hz is therefore (in absolute followed by dB)

$$n = kTB \qquad (3.42)$$

$$N = 10\log_{10}T - 228.6 + 10\log_{10}B \qquad (3.43)$$

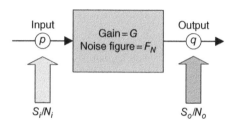

Figure 3.29   Receive component's impact on the S/N

The total system temperature will depend upon external factors (such as those listed previously) and also contributions from equipment. Consider a receiver component in Figure 3.29: it will impact the S/N ratio between input S_I/N_I and output S_O/N_O in two ways:

1. G = the gain of the component (positive for an amplifier, negative for feed loss)
2. F_N = the **noise figure** of the component, identifying how the noise is increased.

Note that the noise figure is defined here in decibels and the absolute equivalent, f_n, is called the noise factor, where the conversion is

$$F_N = 10\log_{10}f_n \tag{3.44}$$

As previously stated, lower case is in general used for values in absolute and upper case for decibel, apart from bandwidth and temperature in kelvin that are always upper case.

Example 3.11
At the input to a receiver, a carrier with bandwidth 10 MHz has $S_I = -124$ dBW/MHz and $N_I = -144$ dBW/MHz so that the $S_I/N_I = 20$ dB. The component amplifies the signal (and input noise) by 25 dB but also introduces 5 dB of additional noise, so that the output $S_O = -99$ dBW/MHz and $N_O = -114$ dBW/MHz with $S_O/N_O = 15$ dB. The two signals are shown in Figure 3.30.

The noise factor f_n is defined in absolute as the ratio of the S/N at the input to the S/N at the output, namely,

$$f_n = \frac{s_i/n_i}{s_o/n_o} \tag{3.45}$$

The receiver component will amplify, with gain g, the inputs, both signal and noise, and introduce additional noise n_a, so that

$$f_n = \frac{s_i/n_i}{gs_i/(n_a + gn_i)} \tag{3.46}$$

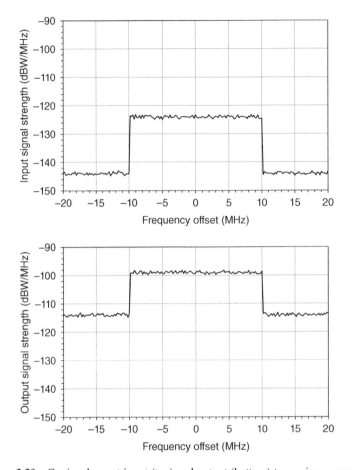

Figure 3.30 Carrier shape at input (top) and output (bottom) to receive component

$$f_n = \frac{n_a + gn_i}{gn_i} \tag{3.47}$$

For terrestrial systems, the input noise temperature will be similar to the reference temperature defined as

$$T_0 = 290 \, \text{K} \tag{3.48}$$

Using Equation 3.42 the input noise for terrestrial systems is therefore

$$n_i = kT_oB \tag{3.49}$$

Hence

$$f_n = \frac{n_a + gkT_oB}{gkT_oB} \tag{3.50}$$

$$f_n = 1 + \frac{n_a/gkB}{T_o} \tag{3.51}$$

This can be simplified using the effective temperature T_e defined as

$$T_e = \frac{n_a}{gkB} \tag{3.52}$$

So

$$T_e = T_o(f_n - 1) \tag{3.53}$$

Or

$$f_n = 1 + \frac{T_e}{T_0} \tag{3.54}$$

What is required is the total noise taking into account the input noise and noise of the receiver but measured at the input to the receiver (point p in Fig. 3.29). This will be the value that, when amplified by the receiver gain g, is the output noise n_o. This can be calculated as follows:

$$n = \frac{n_o}{g} \tag{3.55}$$

$$n = \frac{n_a + gn_i}{g} \tag{3.56}$$

$$n = \frac{T_e gkB + gT_o kB}{g} \tag{3.57}$$

$$n = (T_e + T_o)kB \tag{3.58}$$

And therefore (in absolute followed by dB)

$$n = f_n T_0 kB \tag{3.59}$$

$$N = F_N + 10\log_{10}T_0 - 228.6 + 10\log_{10}B \tag{3.60}$$

For other systems it might not be appropriate to assume that the input noise temperature was the reference temperature. In particular, satellite Earth stations pointing at space can have much lower temperatures, for example, 100 K or below. Radio astronomy receivers tend to use cooling to reduce the noise as much as possible, sometimes to temperatures only a few kelvin above absolute zero, and with antennas that point towards deep space, the results are very low total noise temperatures.

For these systems (i.e. satellite and other non-terrestrial services), it is more common to define the receiver noise via a total noise temperature, T, so that (in absolute followed by dB)

$$n = kTB \tag{3.61}$$

$$N = 10\log_{10}T - 228.6 + 10\log_{10}B \tag{3.62}$$

In many receiving systems there are multiple components that contribute to the total system noise, including the antenna, the line feed and the receiver, and so it is necessary to develop models of each of these to calculate the total noise.

The generic model of cascading elements can be used to calculate the total effective noise as in (Rappaport, 1996)

$$T = T_1 + \frac{T_2}{g_1} + \frac{T_3}{g_1 g_2} + \cdots \tag{3.63}$$

These can be converted into noise factors using Equation 3.54 so that

$$f = f_1 + \frac{f_2 - 1}{g_1} + \frac{f_3 - 1}{g_1 g_2} + \cdots \tag{3.64}$$

A generic model of noise at a receive station considers a number of components, in particular the antenna, a feed and a receiver, as shown in Figure 3.31. There are also two reference points:

- p: just after the antenna, before any feed losses
- q: after any feed losses, at the input to the receiver.

These have different noise temperatures and it is important to be clear which reference point is being used for calculations.

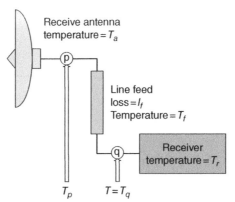

Figure 3.31   Components in receiver noise calculation

For passive components such as line feeds, the noise figure is equal to the device loss (Rappaport 1996). Hence the effective noise temperature T can be calculated from its loss l_f and temperature T_f using

$$T = \left(l_f - 1 \right) T_f \tag{3.65}$$

Hence using Equations 3.63 and 3.65, the temperature T_p can be calculated as

$$T_p = T_a + \left(l_f - 1 \right) T_f + \frac{T_r}{g_r} \tag{3.66}$$

As $g_r = 1/l_f$, we can calculate $T = T_q$ as

$$T = T_q = \frac{T_p}{l_f} \qquad (3.67)$$

Therefore (Bousquet and Maral, 1986)

$$T = \frac{T_a}{l_f} + T_f\left(1 - \frac{1}{l_f}\right) + T_r \qquad (3.68)$$

In practice during most interference analysis studies, it is not possible to identify and agree values for these four parameters (T_a, l_f, T_f, T_r). It is usually easier to agree on a single value, the total noise temperature, T. In some cases this can be used together with a value for the line feed loss.

If the line feed loss is so low it can be assumed to be zero in dB (i.e. 1 in absolute) and the antenna temperature is the same as room temperature, T_0, then the total noise temperature can be simplified to

$$T \cong T_0 + 0 + (f_r - 1)T_0 = f_r T_0 \qquad (3.69)$$

This is an alternative derivation of Equation 3.60.

A generic and transparent approach to receiver noise is to define the total temperature in Kelvin and (where necessary) any feeder loss in dB. If the noise figure is given, it should be assumed to be for the total noise rather than any component unless clearly specified and the conversion to noise power explicitly given.

Example 3.12

A receive satellite Earth station has antenna temperature $T_a = 50$K, feed temperature $T_f = 290$K, feed loss $L_f = 1$ dB and receiver temperature also $T_r = 290$K. From Equation 3.68 it will have a total noise temperature at the input to the receiver $T = 389.4$K. For a bandwidth of $B = 30$ MHz, the noise calculated using Equation 3.43 would therefore be -127.9 dBW.

Example 3.13

A mobile system using a bandwidth of $B = 5$ MHz has a noise figure of the receiver of $NF = 6$ dB, cable loss of $L_f = 3$ dB, antenna temperature of $T_a = 290$K and reference temperature of $T_0 = 300$K. What is the total noise just after the antenna?

In this case the input is the receiver noise factor, so first it must be converted into a temperature using Equation 3.53:

$$T_i - (f_i - 1)T_0 = (3.9811 - 1)300 = 894.3$$

We also note that a cable loss of 3 dB ~ 2 absolute and hence equivalent to a gain ~0.5, and so we can calculate the total temperature at point (p) using Equation 3.66:

$$T_p = 290 + (1.9953 - 1)300 + \frac{894.3}{0.5012} = 2373\text{K}$$

With a bandwidth of 5 MHz, the noise is then

$$N = 10\log_{10}2373.0 - 228.6 + 10\log_{10}5e6 = -127.9\,\text{dBW}$$

The total noise just after the antenna is therefore −127.9 dBW.

Section 3.4.5 noted that to achieve a specified *BER* for a given modulation method, it is necessary to have the required *S/N* at the receiver, and so the noise is calculated at point (q), that is, $T = T_q$. To be consistent it will therefore be necessary to modify Equation 3.16 to add an extra term to take account of the line feed loss as in

$$S = P_{tx} + G_{tx} - L_{fs} + G_{rx} - L_f \tag{3.70}$$

In general it is more useful to know the *S/N* at the input to the receiver (where the decoding is undertaken) so it can be compared against the thresholds for the various modulations, rather than just after the antenna, so unless specified the reference point for calculations will be point (q) so that $T = T_q$.

3.7 Antennas

3.7.1 Basic Concepts

The antenna is fundamental to radiocommunications and hence it is very important to model it as effectively as possible. There are a wide range of types of antenna, some of which are shown in Figure 3.32.

For example, antennas could be:

- Parabolic reflectors
- Flat plates containing arrays of elements
- Wire or cables
- Disguised as part of smartphone casing

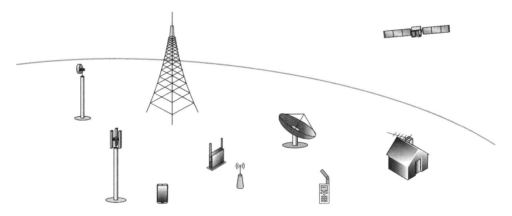

Figure 3.32 Range of systems with different types of antenna

- Bars on top of handset or router
- Rectangular boxes on the side of a base station.

There is a significant amount of electrical engineering that goes into antenna design that is outside the scope of this book, involving reflectors, waveguides and materials. What is important for interference analysis are the antenna's key parameters, namely:

1. **Peak gain**: what is the highest gain value in any direction? This is usually given in dBi. Note there are rare cases in which the peak gain means the gain at boresight and so there could be higher gains in other directions
2. **Gain pattern**: how does the gain vary in different directions? This can be defined as the gain relative to peak in dB. A related parameter is the half-power beamwidth, which is angle over which the gain is within 3 dB of peak
3. **Pointing method**, as described in Section 3.7.12.

This book will in general use these as critical inputs rather than describing how they are derived. The exception is the parabolic or dish antenna, for which it is possible to calculate key parameters from its diameter.

The gain in a specific direction is often defined using the peak gain and the relative gain in that direction via

$$G = G_{peak} + G_{rel} \tag{3.71}$$

The peak gain and the gain pattern of an antenna are characteristics of its far field, excluding the zone in its immediate vicinity, which is considered the near field. Here a receive antenna would have a significant impact on the transmit antenna via electromagnetic effects, while in the far field they can be considered independent.

The limit of the near field is the Fraunhofer distance (Kraus and Marhefka, 2003), which is

$$d = \frac{2D^2}{\lambda} \tag{3.72}$$

In this equation D is the largest element of the antenna and λ is the wavelength, using the same units for both, and additional constraints for the far field are that $d \gg D$ and $d \gg \lambda$. The zone can be quite large, particularly for high frequencies and large diameter antennas.

Example 3.14
The Fraunhofer distance for a Wi-Fi node and radio astronomy observatory are given in Table 3.8.

In general it is better to undertake interference analysis using the methods in this book outside the near field zone, as study within it relates instead to electromagnetic compatibility. However in practice the techniques in this book have been used for almost all distances on the grounds that the Fraunhofer distance is a limit rather than hard boundary and using the antenna gain is likely to be a reasonable assumption in most cases.

Table 3.8   Examples of Fraunhofer distances

Scenario	Wi-Fi node	Observatory
Frequency (GHz)	2.4	43
Antenna size (m)	0.08	10
Wavelength (m)	0.12	0.01
Fraunhofer distance (m)	0.10	28,686.51

As noted previously, the gain is the ratio of the output to input power in a specific direction, which in absolute is a multiple and hence in decibels an additive factor. The antenna does not provide amplification as such; rather it directs the energy (in the transmit case) or absorbs energy (in the receive case) in specific directions. In an ideal antenna all the power provided at its input would be radiated and the key measure of an antenna would be its **directivity**, which is defined as the ratio of the maximum power density to the average value over a sphere measured in the far field.

However in practice some power is lost, for example, due to ohmic losses resulting in heating. The gain, g (in absolute), includes these losses and is related to the directivity d by the antenna ohmic-loss factor k_O via

$$g = k_O d \tag{3.73}$$

The ohmic-loss factor k_O is one of a number of terms that contribute to the antenna efficiency: more information on these terms can be found in specialist antenna references.

The link calculation uses gain (rather than directivity) as it represents the actual and usable performance of an antenna. The total gain of an antenna in absolute when integrated in all directions should be the efficiency, as in Equation 3.74:

$$k_O = \frac{1}{4\pi} \int_{-\pi}^{\pi} \int_{-\pi/2}^{\pi/2} g(\theta, \phi) d\theta d\phi \tag{3.74}$$

In many cases k_O will be close to one, but it should never be greater. However many of the antenna gain patterns described here will integrate to greater than one, in some cases significantly greater. This is due to the need to balance out various different objectives in interference analysis, in particular:

- The need to use parameters of specific equipment types versus the need to use a generic set of parameters that would cover a range of equipment types
- The need to protect the victim by making conservative assumptions versus the need to use parameters that are most likely to occur.

The issues relating to balancing these forces are a theme that will appear in many places in this book.

Possible sources of antenna gain patterns include:

- The service provider should have an antenna model as part of their system design
- The antenna manufacturer should be able to provide the gain pattern as part of their technical sales material
- There are a large number of reference gain patterns in ITU-R and other standardisation bodies' documentation (including CEPT, ETSI and 3GPP). These can be defined by a set of equations or tables and are often parameterised (e.g. by peak gain or the ratio of the antenna diameter to wavelength)
- Basic engineering patterns such as Bessel can also be used but are typically only a last resort as the other approaches are generally preferable due to there being a document trail relevant to that sharing scenario.

3.7.2 Beams and Beamwidths

A key attribute of an antenna is the width of its beam. This is often parameterised via the half-power beamwidth, which is the range of angles within which power of a transmit antenna would be half or more of the peak power (technically the EIRP, but as this ratio is not dependent upon the transmit power, this is really an attribute of the antenna).

Hence the **half-power beamwidth** is defined as the angle over which the gain of the antenna is within 3 dB of the peak gain. There is a trade-off between peak gain and beamwidth, in that wider beams must have lower peak gain and narrow beams have higher peak gains as in

$$G_{peak} = 10 \log_{10} \left[\eta \left(\frac{70\pi}{\theta_{3dB}} \right)^2 \right] \tag{3.75}$$

In this equation G_{peak} is in dBi and θ_{3dB} is in degrees and η is the antenna beam efficiency, a dimensionless number in the range $(0, 1)$ where 1 would be a perfect antenna and zero a perfectly useless antenna. Typically η is in the range $(0.55, 0.75)$ though it can be as high as 0.8. The constant of value 70 relates to the antenna beam versus aperture efficiency as discussed in Section 3.7.5.

For a given η it can be seen that the peak gain and half-power beamwidth are directly connected, as plotted in Figure 3.33.

Note that the half-power beamwidth is the full angle across and it sometimes is useful to consider half the half-power beamwidth, as shown in Figure 3.34.

There can be different half-power beamwidths in different directions, depending upon the symmetries as described in the following section. The figure also shows three of the common components of a gain pattern, namely, the main beam, the side-lobes and the far off-axis range.

3.7.3 Common Gain Pattern Types

To model the wide range of different antenna types, it is often useful to determine whether there are symmetries in the pattern and whether the analysis requires a mask or actual pattern.

The most common symmetries are:

- The same gain in all directions, namely, the isotropic antenna as described in Section 3.7.4
- Gain that is symmetric around an axis, such as the gain pattern of a parabolic antenna around its boresight, as described in Section 3.7.5

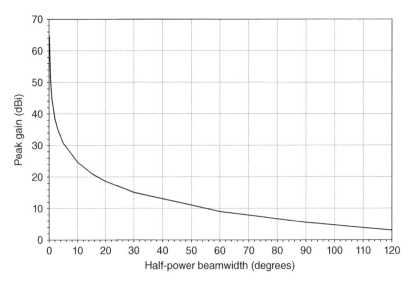

Figure 3.33 Relationship between half-power beamwidth and peak gain for $\eta = 0.6$

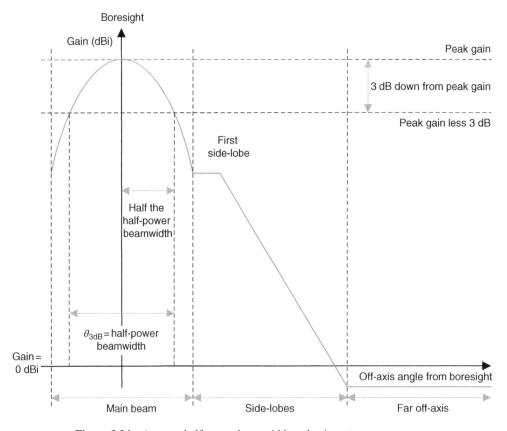

Figure 3.34 Antenna half power beamwidth and gain pattern components

- There can be variations on the boresight symmetric case where the pattern is wider in one direction around the axis of the antenna than others so that its cross section is elliptical. These are described in Section 3.7.6
- Gain varies by azimuth, modelled as being the same for all elevations at a given azimuth as described in Section 3.7.8
- Gain varies by elevation, modelled as being the same for all azimuths at a given elevation. These are sometimes called omnidirectional antennas, as described in Section 3.7.9
- Gain given in azimuth and elevation slices, so that it can be defined along these two axis and the gain in all three dimensions derived, as described in Section 3.7.10
- Gain pattern defined via a three-dimensional table that gives the gain for an array of azimuth and elevation directions, as described in Section 3.7.11.

Some services and geometries have specific methods to define the gain pattern of antennas. An example of this would be GSO satellite systems, which often have their antenna gain patterns defined via contours of (latitude, longitude) as described in Section 7.5.

A factor connected to the gain pattern is the signal's polarization. As this topic is strongly connected to the interference calculation, it is discussed in Section 5.4.

Antenna gain patterns can also be one of the following:

- Theoretical gain pattern, such as a perfect parabolic reflector with Bessel function-based pattern
- Measured gain pattern, which is close to theoretical but will vary slightly due to manufacturing imperfections and the impact of the feeder
- Mask gain pattern, or radiation pattern envelope (RPE), which is the envelope that is unlikely to be exceeded by most actual antennas. This is useful for interference analysis as it ensures the actual gain won't be less than the value used but can lead to a conservative approach to spectrum sharing. They could also be used for calculation of the wanted signal, but can lead to an overly optimistic prediction.

Examples of these three are given in the section on the parabolic dish antenna.
Some commonly used gain patterns for a selection of system types are given in Table 3.9.

Table 3.9 Commonly used gain patterns

System type	Gain pattern
Mobile or handset	Isotropic
Mobile base station	Rec. ITU-R F.1336 or 3GPP TR 36.814
Point-to-point fixed service	Rec. ITU-R F.699 or F.1245
Satellite Earth station	Rec. ITU-R S.465 or S.580
Satellite beam	Rec. ITU-R S.672 or GSO shaped beam
Private mobile radio base station	HCM code
Broadcast fixed receiver	Rec. ITU-R BT.419

3.7.4 Isotropic Gain Pattern

A perfect isotropic gain pattern, which is the same in all directions, should by definition have a gain equal to 0 dBi. However there are times when an interference analysis study can legitimately argue to use other values as long as there is justification and transparency in its application. This can include:

- The antenna itself is close to isotropic but the signal always is attenuated due to an obstruction nearby. For example, mobile applications could include losses due to head obstruction and hence use an isotropic gain = −6 dBi
- The antenna has a low gain (say, around 3 dB) but it would be difficult to identify the direction of the peak gain and it couldn't be controlled in practice. Hence it could be acceptable to use an isotropic gain equal to the peak gain with an explanation as to why that assumption was made
- The antenna could have a large peak gain and highly variable gain pattern (e.g. a parabolic dish), but the analysis is only interested in those moments when the interferer is pointing directly at the victim (or vice versa). For example, a radar station rotates in azimuth but the metric is I/N in any direction so the critical geometry is when it is pointing at the interfering station. Hence a simplification to the model might be to use an isotropic antenna equal to the peak gain.

3.7.5 Parabolic Dish Antennas

One of the classic antenna types is the parabolic dish, which has applications that include satellite Earth stations, radar antennas, point-to-point fixed links and radio astronomy observatories. As well as being widely used, it can illustrate some of the common concepts of antenna gain patterns.

The peak gain of a parabolic antenna can be derived from its area and the ratio of that to the effective area of an isotropic antenna. The effective area of a circular dish antenna with diameter D and efficiency η is given by

$$a_e = \eta.\pi \left(\frac{D}{2}\right)^2 \tag{3.76}$$

The gain is the ratio of this area to the effective area of an isotropic antenna given in Equation 3.6; hence

$$g = \eta \frac{\pi D^2/4}{\lambda^2/4\pi} = \eta \left(\frac{\pi D}{\lambda}\right)^2 \tag{3.77}$$

This is more often given in decibels rather than absolute, so

$$G_{peak} = 10\log_{10}\left[\eta\left(\frac{\pi D}{\lambda}\right)^2\right] \tag{3.78}$$

Note how the peak gain increases with dish diameter and frequency and how in consequence the beamwidth will decrease as described in Section 3.7.2.

For systems such as fixed service links and satellite Earth stations that use parabolic dish antenna, a key design decision is the dish diameter. This can involve a trade-off between:

- Larger antennas have higher gains and smaller beamwidths, which increases the wanted signal and usually reduces the interfering signal (in both directions) and in general results in an increase in spectrum efficiency. For radar systems, these two factors improve both range and accuracy
- Larger antennas cost more.

Example 3.15

For a parabolic dish antenna with an efficiency of 0.6 and a diameter of 2.4 m, the peak gain at a frequency of 3.6 GHz would be 36.9 dBi. Figure 3.35 gives plots of how the antenna peak gain varies as the frequency or dish diameter is changed.

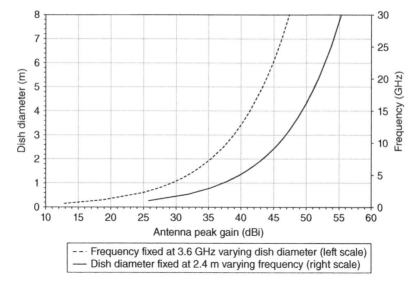

Figure 3.35   Plot of Antenna peak gain vs. frequency or dish diameter

A perfect parabolic dish antenna can be considered equivalent to a circular aperture in an infinite metal plate. Such an aperture would have a gain pattern described by a Bessel function:

$$G_{rel}(\theta) = 20\log_{10}\left[\left(\frac{2j_1(x)}{x}\right)^2\right] \qquad (3.79)$$

where

$$x = \frac{\pi D}{\lambda}\sin\theta \qquad (3.80)$$

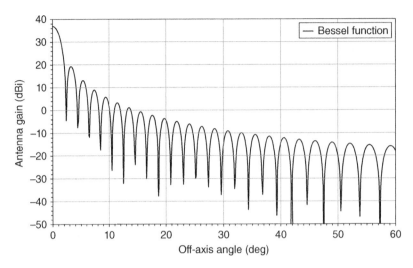

Figure 3.36 Theoretical Bessel gain for 2.4 m diameter dish at 3.6 GHz

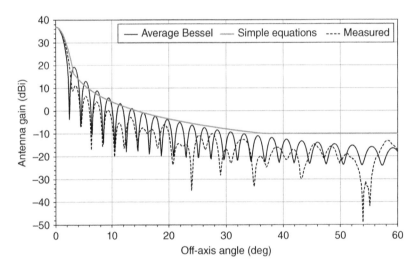

Figure 3.37 Comparison of averaged Bessel against simplified standardisation pattern and measured data

The Bessel function gain pattern is shown in Figure 3.36.

This pattern is unlikely to be realised in practice. For example, the troughs or nulls, which in the figure have a minimum of around −50 dB, in the equations actually go down to minus infinity but have been rounded by the charting software. These nulls could in theory provide significant mitigation benefits if the victim can ensure that its gain pattern has a null towards the interferer.

The location of the nulls is determined by the wavelength, but in practice the radio system will use a range of frequencies and hence there will be an averaging effect that reduces the troughs (see Fig. 3.37). There are also likely to be imperfections and intrusions (such as feeds) that mean the actual gain pattern will not perfectly match the theoretical pattern in Figure 3.36 and will vary depending upon manufacturer and design.

The antenna manufacturer will have to balance the beam efficiency against the aperture efficiency, which influences the speed in which the main lobe rolls off and the level of the first side-lobe. In particular:

- High aperture efficiency, with fast main beam roll-off but higher first side-lobe
- High beam efficiency, with slower main beam roll-off but lower first side-lobe.

This will have a direct impact on the half-power beamwidth. The Bessel function previously has a half-power beamwidth calculated using (Kraus and Marhefka, 2003)

$$\theta_{3dB} = \frac{58}{D/\lambda} \tag{3.81}$$

However, this is equivalent to a high aperture efficiency and a more balanced approach is used in most applications. For example, for GSO services it is the first side-lobe that influences most adjacent satellite interference and so must be controlled. Therefore, according to Bousquet and Maral (1986), the beamwidth should be calculated using

$$\theta_{3dB} = \frac{70}{D/\lambda} \tag{3.82}$$

Generally, the constant used is in the range between these two values.

It is common practice not to use these theoretical patterns but rather an average gain pattern that smooths over the peaks and troughs. Standardisation bodies such as the ITU-R and ETSI generate patterns for two important purposes:

1. Interference studies
2. Type approval.

It should be noted that these can have different priorities: for example, a manufacturer will want the type approval process to create a pattern that it is confident its equipment will meet. However an interference study could want to trade-off likelihood of interference against complexity of antenna design.

A typical standardised gain pattern is described via a set of equations similar to these for a parabolic antenna where θ is the **off-axis angle** from the boresight line:

$$G(\theta) = G_{peak} - 12 \left(\frac{\theta}{\theta_{3dB}} \right)^2 \qquad \theta < 90 \frac{\lambda}{D} \tag{3.83}$$

$$G(\theta) = 29 - 25 \log(\theta) \qquad 90 \frac{\lambda}{D} \leq \theta < 36.4° \tag{3.84}$$

$$G(\theta) = -10 \qquad 36.4° \leq \theta \tag{3.85}$$

The three segments, as shown in Figure 3.34, are:

1. Parabolic main beam which is dependent upon peak gain and beamwidth
2. Side-lobe, which in this case is not dependent upon either the peak gain or beamwidth
3. Constant far off-axis gain.

These elements are often used in standardised gain patterns defined in (say) ITU-R Recommendations.

This pattern is plotted against a Bessel equation gain pattern averaged over a 36 MHz carrier in Figure 3.37, together with the measured gain pattern of a real antenna. Note how the Bessel function has tighter main beam but higher first sidelobe, but the standardisation pattern has a broader main beam (using the standard parabolic roll-off) but lower first sidelobe due to the differences in aperture versus beam efficiency.

The measured data in this case was provided by General Dynamics SATCOM Technologies for one of their 2.4 m C-band antennas measured at 3.625 GHz.

Note how:

- The main lobe is between the theoretical Bessel function and the standardised pattern, showing the balance in the antenna design between aperture and beam efficiency
- The measured data in this case was always below the standardised pattern: if the standardised pattern was used for type approval, the antenna would meet this requirement. Note that some standardised patterns permits the gain to be exceeded for a small percentage of the sidelobe peaks
- If the standardised pattern was being used for interference analysis, then the actual interference using this antenna could be guaranteed to be less than predicted for all off-axis angles. Note that an envelope gain pattern that defines the gain level not to be exceeded could be pessimistic when calculating interference but optimistic for the wanted signal
- The peaks and troughs of the measured data closely align with the theoretical Bessel function near the boresight but less so for the far off-axis region.

Standards for antenna-type approval can define the RPE which the antenna's gain pattern must meet. In many cases it can be advantageous to use tables with gain in dB or dBi against the corresponding angle(s). The gain G at an off-axis angle θ between data points (G_i, θ_i) can be calculated using linear interpolation using

$$G(\theta) = G_1(1 - \lambda_\theta) + G_2\lambda_\theta \tag{3.86}$$

where (θ_1, θ_2) are the off-axis angles from the table that bracket the target off-axis angle θ, (G_1, G_2) the corresponding gains and where

$$\lambda_\theta = \frac{(\theta - \theta_1)}{(\theta_2 - \theta_1)} \tag{3.87}$$

3.7.6   Elliptical Patterns

The standard parabolic dish antenna has a gain pattern that is symmetric around its boresight. If a slice were taken perpendicular to the boresight line then the gain pattern in this cross-section

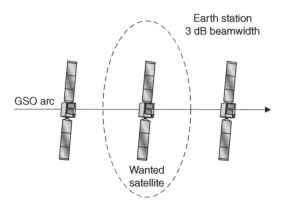

Figure 3.38 GSO ES using elliptical cross-section gain pattern

would be circular. However there are geometries where it can be useful to stretch or compress these circles to make an elliptical pattern.

Take the example of GSO satellite networks: all the satellites are located in a circle around the Earth called the geostationary arc. In the direction of this arc it is critical to manage interference, in particular to adjacent satellites. However perpendicular to the GSO arc there are no satellites and so there is no need to constrain the antenna gain. Therefore one approach to antenna design is for the cross-section of the main beam to be elliptical in shape as in Figure 3.38.

The gain pattern could have the same shape in all directions from the main lobe but be stretched (perpendicular to the GSO arc) or compressed (along the GSO arc). Rather than having a single half-power beamwidth there would be two, one for each of the directions. Equation 3.75 becomes:

$$G_{peak} = 10\log_{10}\left[\eta \frac{(70\pi)^2}{\theta_{3dB,a}\theta_{3dB,b}}\right] \tag{3.88}$$

Here $\theta_{3dB,a}$ and $\theta_{3B,b}$ are the half-power beamwidth in the ellipse major and minor axes respectively as in Figure 3.39. This also shows a critical parameter of elliptical beams, namely, the tilt angle, which gives the orientation of the major axis, typically relative to horizontal.

The tilt, together with the beam centre's azimuth and elevation, complete the three angles necessary to define the pointing direction and orientation of a beam with respect to the containing station.

For some radar stations, it is preferable that the beam is narrow in the horizontal direction, in order resolve the target's azimuth, and acceptable for it to be wider in the vertical direction, and this can be achieved using an elliptical gain pattern with a tilt angle = 90°.

For a mobile base station the opposite is true. In the vertical direction its beam can be narrow as most users will be within 5–10° of its antenna's horizontal plane. However in the horizontal direction it will want to cover a wide range of azimuths, with sector sizes between 60° and 120°, as shown in Figure 3.40.

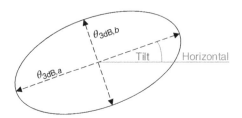

Figure 3.39   Orientation of elliptical beam

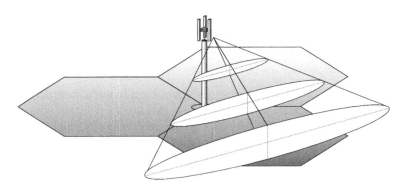

Figure 3.40   Mobile base station with elliptical beam wider horizontally than vertically

Equation 3.82 showed how the half power beamwidth is inversely related to the dish size, so that the larger the dish the smaller the beamwidth. Hence a beam such as that shown in Figure 3.40 that is wider horizontally than vertically will be created by an antenna that is narrower horizontally than it is vertically – such as the vertical box shown in the figure. The converse will be true of the radar example, which would have an antenna wider horizontally than it is vertically.

Mobile base stations often tilt their beams down from the horizontal as shown in Figure 3.41 with elevation angle given by

$$Elevation = -Downtilt \tag{3.89}$$

This can be done in one of two ways:

1. Mechanical downtilt, where the antenna is physically pointed at an angle to the horizontal plane
2. Electronic downtilt, where beamforming methods are used to point the beam downwards.

These produce similar patterns in the forward direction, but differ in the reverse direction. For mechanical downtilt the reverse direction is at 180° to the boresight, while for electronic downtilt it also is tilted, as in Figure 3.42.

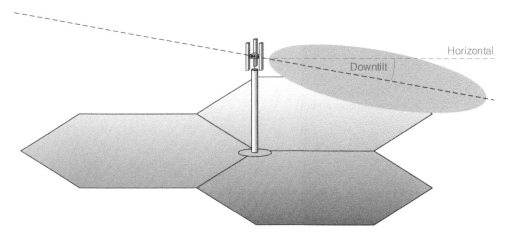

Figure 3.41   Downtilt of mobile phone base station

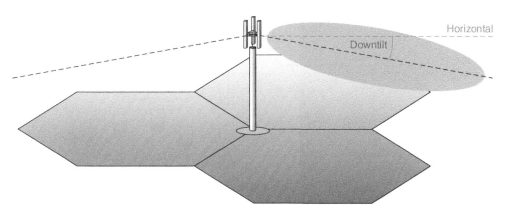

Figure 3.42   Electronic downtilt impact on reverse direction

3.7.7   Phased Array Antennas

An adaptive **phased array antenna**, sometimes called a digital beamforming antenna, is built from multiple elements, each of which is an individual antenna such as a dipole, waveguide or microstrip. Each element can transmit or receive and the signals to/from each are phased to create an overall directional gain pattern that can be steered electronically. The principle is shown in Figure 3.43, where there is an offset in the phase of the signal transmitted by each element which creates a uniform wave front. By varying the phase difference between elements, the beam can be pointed in any direction within the scan angle. In this case the beam has been steered an angle $\theta = 14°$ from the line perpendicular to the array.

Use of phased arrays has a number of advantages, including that there are no moving parts, beams can be repointed near instantaneously, multiple beams can be generated and there is graceful degradation to the failure of any element. However there can be cost and complexity implications.

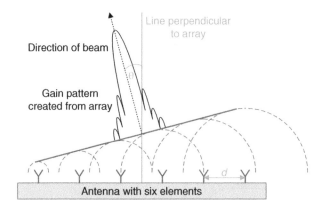

Figure 3.43   Phased array transmit antenna

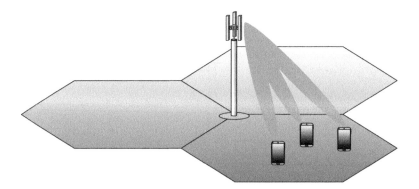

Figure 3.44   Base station antenna using beamforming

There are a wide range of applications that use phased arrays including radar, radio astronomy and remote sensing satellites. Phased arrays could also be used by mobile base stations to improve spectrum efficiency by creating narrow beams that point directly at users, as in Figure 3.44. **Adaptive antennas** can adjust the array phasing in response to the electromagnetic environment to (say) create a null in the direction of an interfering signal.

3.7.8   Azimuth Dependent Antennas

For terrestrial analysis the most important geometry is usually horizontal, and so the most important part of the antenna gain pattern is the slice which defines the gain in each azimuth direction, i.e. Gain(Azimuth). The variation of the gain in the elevation direction is not defined, so by default must be assumed to be the same for all elevations at a particular azimuth.

One of the simplest antennas is the dipole, which has peak gain of 0 dBd = 2.15 dBi and pattern that looks a bit like the cross-section of a doughnut as in Figure 3.45.

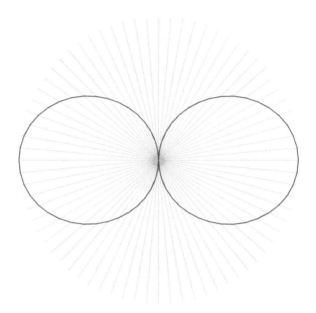

Figure 3.45   Dipole gain pattern

The far field electric pattern for a half-wavelength dipole is given by (Kraus and Marhefka, 2003):

$$e_v = \frac{\cos\left(\frac{\pi}{2}\cos\theta\right)}{\sin\theta} \tag{3.90}$$

Note that to convert from electric field to gain in dB it is necessary to use:

$$G = 20\log_{10}(e_v) \tag{3.91}$$

If this was the gain pattern of an antenna in the azimuth plane then there would be two lobes
(in the figure east and west) and a null (north and south).

There are a wide range of different antennas that could be used for these types of terrestrial
applications and they can be defined either via tables or sets of equations such as the dipole
pattern previously. One particularly useful set of patterns is defined in the Harmonised Calculation Methodology (HCM) (HCM Administrations, 2013) which gives a set of antenna patterns via a format that uses seven alphanumeric characters:

nnnXXmm

where:

- *nnn* is the first numeric parameter, often the beamwidth of the pattern
- *XX* is the code which describes the shape of the pattern
- *mm* is a second numeric parameter, often the far sidelobe level in dB.

Note that an isotropic antenna in HCM format would be *000ND00*. Other patterns are specified via parameterised equations: for example, the CA pattern is given by

$$\varsigma = \sqrt{\frac{(1-a^2)\cos(2\varphi) + \sqrt{(1-a^2)^2 \cos^2(2\varphi) + 4a^2}}{2}} \qquad (3.92)$$

where

$$0 \le a = \frac{nnn}{100} \le 1$$
$$180° \le \vartheta \le 180°$$

Then the gain can be calculated using

$$G_{rel} = 20\log_{10}(\varsigma) \qquad (3.93)$$

Note use of $20\log_{10}$ rather than the standard decibel $10\log_{10}$. There could also be a limit in terms of the front to back ratio (the ratio, often in dB, of the peak gain at boresight to the gain in the opposite direction).

Example 3.16
Two examples of HCM format gain patterns are given in Figure 3.46.

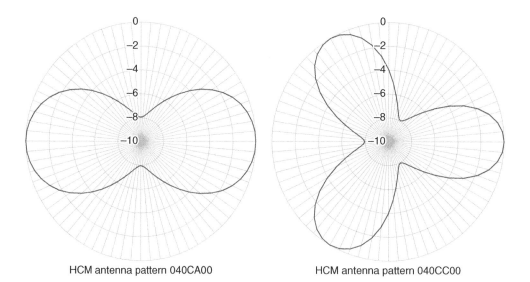

HCM antenna pattern 040CA00   HCM antenna pattern 040CC00

Figure 3.46   Example HCM antenna patterns

3.7.9   Elevation Dependent Antennas

The dipole pattern in Section 3.7.8 could also be rotated so that it gives the pattern in the elevation direction, being the same at all azimuths. This would give a pattern like a horizontal doughnut as per Figure 3.47.

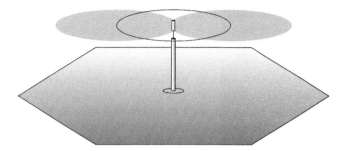

Figure 3.47   Vertically aligned dipole antenna

This is sometimes called omnidirectional in that it gives the same value in all directions though varies in elevation angle. A similar set of HCM codes can be used as for the azimuth pattern in the previous section.

The variation of an antenna's gain pattern in the elevation direction is of particular importance to space to Earth and Earth to space interference paths.

3.7.10   Azimuth and Elevation Slices

In some cases the gain pattern of an antenna is given in three dimensions using two slices:

- The azimuth gain pattern along the elevation = 0 plane, usually with $-180 \le Az \le +180$
- The elevation gain pattern along the azimuth = 0 plane, either in the form $-90 \le El \le +90$ or $-180 \le El \le +180$.

The slices, whether defined via tables or patterns such as the HCM codes, can be combined in a number of ways. The simplest is to just add the values for the gain in the azimuth and elevation directions, as in

$$G(Az, El) = G(Az) + G(El) \tag{3.94}$$

For example, a gain pattern that can be used for base station sectors is given in 3GPP TR 36.814 (3GPP, 2010) and a variant in IEEE 802.16 m Evaluation Methodology Document (IEEE, 2009b) as

$$G(Az) = -min\left[12\left(\frac{Az}{\theta_{3dB,Az}}\right)^2, \ A_m\right] \tag{3.95}$$

$$G(El) = -min\left[12\left(\frac{El - El_{downtilt}}{\theta_{3dB,El}}\right)^2, \ SLA_V\right] \tag{3.96}$$

$$G(Az, El) = -min\{-[G(Az) + G(El)], \ A_m\} \tag{3.97}$$

Suggested values of the constants are $\theta_{3dB,Az} = 70°$, $\theta_{3dB,El} = 10°$, $SLA_V = 20$ dB and $A_m = 25$ dB. The field $El_{downtilt}$ is in the range 6°–11° depending upon the scenario involved.

For some gain tables, simply adding can introduce problems, in particular:

- The gain very close to $El = \pm 90°$ will vary depending upon azimuth, but the reality it will just be a single value, the $G(El = \pm 90°)$
- The far off-axis floor can be significantly overestimated unless a floor is defined (as in Eq. 3.97). For example, if the far off-axis floor is −20 dBi for each table, it will be added twice and give a value = −40 dBi.

These two issues can be resolved by use of a smoothing function as in

$$G(Az, El) = G(Az)(1 - \lambda_{El}) + G(El)\lambda_{El} \tag{3.98}$$

where

$$\lambda_{El} = \frac{|El|}{90} \tag{3.99}$$

While the gain pattern is almost always given in the range [−180, +180] in azimuth, in elevation it can be given in the reduced form [−90, +90]. If the elevation gain is measured it is sometimes given in a more extended range, namely, [−180, +180], where values beyond ±90° are the reverse hemisphere. These require modification to Equation 3.99 to use in the reverse hemisphere:

$$\lambda'_{El} = \frac{180 - |El|}{90} \tag{3.100}$$

If the data provided only covers part of the antenna's field of view then the table could be extended by either using the last value or interpolating with the last two values, though neither approach is ideal.

3.7.11   3D Gain Tables

If there is no symmetry then it is necessary to consider other methods such as use of measured data, which could be defined via an array of gain values for specified (azimuth, elevation) angles. Usually these are defined via a grid, with the following specified:

- Minimum azimuth (degrees)
- Azimuth spacing (degrees)
- Number of azimuth data points
- Minimum elevation (degrees)
- Elevation spacing (degrees)
- Number of elevation data points
- Array of Gain(Number azimuth points, Number elevation points) in dBi.

Between these data points it is usual to calculate the gain using bilinear interpolation:

$$G(\theta) = G_{11}(1 - \lambda_{az})(1 - \lambda_{el}) + G_{12}\lambda_{az}(1 - \lambda_{el}) + G_{21}(1 - \lambda_{az})\lambda_{el} + G_{22}\lambda_{az}\lambda_{el} \tag{3.101}$$

where $(G_{11}, G_{12}, G_{21}, G_{22})$, (Az_1, Az_2) and (El_1, El_2) are the gains, azimuths and elevations from the tables that bracket the target (Az, El) and:

$$\lambda_{Az} = \frac{(Az - Az_1)}{(Az_2 - Az_1)} \tag{3.102}$$

$$\lambda_{El} = \frac{(El - El_1)}{(El_2 - El_1)} \tag{3.103}$$

Outside the defined range in azimuth and elevation the last relevant value can be used, though it is better in practice to define the pattern until the gain reaches a low far off-axis floor value.

3.7.12   Antenna Pointing Methods

The main approaches to defining the pointing of an antenna are:

- Fixed pointing: for example the antenna (azimuth, elevation) might be defined
- Point at another station: a fixed station might be installed so that its boresight with peak gain is targeted at the other end of the link, or a satellite Earth station would be pointed at its target GSO satellite
- Point at the common volume: for point-to-point fixed links using troposcatter, each antenna must be pointed at the location in the sky where the signal would be reflected
- Rotating: an example of this would be a radar station where the antenna rotates 360° of azimuth with defined rotation rate
- Swept broom: sensors on remote sensing satellites can sweep backwards and forwards as the satellite moves forward.

The pointing methods can become quite complicated: for example, consider an antenna at the control station of a non-GSO satellite network. It will point at one of the satellites in that constellation with the one selected depending upon factors such as their positions and which of them are being tracked by other antennas at the control station.

The result in each case should be the same, namely, specific (azimuth, elevation) angles for the time step or simulation sample under consideration.

3.8   Geometry and Dynamics

3.8.1   Geometric Frameworks

Interference analysis is primarily about the power of signals, whether wanted or interfering, and three of the terms of the power Equation 3.16 require information about location and orientation of transmitters and receivers, namely:

1. $G_{tx}(rx)$: gain of the transmit antenna in the direction of the receiver
2. L_{fs}: free space path loss, or more generally, L_p the propagation loss between transmit and receive antennas
3. $G_{rx}(tx)$: gain of the receive antenna in the direction of the transmitter.

To calculate these terms it is necessary to have an underlying geometric framework that defines locations in such a way that distances and angles can be calculated. In addition, for dynamic simulations there could be moving stations in which, for example, satellites follow their orbit and aircraft fly great circle paths.

There are a number of different levels of detail that could be used, and it is worth making sure that the chosen geometric framework is appropriate. For example four common frameworks are given in Table 3.10 together with the type of interference analysis each is most applicable for. In general most interference analysis studies do not require the ellipsoidal Earth model. This is because they are concerned with questions such as:

- What would be the highest level of interference?
- How often would interference occur?
- What would be the length of interference?

These questions can be answered to sufficient accuracy in most cases by a spherical Earth model. The ellipsoidal Earth model could be useful for questions such as:

- At what time would we have to switch off a specific satellite to avoid it causing or suffering harmful interference?

This type of question is more related to mission planning or operations and hence outside the scope of this book.

The boundary for short-range studies is typically between 100 and 1000 m: beyond that it is necessary to consider the curvature of the Earth that will impact distances and angles (and hence antenna gains).

The majority of the examples in this book are derived using an Earth centred inertial (ECI) coordinate frame with positions converted into vectors.

Table 3.10   Geometric frameworks

Geometric framework	Locations defined via	Useful for
Flat Earth	Vector (x, y, h), where position (x, y) is given in a plane and h is height, all typically metres	Short-range studies, for example, within a building or vehicle such as aircraft
Spherical Earth – spherical coordinates	Position on a sphere given in the form of (latitude, longitude, height) in units such as degrees and metres	Terrestrial sharing scenarios involving distances greater than short range
Spherical Earth – vector coordinates	Position in an Earth centred inertial coordinate frame (x, y, z), typically in km assuming Earth spherical	General terrestrial or space sharing scenarios
Ellipsoidal Earth – vector coordinates	Position in an Earth centred inertial coordinate frame (x, y, z), typically in km assuming Earth ellipsoidal	Satellite mission planning and operations

3.8.2 Flat Earth Vectors

In a flat Earth model the position of stations are given by a vector as shown in Figure 3.48.

Here locations are defined with respect to an origin point, with a horizontal (x, y) plane and then height above the plane. As these distances are typically small, a suitable unit might be metres. The directions of the x and y axes could be defined towards the east and north respectively, or relative to a structure such as a building. The vector r_{tx} of the transmit station could therefore be defined as

$$r_{tx} = \begin{pmatrix} x_{tx} \\ y_{tx} \\ h_{tx} \end{pmatrix} \tag{3.104}$$

With a similar notation for the position vector of the receive station, the distance between them can be calculated using

$$d = |r_{tx} - r_{rx}| = \sqrt{(x_{tx} - x_{rx})^2 + (y_{tx} - y_{rx})^2 + (h_{tx} - h_{rx})^2} \tag{3.105}$$

These stations can be in motion: for example the position of each of the stations defined via a starting point, heading and speed in m/s $= v_m$, with the heading defined (say) by angle θ to the y-axis. Then the position vector at time t_s in seconds would be

$$r(t) = r_0 + v_m t_s \begin{pmatrix} \cos\theta \\ \sin\theta \\ 0 \end{pmatrix} \tag{3.106}$$

A variation on the flat Earth model is to use **wrap around geometry**. This is helpful to avoid edge effects when analysing aggregate interference. Consider the scenario in Figure 3.49 which shows a grid of base stations and two handsets:

- One is in the centre of the grid and would receive the full aggregation effect
- The other is at the edge of the grid, and so would appear to receive less interference.

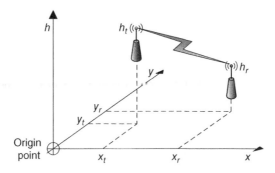

Figure 3.48 Flat Earth vector framework

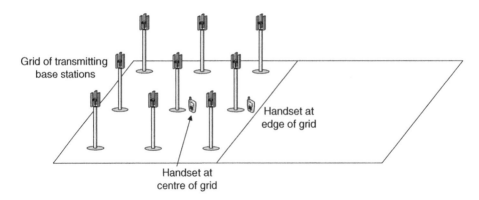

Figure 3.49   Aggregate scenario without wrap around geometry

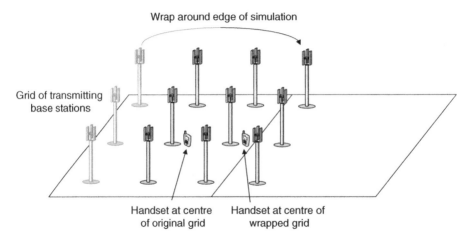

Figure 3.50   Aggregate scenario with wrap around geometry

It might be that this is acceptable as both cases can occur in reality, for example with one handset in the centre of a city while the other is at its edge where buildings give way to fields. However to model high density deployments in large cities in detail would require extremely large simulations: a simpler approach is to wrap around the geometry so that the left hand side connects to the right hand side as in Figure 3.50. In addition the top side would be connected to the bottom side, though this is not shown.

With wrap around geometry the stations are assumed to be located at one of n-positions, offset from their actual positions by integer multiples of the size of the wrap around zone. For example, if the zone had width and height of d_w then the position of a transmitter at (x, y, h) could be defined by

$$\boldsymbol{r}_{tx} = \begin{pmatrix} x_{tx} + id_w \\ y_{tx} + jd_w \\ h_{tx} \end{pmatrix} \qquad (3.107)$$

where (i, j) = one of $\{-1, 0, +1\}$.

In the simplest case there is only one of each station located at the combination of (i, j) that minimises the distance to the victim i.e. the receiver is always assumed to be at the centre. In more complex cases the grid could be extended further so that there are multiple copies of each transmitter.

Wrap around geometry assists in avoiding edge effects in aggregate interference scenarios but introduces complexity and is hard to extend to more general geometric systems such as Earth spherical coordinates.

3.8.3 Earth Spherical Coordinates

3.8.3.1 Definitions

Earth spherical coordinates are the familiar combination of (latitude, longitude, height) as shown in Figure 3.51.

The Earth is defined as a sphere with rotation axis between the North and South Poles. The plane perpendicular to the rotation axis through the centre of the sphere is called the equatorial

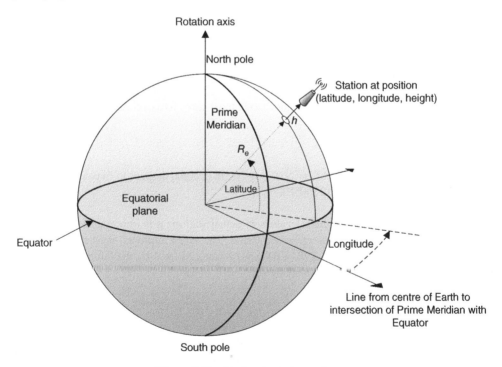

Figure 3.51 Earth spherical coordinates

plane, and the circle of the intersection of the equatorial plane with the sphere is called the equator. The lines between the North Pole and the South Pole along the surface of the sphere are called great circles. One of these is defined as the Prime Meridian where longitude is zero.

The position of a station on the sphere can be defined by two angles:

1. **Latitude**: angle between the line from the centre of the Earth to the station and the equatorial plane
2. **Longitude**: the angle in the equatorial plane between the great circle on which the station is located and the prime meridian.

The position of the station in three dimensions is completed by its height above the sphere.

The (latitude, longitude) angles can be defined in a variety of formats, the most common being decimal degrees and degrees, minutes, seconds (DMS), where a degree is divided into 60 minutes and a minute divided into 60 seconds so that

$$Decimal\ Degrees = Degrees + \frac{Minutes}{60} + \frac{Seconds}{3600} \tag{3.108}$$

The notation for (degrees, minutes, seconds) is (D° M′ S″).

The benefits of using DMS mostly relates to nautical navigation whereby 1 minute of latitude is 1 nautical mile: in most other situations it is easier to use decimal degrees.

The signs of the latitude and longitude are given as follows:

- Latitude is positive for points north of the equator and negative for points south of the equator
- Longitude is positive east of the Prime Meridian and negative west of the Prime Meridian. Sometimes points west of the Prime Meridian are described using a positive number and then 'W' to identify the longitude is westerly i.e. a negative number.

Note that $0.00001°$ is about 1 m in latitude, so in most cases five decimal places is sufficient accuracy for station locations.

Example 3.17

London's position in DMS is $51°30′26″$N $0°7′39″$W. In decimal degrees to five decimal places this would be ($51.50722°$N, $-0.12750°$E).

When using the (latitude, longitude) values of actual stations it is important to ensure that they are all defined using the same reference frame. The safest approach is to ensure that all are in the World Geodetic System 1984 (WGS84) format, as used by the Global Positioning System (GPS).

3.8.3.2 Distance and Azimuth

The distance between two points on the surface of the Earth and the azimuth (i.e. direction from North) of one station as seen by another can be calculated using spherical geometry as shown in Figure 3.52.

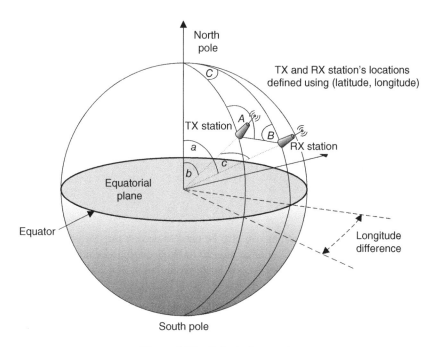

Figure 3.52   Spherical geometry

Given TX and RX stations with known (lat, long) coordinates, the first stage is to create a spherical triangle ABC on the surface of the Earth with angles at the centre of the Earth a, b, c calculated using

$$C = long_{RX} - long_{TX} \qquad (3.109)$$

$$a = \frac{\pi}{2} - lat_{RX} \qquad (3.110)$$

$$b = \frac{\pi}{2} - lat_{TX} \qquad (3.111)$$

Then the spherical trigonometry equations can be used to calculate c, A, B:

$$\cos c = \cos a \cos b + \sin a \sin b \cos C \qquad (3.112)$$

$$\cos A = \frac{\cos a - \cos b \cos c}{\sin b \sin c} \qquad (3.113)$$

$$\cos B = \frac{\cos b - \cos c \cos a}{\sin c \sin a} \qquad (3.114)$$

From these the various distances and angles can be calculated. For example, the great circle distance d between the two stations can be calculated using the radius of the Earth R_e as

$$d = R_e c \qquad (3.115)$$

Note that here c is in radians.

In this case, where the longitude of the RX is greater than that of the TX, the azimuth of the RX station as seen by the TX station is simply angle A while the azimuth of the TX station as seen by the RX station is $-B$. Note that if the longitude of the TX was greater than that of the RX the signs would be the other way round.

Example 3.18
For a transmitter in London at (51.50722°N, −0.1275°E) and receiver in Paris at (48.85667°N, 2.35083°E) the great circle distance between them is about 343.9 km while the azimuth at the transmitter towards the receiver is 148.1° from north.

3.8.3.3 Great Circle Motion

These spherical geometry calculations can also be used to predict the position of moving stations that are on **great circle** paths – such as aircraft or ships on transoceanic voyages. Given a start position, a heading azimuth, speed v_{km} in km/hour and time of travel t_h in hours it is possible to calculate the final position as follows:

$$c = \frac{v_{km} t_h}{R_e} \tag{3.116}$$

$$b = \frac{\pi}{2} - lat_{tx} \tag{3.117}$$

$$A = Azimuth \tag{3.118}$$

Then
$$\cos a = \cos b \cos c + \sin b \sin c \cos A \tag{3.119}$$

$$\cos B = \frac{\cos b - \cos c \cos a}{\sin c \sin a} \tag{3.120}$$

$$\cos C = \frac{\cos c - \cos a \cos b}{\sin a \sin b} \tag{3.121}$$

Hence
$$long_{rx} = C + long_{tx} \tag{3.122}$$

$$lat_{rx} = \frac{\pi}{2} - a \tag{3.123}$$

Example 3.19
An aircraft over London at (51.50722°N, −0.1275°E) flies at a speed of 800 km/h with heading 148.1° flies on a great circle. After 0.43 hours (25 minutes 48 seconds) it will reach a position of (48.85638°N, 2.35400°E), somewhere over Paris.

3.8.3.4 Elevation Angle

The azimuth angle is complemented by the elevation angle which can be calculated using the geometry in Figure 3.53.

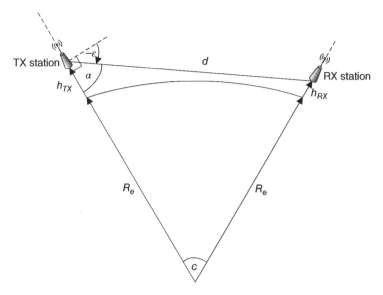

Figure 3.53 Calculation of elevation angle

From a triangle created with the centre of the Earth (radius R_e) and the two stations the elevation angle ε can be calculated as follows:

$$d^2 = (R_e + h_{tx})^2 + (R_e + h_{rx})^2 - 2(R_e + h_{tx})(R_e + h_{rx})\cos c \tag{3.124}$$

$$\sin\alpha = \frac{R_e + h_{rx}}{d}\sin c \tag{3.125}$$

Then

$$\varepsilon = \alpha - \frac{\pi}{2} \tag{3.126}$$

Example 3.20
For a transmitter at a height of 100 m in London the elevation of a receiver at a height of 10 m in Paris, such that $c = 3.1°$, would be $\varepsilon = -1.56°$.

A special case is the calculation of the elevation angle ε_0 of the horizon of a spherical smooth Earth (i.e. without terrain) which is dependent upon the station height using

$$\cos\varepsilon_0 = -\frac{R_e}{R_e + h} \tag{3.127}$$

Example 3.21
For a transmitter at a height of 100 m above a smooth spherical Earth the horizon elevation angle would be $\varepsilon = -0.32°$.

3.8.4 ECI Vector Coordinates

3.8.4.1 Definition

The Earth centred spherical coordinates are useful for terrestrial applications but become increasingly cumbersome for modelling scenarios that include space applications. For this reason it is usually more helpful to use a generic vector based framework with origin at the centre of the Earth. This framework is inertial, i.e. it does not rotate with the Earth and so is called the **Earth centred inertial** (ECI) vector framework. Positions are then defined by three values, the (x, y, z) coordinates within this frame as in Figure 3.54.

The z axis is aligned with the Earth's rotation axis while the xy plane is the same as the equatorial plane. The key question is then, where is the x axis pointing? There are two approaches that are frequently used in simulations:

1. Set it to coincide with the Prime Meridian (longitude = 0) at the start of the simulation. This is helpful if all stations (including satellites) can be defined at the start of the simulation with respect to the Earth
2. Align the x axis with the astronomical reference point called the vernal equinox direction which is fixed in space and is often used to define satellite orbits that are not fixed with respect to the Earth.

In the second case the Prime Meridian will have an offset at the start of the simulation from the x axis, θ_g, which can be extracted from astronomical references and depends upon the

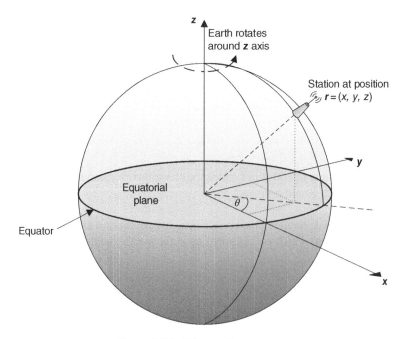

Figure 3.54 ECI coordinate system

assumed reference time. However for simplicity in the following examples, the first case will be used so that $\theta_g = 0$.

The Earth rotates around its axis at an average rotation rate of $\omega_e = 0.004178074°/s$ so a station at a fixed location with longitude λ the angle θ in Figure 3.54 will increase in $t =$ time in seconds since the simulation start according to

$$\theta(t) = \lambda + \omega_e t \tag{3.128}$$

3.8.4.2 Conversion of Spherical Coordinates to ECI

If the Earth is modelled as a sphere then a position in latitude, longitude, height in metres = (φ, λ, h_m) can be converted to an ECI vector using

$$r = R_e + \frac{h_m}{1000} \tag{3.129}$$

$$r = \begin{pmatrix} r\cos\theta(t)\cos\vartheta \\ r\sin\theta(t)\cos\vartheta \\ r\sin\vartheta \end{pmatrix} \tag{3.130}$$

The process can be reversed from ECI to position using

$$h_m = 1000(r - R_e) \tag{3.131}$$

$$\sin\vartheta = \frac{z}{r} \tag{3.132}$$

$$\tan\theta(t) = \frac{y}{x} \tag{3.133}$$

Example 3.22
For a transmitter in London at (51.50722°N, −0.1275°E) and receiver in Paris at (48.85667°N, 2.35083°E), both with height = 10 m, the ECI vectors at $t = 0$ assuming $\theta_g = 0$ would be

TX station (London) = (3969.85, −8.83, 4992.09)
RX station (Paris) = (4192.94, 172.13, 4803.17)

3.8.4.3 Orbit Parameters

To work out the position of a satellite in ECI coordinates its first necessary to consider the parameters used to describe its orbit. For most interference analysis studies it is sufficient to use the classical orbit elements shown in Figures 3.55 and 3.56 (Bate et al., 1971).

The position of the satellite at a specified time can be defined via multiple combinations of different parameters, but one common set are:

$a =$ semi-major axis, which defines the size of the orbit
$e =$ eccentricity, which defines the shape of the orbit
$i =$ inclination, which specifies the angle between the equatorial plane and the orbital plane

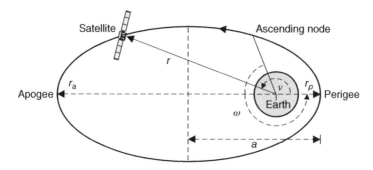

Figure 3.55   Orbit shape and position of satellite in orbit

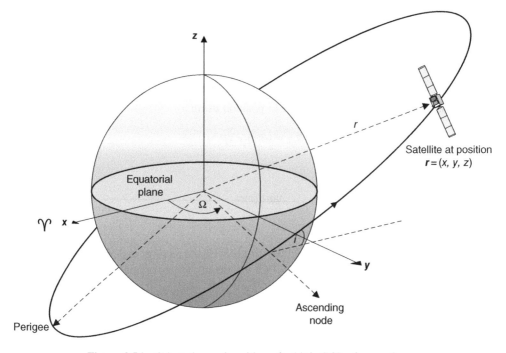

Figure 3.56   Orientation and position of orbit in ECI reference frame

Ω = longitude of the ascending node, i.e. the angle at the centre of the Earth in the equatorial plane between the x axis (towards the vernal equinox, Υ) and the line towards the ascending node, which is where the plane of the orbit and the equatorial plane intersect and the satellite moves from the southern hemisphere to the northern hemisphere

ω = argument of perigee, i.e. the angle at the centre of the Earth in the plane of the orbit between the line to the ascending node and the line to the perigee point

ν = true anomaly, i.e. the angle at the centre of the Earth in the plane of the orbit between the line to the perigee point and the line to the satellite.

These parameters are usually either defined for a specific time or when the satellite crosses the ascending node.

There are a range of orbit shapes, such as circular, elliptical, parabolic and hyperbolic depending upon the eccentricity as follows:

$e = 0 \Rightarrow$ circular orbit
$0 < e < 1 \Rightarrow$ elliptical orbit
$e = 1 \Rightarrow$ parabolic orbit
$e > 1 \Rightarrow$ hyperbolic orbit.

The most common orbit types for interference analysis are circular and elliptical. In practice because of launch errors and perturbations, no orbit is exactly circular though many are described, and can be modelled, as such. Similarly no orbit is exactly parabolic, so satellites launched into deep space can be modelled as being on hyperbolic escape trajectories.

The perigee is the closest point of the orbit to the Earth and the apogee the furthest point. Note that orbits around other celestial bodies use alternative names – for example around the sun the nearest point would be called the perihelion.

For circular orbits there is no perigee and hence terms such as the argument of perigee are not defined. Often these can either be assumed to be zero or defined relative to an arbitrary point on the orbit. Similarly equatorial orbits with $i = 0$ will have no defined longitude of ascending node.

The ITU requires in RR Appendix 4 (ITU-R, 2012a) that administrations file a satellite's phase angle (PA) which is defined in Rec. ITU-R S.1503 (ITU-R, 2013p) as

$$PA = \omega + v \tag{3.134}$$

The semi-major axis and eccentricity are connected to the radius distance at perigee and apogee using

$$a = \frac{r_a + r_p}{2} \tag{3.135}$$

$$e = \frac{r_a - r_p}{r_a + r_p} \tag{3.136}$$

Or:

$$r_a = a(1 + e) \tag{3.137}$$

$$r_p = a(1 - e) \tag{3.138}$$

The height of an orbit is often given instead of orbit radius (in particular for circular orbits) and the conversion is simply:

$$h = r - R_e \tag{3.139}$$

The orbit radius at any point of an elliptical orbit can be calculated using

$$r = \frac{a(1 - e^2)}{1 + e \cos v} \tag{3.140}$$

From these parameters it is necessary to first work out where the satellite will be at time t and then convert that position into ECI vectors.

3.8.4.4 Orbit Prediction

The motion of a satellite around the Earth is dependent upon a range of factors including:

1. Its position and velocity at a specific time
2. The gravitational field
3. Atmospheric drag
4. Solar radiation pressure.

The gravitational field alone can be very complex, taking into account the non-spherical nature of the Earth and the gravitational fields of other celestial bodies such as the Sun, Moon and planets.

There is a choice to be made as to how accurate an orbit prediction model is needed when undertaking interference analysis that includes satellites. The most common models are:

1. Point mass, which only includes the gravitational effects of the Earth and assumes its field can be modelled as if all its mass were located at its central point
2. Point mass plus J_2, which extends the point mass model to take account of the ellipsoidal shape of the Earth
3. Simplified General Perturbations 4 (SGP-4): these are more advanced orbit models that include additional gravitational effects and perturbations, in particular for low Earth orbits (Vallado et al., 2006).

There are also high fidelity models that can predict to even greater accuracy.

In general, as noted previously, interference analysis differs from satellite mission planning as the types of questions it asks relate to levels of interference, likelihood of interference and interference event lengths, all of which can be calculated with sufficient accuracy using one of the first two types of orbit propagators.

The point mass model prediction method has alternative approaches, some of which can handle all orbit types. The approach described here is suitable for circular and most elliptical orbits apart from those that are near parabolic. In a point mass model, the orbit plane is fixed in inertial space and so the prediction problem is to identify what the position of the satellite is within the plane, in particular to calculate the true anomaly $= \nu$. This can be calculated from the mean motion n which is constant and relates to Kepler's law of equal areas, in that the area swept out on a satellite's ellipse in a given time is always the same.

The mean motion n can be calculated using

$$n = \sqrt{\frac{\mu}{a^3}} \tag{3.141}$$

Here μ is the gravitational constant of Earth $= 398600.4418 \text{ km}^3/\text{s}^2$. The period of an orbit is therefore

$$P = 2\pi \sqrt{\frac{a^3}{\mu}} \tag{3.142}$$

Note how the orbit period is only dependent upon the semi-major axis, not on other parameters such as the orbit's eccentricity.

The change in the mean anomaly from its initial value of M_0 over time t can then be calculated using

$$M = M_0 + nt \tag{3.143}$$

It can be shown that this is connected to the true anomaly v via the eccentric anomaly E using

$$M = E - e \sin E \tag{3.144}$$

$$\cos v = \frac{e - \cos E}{e \cos E - 1} \tag{3.145}$$

This can be reversed using

$$\cos E = \frac{e - \cos v}{1 + \cos v} \tag{3.146}$$

So an algorithm to calculate the position of a satellite at time $= t$ could be:

Given $(a, e, i, \Omega, \omega, v_0)$ at $t = 0$:

1. Use (e, v_0) to calculate the initial eccentric anomaly E_0 using Equation 3.146
2. Use (e, E_0) to calculate the initial mean anomaly M_0 using Equation 3.144
3. Use (a) to calculate the mean motion n from Equation 3.141
4. Use (M_0, n, t) to calculate M using Equation 3.143
5. Use (M, e) to calculate E using Equation 3.144 and iteration
6. Use (E, e) to calculate v using Equation 3.145.

Steps 1–3 need only be calculated once, steps 4–6 need to be repeated for each time step. The velocity can be calculated using

$$v^2 = \mu \left(\frac{2}{r} + \frac{1}{a} \right) \tag{3.147}$$

For many studies a point mass orbit predictor is sufficient, but some require additional accuracy. For example, the methodology described in Section 7.6 to calculate equivalent power flux density (EPFD) from non-GSO satellite systems in Rec. ITU-R S.1503-2 (ITU-R, 2013p) extends the point mass model to include J_2 terms. The affect of the J_2 terms are to rotate the orbit plane around the equator and the perigee within the orbit plane.

The unperturbed mean motion term is from Equation 3.141:

$$n_0 = \sqrt{\frac{\mu}{a^3}} \tag{3.148}$$

It is replaced by the perturbed mean motion:

$$\bar{n} = n_0 \left(1 + \frac{3}{2} \frac{J_2 R_e^2}{p^2} \left(1 - \frac{3}{2} \sin^2(i) \right) (1 - e^2)^{1/2} \right) \tag{3.149}$$

where $J_2 = 0.001082636$ and:

$$p = a(1 - e^2)$$ (3.150)

So now:

$$M = M_0 + \bar{n}t$$ (3.151)

The argument of perigee and the longitude of the ascending node are not constant but change in time according to

$$\omega = \omega_0 + \omega_r t$$ (3.152)

$$\Omega = \Omega_0 + \Omega_r t$$ (3.153)

where ω_0, Ω_0 are the initial values and ω_r, Ω_r are the rates of change respectively, given by

$$\omega_r = \frac{3}{2} \frac{J_2 R_e^2}{p^2} \bar{n} \left(2 - \frac{5}{2} \sin^2(i) \right)$$ (3.154)

$$\Omega_r = -\frac{3}{2} \frac{J_2 R_e^2}{p^2} \bar{n} \cos(i)$$ (3.155)

The revised algorithm to calculate the position of the satellite at time t is then

Given $(a, e, i, \Omega_0, \omega_0, \nu_0)$ at $t = 0$:

1. Use (e, ν_0) to calculate the initial eccentric anomaly E_0 using Equation 3.146
2. Use (e, E_0) to calculate the initial mean anomaly M_0 using Equation 3.144
3. Use (a) to calculate the mean motion n_0 from Equation 3.148
4. Use (n_0, a, i, e) to calculate $\bar{n}$ using Equation 3.149
5. Use $(\bar{n}, a, i, e)$ to calculate ω_r using Equation 3.154
6. Use $(\bar{n}, a, i, e)$ to calculate Ω_r using Equation 3.155
7. Use $(M_0, \bar{n}, t)$ to calculate M using Equation 3.151
8. Use (M, e) to calculate E using Equation 3.144 and iteration
9. Use (E, e) to calculate ν using Equation 3.145
10. Use (ω_0, ω_r, t) to calculate ω using Equation 3.152
11. Use (Ω_0, Ω_r, t) to calculate Ω using Equation 3.153.

Steps 1–6 need only be calculated once, steps 7–11 need to be repeated for each time step. The rate of precession of the perigee in Equation 3.154 becomes zero for an orbit inclination given by

$$\sin^2(i) = \frac{4}{5}$$ (3.156)

This gives an inclination of around $i = 63.43°$ and is often used for elliptical orbit systems such as the Molniya constellation, as it will be stable without the perigee drifting so that the apogee remains fixed. There will continue to be precession of the orbit around the equator due to the terms in Equation 3.155, but these can be countered by adjusting the semi-major axis so that the orbit period is not precisely the same as an exact fraction of that of the Earth, with the difference being just enough to keep the orbit's ground track fixed.

To be able to calculate ground tracks we need to be able to convert the satellite position from orbit elements into a vector, as described in the following section.

3.8.4.5 Orbit Conversion to ECI

As with the orbit prediction algorithm there are multiple ways to convert the position of a satellite within its orbit into ECI vectors, but one approach starts by defining vectors P and Q within the orbit plane as in Figure 3.57.

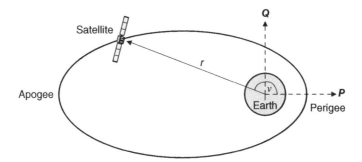

Figure 3.57 Orbit (P, Q) vectors

Within this coordinate system the position of the satellite is

$$r = r\cos\nu P + r\sin\nu Q = pP + qQ \tag{3.157}$$

Note that the orbit radius is given by Equation 3.140.

It is then necessary to convert these into ECI coordinates and one method (Bate et al., 1971) is to define a rotation matrix such that

$$\begin{bmatrix} x \\ y \\ z \end{bmatrix} = \begin{bmatrix} R_{11} & R_{12} & R_{13} \\ R_{21} & R_{22} & R_{23} \\ R_{31} & R_{32} & R_{33} \end{bmatrix} \begin{bmatrix} p \\ q \\ 0 \end{bmatrix} \tag{3.158}$$

The elements of rotation matrix R are derived from the angles that specify the orientation of the orbit plane with respect to the ECI frame, namely, $\{\Omega, \omega, i\}$ such that

$$R_{11} = \cos\Omega\cos\omega - \sin\Omega\sin\omega\cos i \tag{3.159}$$

$$R_{12} = -\cos\Omega\sin\omega - \sin\Omega\cos\omega\cos i \tag{3.160}$$

$$R_{13} = \sin\Omega\sin i \tag{3.161}$$

$$R_{21} = \sin\Omega\cos\omega + \cos\Omega\sin\omega\cos i \tag{3.162}$$

$$R_{22} = -\sin\Omega\sin\omega + \cos\Omega\cos\omega\cos i \tag{3.163}$$

$$R_{23} = -\cos\Omega\sin i \tag{3.164}$$

$$R_{31} = \sin\omega\sin i \tag{3.165}$$

$$R_{32} = \cos\omega\sin i \tag{3.166}$$

$$R_{33} = \cos i \tag{3.167}$$

There can be shortcuts – for example note that terms (R_{13}, R_{23}, R_{33}) are not used. These ECI vectors can then be converted into (latitude, longitude) pairs to show the motion on the satellite as a **ground track** using Equations 3.131, 3.132 and 3.133.

A similar approach can be used for the velocity vector:

$$v = \sqrt{\frac{\mu}{p}}[-\sin\nu\boldsymbol{P} + (e + \cos\nu)\boldsymbol{Q}] \tag{3.168}$$

Example 3.23

The International Space Station (ISS) has the following orbit parameters:

Perigee height: 418 km
Apogee height: 423 km
Inclination angle: 51.64°

An example of this orbit's ground track is shown in Figure 3.58.

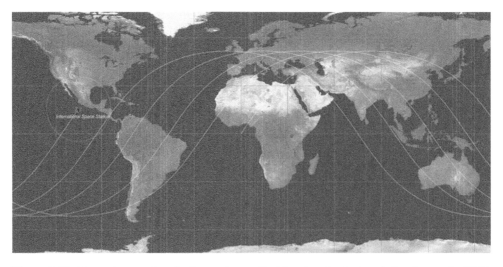

Figure 3.58 Example ground track. *Screenshot credit: Visualyse Professional. Overlay credit: NASA Visible Earth*

3.8.4.6 Geostationary and Repeating Orbits

The ISS is in low Earth orbit (LEO) with a height of 418 km, which means it takes around 93 minutes to circle the planet. It can sometimes be seen as a bright star rising just after sunset taking a few minutes to cross the sky before blinking out.

This motion requires radio systems that communicate with the ISS with directional antennas to track the space station, using pointing mechanisms and logic, which increases the cost. However as noted by Arthur C. Clarke in 1945 (Clarke, 1945), if a satellite in circular equatorial orbit has angular velocity equal to that of the Earth it will appear to an observer on the ground to be fixed.

The Earth takes 1 day (86,400 seconds) to return to the same position with respect to the sun, but it has moved forward on its orbit and will be in the same position with respect to an inertial frame (in particular the vernal equinox) after:

$$T_{sideral} = 23 \text{ hours } 56 \text{ minutes } 4.091 \text{ seconds} = 86,164.0916 \text{ seconds}$$

Using this as the orbit period in Equation 3.142 we get an orbit radius and height of

$$R_{gso} = 42,164.17 \text{ km}$$
$$h_{gso} = 35,786.1 \text{ km}$$

A satellite with circular orbit with these parameters would be called geosynchronous, in that its orbit would on average keep track with the Earth. To be in a geostationary satellite orbit (GSO) it also must be equatorial, i.e. with inclination $i = 0$. A quick way to calculate the ECI coordinates of a GSO satellite is to use the same approach as for terrestrial stations but with a height = h_{GSO}.

If a satellite has a near geosynchronous circular orbit with a small but non-zero inclination it will appear to move slowly in a figure of eight as seen by the Earth. This can be useful for satellites that want to provide occasional coverage to Polar Regions.

It is also possible to change the orbit's eccentricity while maintaining a repeating ground track, though it will no longer be a figure of eight. For example, there is a class of orbit called highly elliptical orbit (HEO) that has a period which is a multiple or fraction of that of the Earth but with non-zero inclination and eccentricity.

Example 3.24

The Sirius satellite radio network uses a constellation of three satellites in HEO using the parameters in Table 3.11 (International Launch Service, 2000). This orbit was selected as it

Table 3.11 Orbit parameters of the Sirius satellite radio constellation

Inclination	63.4°
Eccentricity	0.2684
Semi-major axis	42,164
Argument of perigee	270.0°
Apogee longitude	96.0°W

gives much higher minimum elevation angles for listeners across North America than could be achieved using a satellite in GSO.

The argument of perigee = 270° ensures that the apogee is over the northern hemisphere. This is important as the satellite travels slower when at apogee than perigee so the it will 'hang' there for longer. Note how the inclination angle was selected to be the value calculated by Equation 3.156.

The ground track of the Sirius satellite constellation is shown in Figure 3.59 together with a GSO satellite (at a fixed location), and inclined GSO with a figure of eight track and field of view that can cover the North Pole.

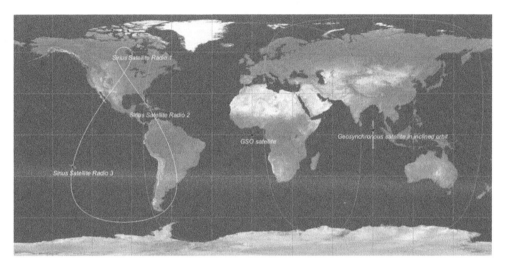

Figure 3.59 Examples of geosynchronous orbits. *Screenshot credit: Visualyse Professional. Overlay credit: NASA Visible Earth*

3.8.5 Ellipsoidal Earth and Orbit Models

For almost all interference analysis work the spherical Earth ECI approach is sufficient but in some cases additional accuracy could be required – such as identification of exactly what time a spacecraft must cease emissions to avoid harmful interference.

As noted previously, there are more advanced satellite orbit prediction models that could be used such as SGP-4 but it can also be useful to model the Earth as an ellipsoid as shown in Figure 3.60 where L = geodetic latitude (usually used) and L' = geocentric latitude.

This model defines the Earth's shape via the parameters of the cross-sectional ellipse. The values for the most commonly used model, WGS84, are given in Table 3.12.

The flattening and eccentricity of the ellipse can be calculated from the semi-major and semi-minor axis using

$$f = \frac{a-b}{a} \tag{3.169}$$

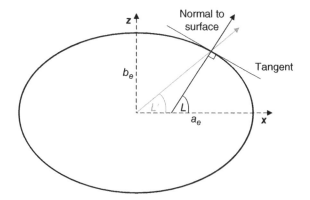

Figure 3.60   Ellipsoidal Earth model

Table 3.12   WGS 84 ellipsoid parameters

a_e = semi-major axis	6378.137 km
b_e = semi-minor axis	6356.752314245 km
$1/f_e$ = inverse flattening	298.257223563

$$e^2 = 1 - \frac{b^2}{a^2} \tag{3.170}$$

From these the ECI vector in the xz plane can be calculated using

$$x = \left| \frac{a_e}{\sqrt{1 - e^2 \sin^2 L}} + h \right| \cos L \tag{3.171}$$

$$z = \left| \frac{a_e(1 - e^2)}{\sqrt{1 - e^2 \sin^2 L}} + h \right| \sin L \tag{3.172}$$

Using the longitude, the full ECI vector can be created.

Note there are a large number of reference geoids, in particular those optimised for specific countries. For example, in the UK the Ordnance Survey's OSGB uses the Airy geoid. If a position is given in (latitude, longitude) it is important to check the reference geoid and if necessary convert to a common frame such as WGS84.

3.8.6   Delay and Doppler

The distances involved in satellite networks can lead to delays in a communications link and hence increased latency for two way communications. In addition due to the relative speeds there can be changes in the measured frequency, the Doppler effect. These can be calculated using

$$\Delta t = \frac{d}{c} \tag{3.173}$$

$$\Delta f = \frac{\Delta v}{c} f_0 \tag{3.174}$$

where d is the path length, Δv the relative velocity, c the speed of light and f_0 the system's stationary frequency.

It is not common for these to be major issues for interference analysis but they are key for system design – in particular for the selection of satellite constellation orbit. For these cases where there is both an uplink and downlink the delay is effectively doubled. Delay is also a significant issue for single frequency networks (SFN), used for terrestrial digital broadcasting, as discussed in Section 7.3.

Example 3.25
Round trip delays for three satellite networks listed in Rec. ITU-R M.1184 (ITU-R, 2003a) are given in Table 3.13 using an elevation angle of 20°. It can be seen that using GSO would introduce a significant round trip delay to interactive communications.

Table 3.13 Round trip delays for example satellite networks

System	LEO-D	LEO-F	GSO
Height (km)	1414	10,355	35,786
Elevation angle (degrees)	20	20	20
Round trip delay (mS)	18.7	111.6	281.3

3.9   Calculation of Angles

The previous section described how to calculate the position of stations using a common ECI set of vectors. To calculate the gain at a transmit or receive station it is necessary to convert these vectors into angles, in particular the azimuth and elevation.

3.9.1   Azimuth and Elevation

Azimuth and elevation can be calculated from vector components as shown in Figure 3.61. Note the use of capital letters (X, Y, Z) to identify the vector is relative to the station's reference frame and not the ECI (x, y, z).

The equations to convert from vector (X, Y, Z) of length R to (Az, El) are

$$Az = \tan^{-1}\frac{X}{Y} \tag{3.175}$$

$$El = \sin^{-1}\frac{Z}{R} \tag{3.176}$$

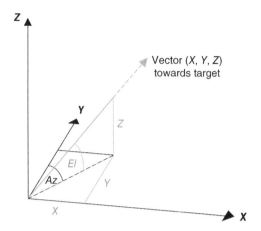

Figure 3.61 Calculation of azimuth and elevation

The equations to convert from (Az, El) to a unit vector (X, Y, Z) are

$$X = \sin Az \cos El \tag{3.177}$$

$$Y = \cos Az \cos El \tag{3.178}$$

$$Z = \sin El \tag{3.179}$$

The key question is then how the reference frame created by the XYZ axes is orientated, and this can depend upon the station type, as described in the following sections.

3.9.2 Terrestrial

The simplest reference frame is for terrestrial systems on the surface of the Earth where azimuth = 0 is due North and elevation = 0 is the horizontal plane perpendicular to the vertical, as shown in Figure 3.62.

This reference frame is identified by the three right angled vectors heading North, East and towards the zenith, hence NEZ. The NE plane is horizontal and at right angles to the Zenith vector.

Another reference frame is for moving stations, where azimuth = 0 is straight ahead, sometimes described as 12 o'clock. Here the three reference directions are:

1. **Y**: in the direction of motion
2. **Z**: towards the zenith
3. **X**: at right angles to **ZY** to complete the **XYZ** coordinate system.

3.9.3 Satellite

Just as there are multiple coordinate systems for terrestrial systems there are also alternatives for satellites. For GSO systems the coordinate frame can be defined by the north and east directions, with the third axis aligned with the sub-satellite point, as in Figure 3.63.

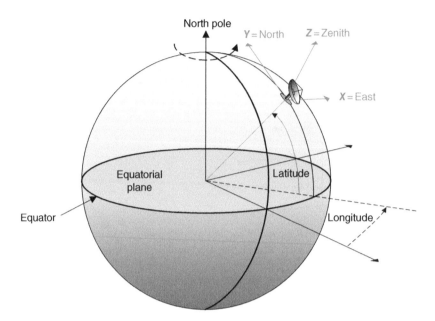

Figure 3.62   North East Zenith coordinate system

For non-GSO systems the coordinate system is usually referenced to the direction of motion as in Figure 3.64, though the PFD mask used to calculate EPFD, as described in Section 7.6.3, uses the reference frame in Figure 3.63 for non-GSO satellite systems.

Another way to consider the non-GSO reference frame is to consider the view from the satellite looking down at the sub-satellite point with the direction of motion at the top as in Figure 3.65.

3.9.4   Angles in the Antenna Frame

To get the angles used for the gain calculation in the antenna frame from ECI coordinates requires a final stage in which the (Az, El) angles of the antenna boresight are taken into account. As can be seen in Figure 3.66, there will be two sets of (Az, El) to consider, one for the antenna boresight and another for the off-axis direction under consideration.

There are a number of methods to calculate off-axis angles as seen by the antenna including spherical geometry and vector rotations. In addition the result can be derived in (Az, El) angles or (θ, ϕ) angles described in the following section.

The two pairs of (Az, El) angles are very similar to (latitude, longitude), with

Azimuth ~ Longitude
Elevation ~ Latitude

Therefore the calculation of the angle between them is very similar to the spherical geometry used in Section 3.8.3 using

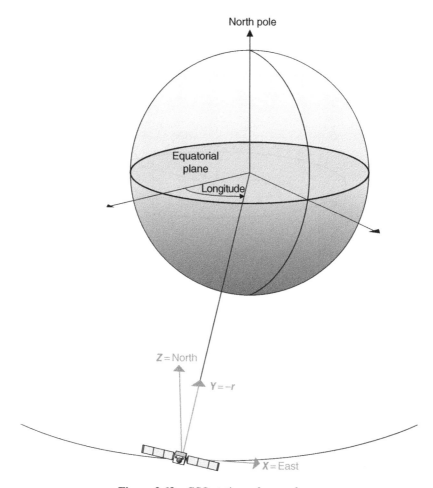

Figure 3.63   GSO station reference frame

$$C = Az_0 - Az \tag{3.180}$$

$$a = \frac{\pi}{2} - El \tag{3.181}$$

$$b = \frac{\pi}{2} - El_0 \tag{3.182}$$

The spherical trigonometry equations can then be used to calculate c, A, B. In particular the off-axis angle c can be derived using

$$\cos c = \cos a \cos b + \sin a \sin b \cos C \tag{3.183}$$

3.9.5   Off-Axis Angle from ECI Vectors

The off-axis angle can be calculated directly from the ECI vectors. Given three stations, A, B, C, where the antenna at station B is pointing directly at station A, then the off-axis angle at

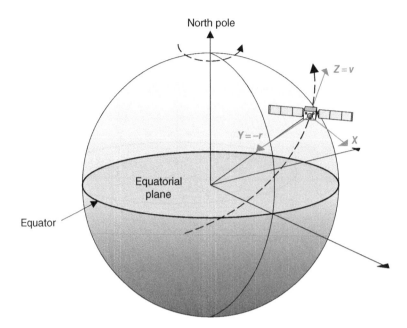

Figure 3.64 Non-GSO satellite reference frame

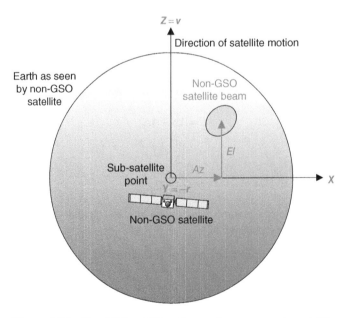

Figure 3.65 Non-GSO satellite reference frame as seen by satellite

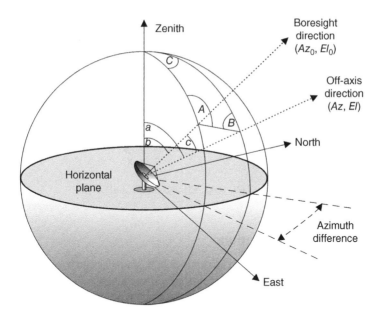

Figure 3.66 Boresight and Off-axis direction pointing angles

station B towards station C is the angle ABC. The can be calculated by first deriving the two vectors r_{BA} and r_{BC}:

$$r_{BA} = r_A - r_B \tag{3.184}$$

$$r_{BC} = r_C - r_B \tag{3.185}$$

Then the off-axis angle θ can be calculated using the dot product of the two:

$$\cos\theta = \frac{r_{BA} \cdot r_{BC}}{|r_{BA}||r_{BC}|} \tag{3.186}$$

3.9.6 Theta Phi Coordinates

As well as the (Az, El) coordinates another way to define the pointing angles is using (θ, ϕ) as shown in Figure 3.67 and documented in Rec. ITU-R BO.1443 (ITU R, 2014c). Note that ϕ is effectively the off-axis angle.

It is possible to convert between these two sets of angles using

$$\cos\phi = \cos Az \cos El \tag{3.187}$$

$$\sin\theta = \frac{\sin El}{\sin Az} \tag{3.188}$$

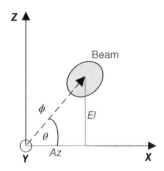

Figure 3.67   (Az, El) and (θ, ϕ) angles

3.10   Statistics and Distributions

Interference analysis involves significant amounts of both statistics and use of distributions so it is worth having a good understanding of both topics. For statistics two key concepts are the mean and standard deviation. The mean or average given a set of N values of variable x is

$$\mu = \bar{x} = \frac{1}{N}\sum_{i=1}^{i=N} x_i \tag{3.189}$$

The standard deviation σ is a representation of how much variation there is around the mean value and can be calculated using

$$\sigma^2 = \frac{1}{N-1}\sum_{i=1}^{i=N}(x_i - \bar{x})^2 \tag{3.190}$$

This is the standard deviation of a sample as the mean was calculated rather than being known, as would be the case for a population, in which case the $(N-1)$ would be replaced by N.

Example 3.26
For the data set = {3, 2, 4, 7, 3} the mean = 3.8 and sample standard deviation = 1.92

Given two independent variables (x, y), then the standard deviation of their sum can be calculated using

$$\sigma^2(x,y) = \sigma^2(x) + \sigma^2(y) \tag{3.191}$$

This is useful in propagation modelling when there are two independent fading mechanisms.
Note that the mean and standard deviation could be calculated in either absolute or in decibel scales and the results are likely to be different. When calculating power levels it is often better to average in absolute e.g. watts rather than dBW.

There are a number of distributions that are useful in many aspects of interference analysis, and the most common are:

- Normal
- Linear or uniform
- Triangular
- Rayleigh
- Rician.

For each of these distributions there is a **probability density function** (PDF) and **cumulative distribution function** (CDF). The PDF can be integrated into the CDF using

$$CDF(X) = \int_{-\infty}^{X} PDF(x)dx \tag{3.192}$$

Typically the data to generate PDFs and CDFs are given in quantised form as a **histogram** H (i) where $i = \{0 \dots n\}$ and each value i can be mapped to a data value x using

$$x(i) = x_{min} + i.x_{BinSize} \tag{3.193}$$

The bin relating to data value x is then

$$i(x) = Round \left[\frac{x - x_{min}}{x_{BinSize}} \right] \tag{3.194}$$

Care is required with the rounding direction and it is better to use a small bin size to reduce rounding errors.

Then the CDF can be generated from the histogram as a percentage using

$$CDF(X) = 100.\frac{\sum_{i=0}^{i(X)} H(i)}{\sum_{i=0}^{i=n} H(i)} \tag{3.195}$$

An important principle is that using a random number generator, the CDF can be sampled in reverse (i.e. from probability to value rather than from value to probability) in order to generate a sequence with distribution which matches the PDF. This is required for Monte Carlo analysis, as described in Section 6.9.

The **normal distribution** (sometimes called Gaussian) is particularly important for statistics and modelling radio systems with PDF and CDF given by

$$PDF(x) = \frac{1}{\sigma\sqrt{2\pi}} e^{-\frac{(x-\mu)^2}{2\sigma^2}} \tag{3.196}$$

$$CDF(x) = \frac{1}{2}\left[1 + erf\left(\frac{x-\mu}{\sigma\sqrt{2}} \right) \right] \tag{3.197}$$

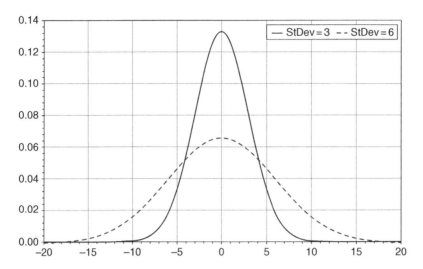

Figure 3.68 Example PDFs of the normal distribution with $\mu = 0$

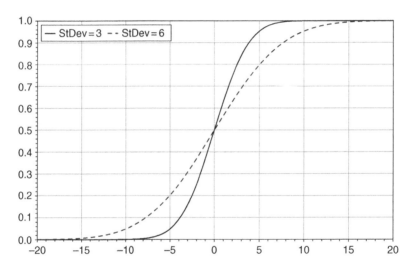

Figure 3.69 Example CDFs of the normal distribution with $\mu = 0$

The complement of the normal CDF is the Q function, which is particularly useful as it gives the probability that a value will be exceeded:

$$Q(x) = 1 - CDF(x) \qquad (3.198)$$

The normal distribution has a bell-like shape as in Figure 3.68 which shows the PDF for a normal distribution with mean 0 and standard deviation of 3 or 6. The width of the bell varies depends upon the standard deviation and the centre point is the mean. The CDF is created by integrating the PDF which for this example is shown in Figure 3.69.

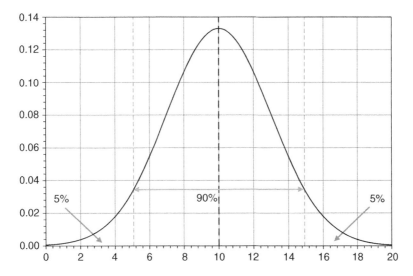

Figure 3.70   Ninety percent likelihood range for normal distribution $\mu = 10$, $\sigma = 3$

Table 3.14   Confidence interval factors

p = confidence interval $[\mu - z\sigma, \mu + z\sigma]$	z
0.8	1.282
0.9	1.645
0.95	1.960
0.99	2.576

The wider the PDF curve the more variation there is in the data and, in particular for outputs, the greater uncertainty in the results. A confidence interval can be generated by specifying a required percentage likelihood and then identifying the range of values involved. For example Figure 3.70 shows the 90% likelihood range for the case where the mean is 10 and the standard deviation 3.

Sometimes distributions are described as 'log-normal'. What is usually meant is that the distribution is applied to the variables in decibel form i.e. $10\log_{10}(x)$ rather than absolute. An example would be the log-normal location variability described in Section 4.3.5.

Example 3.27
A normal distribution has mean $\mu = 0$ and standard deviation $\sigma = 5.5$ dB. The 95% confidence interval is the central portion excluding the 2.5% at either end. This is about 1.96 times the standard deviation, so the 95% confidence range would be ±10.8 dB.

In general the confidence interval of variable X can be calculated to be the range $[X_{Min}, X_{Max}]$ where

$$X_{Min} = \mu - z\sigma \tag{3.199}$$

$$X_{Max} = \mu + z\sigma \tag{3.200}$$

Table 3.15   Likelihood not exceeded factors

p = likelihood $\mu + z\sigma$ exceeded	z
0.9	1.282
0.95	1.645
0.975	1.960
0.995	2.576

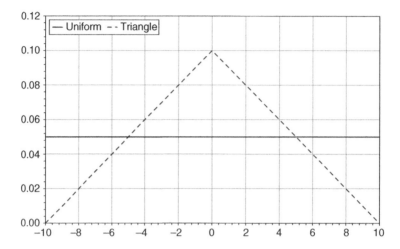

Figure 3.71   PDF of uniform and triangular distributions

The value of z to use depends upon the confidence level required, and some of the standard values are given in Table 3.14.

If the statistic of interest is not the confidence interval but the level not exceeded for a certain likelihood, then it is necessary to consider the factors in Table 3.15.

Simpler than the normal distribution are the uniform (or flat) and triangular distributions which are shown as PDFs and CDFs in Figures 3.71 and 3.72. The distributions can be defined either via the minimum and maximum or by mean value and range either side. The values used in these figures are:

Either:	min $= -10$	max $= +10$, often written as $[-10, 10]$
Or:	mean $= 0$	range $= \pm 10$

The PDF shows how likely each value is: for a flat distribution all values in the range are equally likely. The triangular distribution is more likely to be around the mean than the ends and can be created from the summation of two independent random variables each defined by uniform distributions.

Example 3.28

If the transmit power is defined by a uniform distribution $[-10, 0]$ dBW and the transmit gain is another uniform distribution $[0, 10]$ dBi independent of the transmit power, then the EIRP will have triangular distribution mean 0 dBW range ± 10 dB.

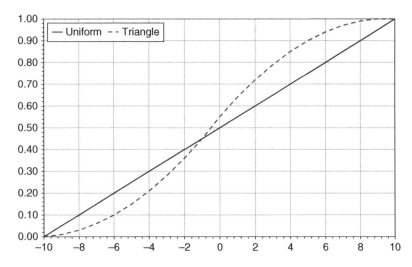

Figure 3.72 CDF of uniform and triangular distributions

As with any distribution, the uniform and triangular CDFs can be reversed using a uniform random number to generate a sequence of values that matches the required PDF. It is important that the pseudorandom generator has a long repeat period and able to create numbers close to 0 and 1 with a flat probability density.

Example 3.29
For an EIRP with triangular distribution [−10, +10] (as in Example 3.28 and Fig. 3.72) with random number $r = 0.28$ the resulting of EIRP $= −3$ dBW.

Signal fading is sometimes described using Rayleigh or Rician distributions. The Rayleigh distribution is defined via a standard deviation σ using

$$PDF(x) = \frac{x}{\sigma^2} e^{-\frac{x^2}{2\sigma^2}}$$ (3.201)

$$CDF(x) = 1 - e^{-\frac{x^2}{2\sigma^2}}$$ (3.202)

The PDF for the Rayleigh distribution is shown in Figure 3.73 for $\sigma = 3$ and 6.

These distributions are based upon mathematical concepts but it is possible to be generic and use a table of values via a CDF in the format:

[Value, Probability that Value is exceeded]

The table can then be sampled during Monte Carlo analysis as described in Section 6.9, which also covers the reliability of statistics and tests for significance.

3.11   Link Budgets and Metrics

The fundamental tool of interference analysis is the **link budget** which is the calculation of the strength of a received signal taking into account the transmit power and the various gains and

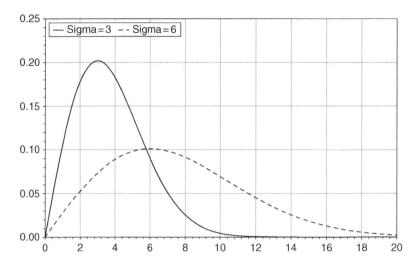

Figure 3.73   Example Rayleigh distribution PDFs

losses involved. The gain could be due to an antenna or amplifier and losses to a range of factors including propagation, polarisation and line feed.

The link budget can be used to calculate either C = wanted (victim) or I = interfering signal with equations similar to these:

$$C = P_{tx} + G_{tx} - L_p + G_{rx} - L_f \qquad (3.203)$$

$$I = P'_{tx} + G'_{tx} - L'_p + G'_{rx} - L_f \qquad (3.204)$$

Here the feed loss at the receive station has been included so the reference point is the input to the receiver, as discussed in Section 3.6. Note that there could be losses at the transmitter too but here the transmit power is defined at the input to the antenna and hence after any such losses: this is shown graphically in Figure 3.74.

For many spectrum management tasks, including licensing and interference analysis, it is the **total radiated power** (TRP) that is of interest and hence the measurement point at the transmit side is at the input to the antenna. However the user's quality of service is driven by the signals at the input to the receiver, hence that is the measurement point at the receive side.

Other terms could be added and it should be noted that the free space term L_{fs} from Equation 3.16 has been replaced by L_p which is the total propagation losses calculated using the relevant propagation model, as discussed in Chapter 4.

The units of the transmit power P_{tx} determines the units of the signal, whether C or I. Usually it is one of dBW or dBm, i.e. either decibel (watts) or decibel (milliwatts) respectively. The conversion between the two is simply:

$$dBm = dBW + 30 \qquad (3.205)$$

$$dBW = dBm - 30 \qquad (3.206)$$

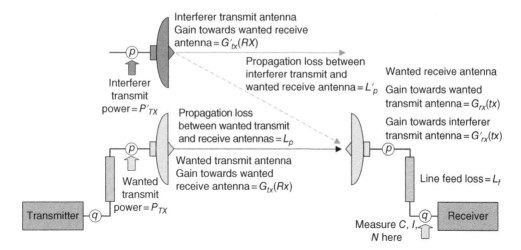

Figure 3.74 Wanted and interferer links

It can be useful to consider the power density in dBW/Hz by subtracting in dB the relevant bandwidth B in Hz using

$$C_0 = C - 10\log_{10}B \qquad (3.207)$$

$$I_0 = I - 10\log_{10}B \qquad (3.208)$$

The other key attribute of the link is its noise which can be calculated from the noise figure or temperature and bandwidth using (from Section 3.6) one of

$$N = F_N + 10\log_{10}T_0 - 228.6 + 10\log_{10}B \qquad (3.209)$$

$$N = 10\log_{10}T - 228.6 + 10\log_{10}B \qquad (3.210)$$

Note that the result of this equation is the noise given in units of dBW, as Boltzmann's constant in decibel form is -228.6 dBW/K/Hz.

From these three attributes various metrics can be created, including:

- C/I: the ratio of the wanted to interfering signal
- C/N: the ratio of the wanted signal to noise
- $C/(N+I)$: the ratio of the wanted signal to noise plus interference
- I/N: the ratio of the interfering signal to noise.

These are usually given in decibels so that

$$\left(\frac{C}{I}\right) = C - I \qquad (3.211)$$

$$\left(\frac{C}{N}\right) = C - N \tag{3.212}$$

$$\left(\frac{I}{N}\right) = I - N \tag{3.213}$$

There are also versions of these ratios normalised to one hertz as in

$$\left(\frac{C_0}{N_0}\right) = C_0 - N_0 \tag{3.214}$$

$$\left(\frac{I_0}{N_0}\right) = I_0 - N_0 \tag{3.215}$$

As the bandwidth for the wanted system's carrier and noise are likely to be similar, if not identical, Equations 3.212 and 3.214 will give equally similar results. However the interferer could be using a different bandwidth to the victim's noise bandwidth, in which case the results of the calculations using Equations 3.213 and 3.215 would be different. This is discussed further in Section 5.2.

In general it is considered acceptable practice to write $C/I = C - I$ and assume that the context makes it clear that this must be a decibel calculation.

To calculate the $C/(N + I)$, it is necessary to add the I and the N in absolute; hence

$$\left(\frac{C}{N+I}\right) = C - 10\log_{10}\left[10^{N/10} + 10^{I/10}\right] \tag{3.216}$$

An important quantity is the energy per bit which is related to the wanted signal and data rate via

$$E_b = C - 10\log_{10}R_d \tag{3.217}$$

Note that here C is assumed to be in dB (watts) so that the energy per bit is in dB (joules) as watts = joules/second and the rate is in bits/second. Hence the E_b/N_0 is

$$\left(\frac{E_b}{N_0}\right) = (C - 10\log_{10}R_d) - (N - 10\log_{10}B) \tag{3.218}$$

Example 3.30
The link budget for a satellite to Earth link is given in Table 3.16 based on the parameters for the scenario in Example 3.33.

Note there is no single industry-wide agreed way to present a link budget, but some organisations do use a standardised template. The contents of the link budget will vary depending upon circumstances and what information is available and required.

Satellite interference is often calculated in terms of the percentage in absolute of the ratio between the change in temperature, ΔT or DT, to T:

Table 3.16 Example link budget

Field	Symbol(s)	Value	Calculation
Transmit frequency (GHz)	f	3.6	
Bandwidth (MHz)	B	30	
Transmit power (dBW)	P_{tx}	−10	
Transmit peak gain (dBi)	G_{tx}	36.9	
Transmit EIRP (dBW)	$EIRP$	26.9	$P_{tx} + G_{tx}$
Distance (km)	d	38,927.2	
Free space path loss (dB)	L_{fs}	195.4	From Eq. 3.13
Receive peak gain (dBi)	G_{RX}	36.9	
Feed loss (dB)	L_f	1	
Receive signal (dBW)	C	−112.5	From Eq. 3.203
Receive temperature (K)	T	300	
Receive noise (dBW)	N	−129.1	From Eq. 3.210
C/N (dB)	C/N	16.5	$C - N$

$$\frac{DT}{T} = 100.10^{\left(\frac{I/N}{10}\right)} \tag{3.219}$$

Or:

$$\frac{I}{N} = 10\log_{10}\left(\frac{DT/T}{100}\right) \tag{3.220}$$

Example 3.31
A satellite system has a $DT/T = 8\%$. In I/N this is $10\log_{10}(0.08) = -10.97$ dB.

If there are multiple interferers then the power should be aggregated in absolute (i.e. in watts rather than dBW) and then, if necessary, converted back to dBW. The **aggregate interference** can therefore be calculated using

$$I_{agg} = 10\log_{10}\sum_i 10^{I_i/10} \tag{3.221}$$

Just as there is a single entry I_i/N, C/I_i, etc., there can be an I_{agg}/N, C/I_{agg}, etc., which can have separate thresholds. For some scenarios it is important to consider both the single entry interference and aggregate as discussed in Section 5.9.

For regulatory analysis it is often useful to know the *PFD* which, as noted, in absolute is

$$pfd = \frac{eirp}{4\pi r^2} \tag{3.222}$$

Sometimes it is useful to include the receive relative gain in the *PFD* calculations, which results in a metric called the equivalent power flux density (*EPFD*):

$$epfd = \frac{eirp}{4\pi r^2} \cdot g_{rx,rel} \tag{3.223}$$

An alternative way of defining the signal strength of a radio system is by the magnitude of the electric field. The **field strength** E generated is similar to the *PFD* but gives the power in dBμV/m. The average power is determined by the square of the voltage, so decibel calculations relating to the electric field include a $20\log_{10}$ term.

The *PFD* and field strength E are related via the impedance of free space $Z_0 = 119.9169832\pi$ ohms, often rounded to 120π ohms. The conversion from a field strength e_v in volts/m to *pfd* in W/m^2 is then:

$$pfd = \frac{e_v^2}{Z_0}$$

(3.224)

So a field strength $E_{\mu V}$ in dBμV/m can be converted to a *PFD* in dBW/m^2 using:

$$PFD = E_{\mu V} - 145.8$$

(3.225)

The *PFD* can be converted to and from the signal at an isotropic receiver simply using Equation 3.6:

$$S = PFD + A_{e,i}$$

(3.226)

where:

$$A_{e,i} = 10\log_{10}\left(\frac{\lambda^2}{4\pi}\right)$$

(3.227)

A useful metric of a satellite receiver is its *G/T*, which is the ratio of the peak gain to its temperature in units of dB/K. A higher value is better as the received wanted signal to noise is greater, as in

$$\frac{C}{N_0} = P_{tx} + G_{tx} - L_p - L_f + \frac{G}{T} - k$$

(3.228)

3.12 Spectrum Efficiency and Requirements

Most spectrum management regulations highlight the need for efficient use of the radio spectrum, but what does that mean? The simplest metric, described in Section 3.4.5, is the bandwidth efficiency of the carrier, namely, the ratio of the data rate R_d (bps) to the bandwidth B (Hz):

$$e_C = \frac{R_d}{B}$$

(3.229)

This metric increases for higher-order modulations which require greater transmit power and hence can lead to interference and/or range issues. Having a high efficiency of the carrier is generally beneficial for one user but the radio spectrum is a shared resource and it is therefore useful to consider the aggregate data rate that can be transmitted to all users within a certain area A.

$$e_A = \frac{\sum R_d}{A.B}$$

(3.230)

This is in units of bits/s/Hz/km^2.

The spectrum efficiency in terms of data rate per area is commonly used for mobile systems, and it is almost always possible to increase this metric by one of:

- Increasing the number of sectors per base station
- Increasing the number of base stations.

However this metric is less useful for other applications. In particular, for broadcasting it could be argued that the most efficient radio system is the one which can provide the greatest coverage from a single transmitter, so that

$$e_T = \frac{A.R_d}{B} \tag{3.231}$$

These technical measures of efficiency are not sufficient on their own as it is also necessary to consider cost implications. The metric e_A can be optimised by having a very large number of small cells, but this would be very expensive, and there would be resistance on environmental grounds to the deployment of large numbers of base stations.

For many mobile operators there is therefore a trade-off between:

- The cost of spectrum, for example bought at auction
- The cost of installing base stations including associated infrastructure.

Mobile operators are therefore able to put a value on spectrum, taking into account their income stream and costs of base stations, backhaul, etc. This can also be used to identify a requirement for spectrum using traffic models to identify demand and deployment models to identify the best way to meet that.

These sorts of trade-offs occur for other radio services. For example, the GSO satellite arc is considered another resource along with the radio spectrum. One measure of how efficiently it can be used is the minimum separate distance between co-frequency co-polarization satellites providing co-located services. For a given C/I this will be a function of the dish diameter at the Earth station.

Example 3.32
Assuming the threshold to protect a BPSK link is a $C/I = 22.7$ dB, the dish diameter required at $f = 3.6$ GHz and $f = 12$ GHz for a co-frequency, co-polar, co-located satellite service assuming clear air (i.e. free space path loss) can be estimated as depending upon the separation on the GSO arc between wanted and interfering satellites according to Figure 3.75.

It can be seen that the dish diameter required increases dramatically as the GSO arc separation decreases. Therefore the most efficient use of the GSO arc would also be the most expensive in terms of ground segment cost.

For both satellite and terrestrial systems there can be clear trade-offs between efficient use of a resource (spectrum, GSO arc) and total system costs including equipment and deployment. It is not always sufficient to use technical metrics such as bits/s/Hz or bits/s/Hz/km^2. In consequence, this complicates any analysis of spectrum requirements.

There may also be other metrics that need to be considered, for example power consumption. There can be requirements to reduce the environmental impact of radio systems and their total

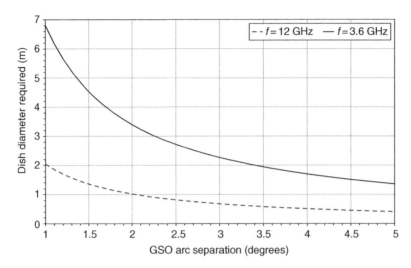

Figure 3.75   Dish diameter vs. GSO arc separation angle

CO_2 emissions as part of wider societal goals. An example of this was in the transition from analogue to digital terrestrial TV. While this resulted in an increase in spectrum efficiency, with more TV channels transmitted within the same bandwidth and/or spectrum being released for other applications, there was an increase in total power consumption due to the greater complexity of the receivers. In the UK the extra power required was estimated to be 31 MWh/day (Scientific Generics Limited, 2005).

The need to balance these various requirements is why regulators use a combination of tools to manage the radio spectrum, including auctions, licensing and beauty-contests. There are strong connections between interference issues as discussed in this book and spectrum economics, though this field remains under-researched, and would benefit from further study.

One approach that connects the concepts of spectrum efficiency, interference analysis and economic models is the N-Systems methodology, which uses interference analysis to determine the number of stations of a particular type (equipment and service) that could be deployed in a given reference area before reaching a limit due to interference. The density of deployment is another metric of spectrum efficiency, and it is possible to determine the decrease in deployment density of one system due to the introduction of systems of another type of service. This could be used to determine a per station spectrum opportunity cost that could be included in a licence fee, together with any fixed costs. This approach is described further in Section 7.8.

3.13   Worked Example

Example 3.33   In this example scenario there are two GSO satellite systems as shown in Figure 3.76:

- Victim: GSO satellite at 30°E transmitting to an Earth station in Paris
- Interferer: GSO satellite at 25°E transmitting to an Earth station in London.

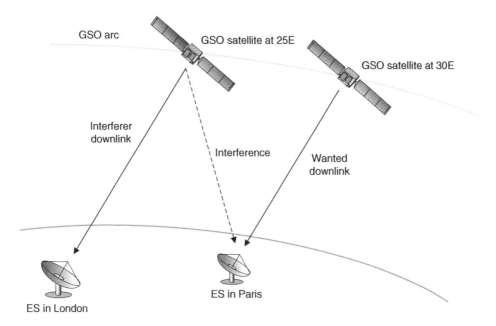

Figure 3.76   Interference scenario

A satellite scenario has been chosen for its simplicity as many terrestrial scenarios would require propagation models more complex than free space path loss: this scenario is, however, topologically very similar to that between two fixed links.

Both satellites are transmitting at 3.6 GHz with $P_{tx} = 10$ dBW and a bandwidth of 30 MHz. All antennas are assumed to be 2.4 m diameter parabolic dishes with efficiency = 0.6, line feed loss of 1 dB and have a total noise temperature = 300 K.

Assuming that all beams have the standardised gain pattern described in Section 3.7.5, and that the GSO satellite and ES's beams are pointing directly at each other, what would be the link attributes $\{C/N, C/I, I/N, DT/T, C/(N + I)\}$ assuming free space path loss?

The following resources are available:

Resource 3.1   Spreadsheet 'Chapter 3 Calculations.xlsx' contains the calculations in this example

Resource 3.2   Visualyse Professional simulation file 'GSO Example.SIM' has been configured to match this example.

The starting point is to convert the locations of the satellites and Earth stations into a common geometric format, and it would be appropriate in this case to use ECI with a spherical Earth model. Using Section 3.8.4 we can calculate the vectors to be as shown in Tables 3.17 and 3.18.

From these vectors it is possible to calculate off-axis angles at the victim ES towards the interfering satellite, and the interfering satellite towards the victim ES using the method in Section 3.9.5. This is shown in Table 3.19, which also shows the relative gains, assuming the standardised gain pattern described in Section 3.7.5.

Table 3.17 GSO satellite ECI vectors

Satellite	GSO at 25E	GSO at 30E
x (km)	38,213.7	36,515.2
y (km)	17,819.3	21,082.1
z (km)	0.0	0.0

Table 3.18 ES ECI vectors

Earth station	London	Paris
x (km)	3,969.85	4,192.94
y (km)	−8.83	172.13
z (km)	4,992.09	4,803.17

Table 3.19 Interfering path angles and relative gains

Off-axis angles and relative gains	Off-axis angle	Relative gain
At Paris ES towards satellite at 25°E	5.4°	−26.2 dB
At satellite at 25°E towards Paris ES	0.2°	−0.1 dB

Table 3.20 Distances and free space path loss from Paris ES to two GSO satellites

GSO satellite	GSO at 25E	GSO at 30E
Distance (km)	39,100.8	38,794.1
Free space path loss (dB)	195.4	195.4

The distances are simply the magnitude of the vectors from the Paris ES to the two satellites. Putting the distances and a frequency = 3.6 GHz into Equation 3.13 it is possible to calculate the free space path loss of the wanted and interfering paths as in Table 3.20.

The peak gain can be calculated using Equation 3.78 to be 36.9 dBi.

From the aforementioned, the wanted and interfering link budgets can be calculated as in Table 3.21. The noise can be calculated from Equation 3.43 as $N = -129.1$ dBW. Hence the link attributes are those given in Table 3.22.

But what do these numbers mean? Are they evidence that the interference is acceptable or harmful? The topic of thresholds will be discussed in general in Sections 5.9 and 5.10, then in more detail for selected services in Chapter 7.

3.14 Further Reading and Next Steps

In this chapter we have developed a model of the link calculation from first principles to the link budget and associated attributes. This has included a description of key concepts such as carrier modulation, access methods, noise, antennas and geometry.

Table 3.21 Wanted and interfering link budgets

Signal	Wanted	Interfering
Frequency (MHz)	3 600	3 600
Transmit power (dBW)	10	10
Transmit peak gain (dBi)	36.9	36.9
Transmit off-axis angle (deg)	0.000	0.227
Transmit relative gain (dB)	0.0	−0.1
Distance (km)	38,927.2	39,100.8
Free space path loss (dB)	195.4	195.4
Receive peak gain (dBi)	36.9	36.9
Receive off-axis angle (deg)	0.0	5.4
Receive relative gain (dB)	0.0	−26.2
Receive losses (dB)	1.0	1.0
Received signal (dBW)	−112.5	−138.9

Table 3.22 Example link attributes

C/N (dB)	16.5
I/N (dB)	−9.9
DT/T (%)	10.3
C/I (dB)	26.4
$C/(N+I)$ (dB)	16.1

The next step for interference analysis is to go into more details of the link calculation, in particular:

- Developing the propagation model beyond free space, as described in Chapter 4
- Developing the interference calculation to include factors such as carrier shape, polarization, traffic models and aggregation effects, as in Chapter 5
- Different methodologies that can be used to analyse interference, including static, dynamic and Monte Carlo, as described in Chapter 6
- Specific algorithms designed for particular services and sharing scenarios are described in Chapter 7.

This chapter has touched on issues relating to carrier and antenna engineering, looking at those aspects relevant to interference analysis. Further information can be found in the books referenced, in particular (Bousquet and Maral, 1986; Kraus and Marhefka, 2003; Rappaport, 1996).

4

Propagation Models

The defining characteristic of radio systems is that signals are transmitted wirelessly. Between the transmit (TX) and receive (RX) antennas, the signal strength will be altered, typically reduced, by the environment through which the electromagnetic wave propagates. The propagation model is the algorithm to calculate this reduction in signal strength.

While there is a theoretical model of electromagnetic waves, defined via Maxwell's equations, this is not typically used in interference analysis due to its computational complexity. Instead there are a range of models, dozens in total, and a key question for any interference analysis is which one(s) should be selected.

The objective of this chapter is to review the main propagation models and identify for each (where applicable):

- A reference that specifies the model
- The motivation behind it
- What factors are included and which are excluded
- The range of applicability, such as geometry, heights and frequencies
- Examples of the predicted propagation loss.

Key to these models is a definition of the propagation environment, which can include geoclimatic parameters plus terrain, surface and land use databases, which are discussed in Section 4.2. As there are significant differences between terrestrial propagation models and those for Earth to/from space, these are handled separately, in Sections 4.3 and 4.4, respectively. There are also discussions of models for aeronautical applications and attenuations due to buildings, floors and other obstructions.

Interference Analysis: Modelling Radio Systems for Spectrum Management, First Edition. John Pahl.
© 2016 John Wiley & Sons, Ltd. Published 2016 by John Wiley & Sons, Ltd.
Companion website: www.wiley.com/go/pahl1015

The following section provides an overview of the main types of propagation models and concepts, while the chapter ends with a discussion of the selection of the most appropriate model when undertaking interference analysis.

The following resource is available:

Resource 4.1   Spreadsheet 'Chapter 4 Examples.xlsx' contains the calculations used in the derivation of diffraction loss, Earth size, P.452 clutter loss, dual slope loss model and location variability including the standard deviation, σ.

4.1   Overview

The selection of propagation model(s) can have a dramatic impact on the resulting predicted levels of interference, to the extent of converting a scenario with harmful interference into one with no significant interference and vice versa. Every interference study requires one or more propagation models to be used, and the selection must be documented and justified.

There has been decades of research into propagation models done within the remit of the ITU-R in Study Group 3 and also by organisations, including regulators like Ofcom and research organisations such as NASA and the Rutherford Appleton Laboratory (RAL). The models are developed based on defining a scope (such as range of frequencies, heights, percentages of time, etc.) and then undertaking measurements. A model is then created, sometimes including physical concepts and other times simply curve fitting. The model is typically peer reviewed at (say) ITU-R SG 3 or equivalent before being accepted. Often additional data becomes available, and so the model is updated accordingly.

There are regular updates to many of the propagation models, and the reader is invited to check the ITU-R website for the latest versions. The most recent version of each of the Recommendations should always be selected, except when trying to reproduce the results from earlier studies. The core concepts, however, tend to be stable for many years, and it is very rare for an old model to be deleted.

These models therefore represent the result of a rigorous process, and it is not necessary to be involved in their development to use them in studies. Propagation models gain credibility in passing through this approval process and being accepted as an approved Recommendation. Even greater confidence comes from their use in spectrum management for interference analysis, planning and frequency assignment over years and decades, involving potentially thousands of radio systems.

This confidence means that if a suitable model has been selected, it can be assumed for the purposes of the study as being 'correct'. In reality there could be large differences with an actual link: standard deviations of over 10 dB in the difference between prediction and measurement are not unusual (Ofcom, 2008a). However, the defence is that these models are the best available to undertake that part of the analysis.

Some of these models are very complicated and so will not be described on an equation by equation basis in this book. The focus will instead be on describing the main features or components of each model to allow a greater understanding of what they are trying to achieve and hence provide a guide as to their applicability. Most interference analysis study work is undertaken by software tools that implement these models, and therefore this chapter is targeted at answering how to select and configure the appropriate model or models.

The models are typically targeted at specific issues, segmented by the following:

- Geometry: in particular, is the path terrestrial, between ground and air or between Earth and space?
- What range in distance is the study considering? A model of radio wave propagation within a building will be different from one working to a range of a thousand kilometres.
- What range of percentages of time are of interest?
- Is the analysis aimed at a generic study, or is it specific to a certain location?
- Is the analysis considering the coverage (wanted or interfering) over an area or loss at a specific point?

A key issue is the percentage of time for which a signal level could be exceeded. The atmosphere is continually changing, with variations in temperature and pressure, with rain, snow, sleet, ice and sand effects. Rather than modelling each meteorological phenomenon, some propagation models try to predict the statistical variation in signal strength due to these factors.

The format is often $L_p(p)$, which gives the loss exceeded for $p\%$ of the year, as from Rec. ITU-R P.452 (ITU-R, 2013d):

> p = *required time percentage(s) for which the calculated basic transmission loss is not exceeded*

The propagation loss (a positive number) is taken off from the signal strength, so the percentage of time is as follows:

For propagation loss, p = *time percentage for which the propagation loss* **is not** *exceeded*	⇒	*For signal strength,* p = *time percentage for which the received signal level* **is** *exceeded*

Usually the percentage of time is averaged over a year, though it could alternatively be over the worst month. The critical range of percentages will differ depending upon whether the signal is:

- Wanted: in this case it is decreases in signal strength (i.e. increases in loss) that are of interest, and so the critical region is where p is 50% or higher of the time
- Interfering: in this case it is increases in signal strength (i.e. decrease in loss) that are of interest, and so the critical region is where p is 50% or lower of the time.

In some cases the only term in the link budget that varies in time is the propagation model. This means that the signal strength for a given percentage of time can be calculated by putting that value in the propagation model. If other terms vary then it becomes necessary to undertake more detailed modelling, such as Monte Carlo analysis.

As will be seen later, the percentage of time for which a signal is met or exceeded is important for comparison against interference thresholds. These are sometimes categorised by:

- **Short-term**: interference for less than 1% of time
- **Long-term**: interference for 20–50% of the time.

These terms, therefore, are often applied also to the propagation model percentage, for example, in Rec. ITU-R SM.1448 (ITU-R, 2000b).

Some models do not give an associated percentage of time. This could be because either the modelled phenomena do not vary significantly (e.g. free space path loss) or because it is a prediction of the median, that is, the most likely path loss.

Point-to-area propagation models have an associated percentage of locations, $q\%$, which has similar structure to the percentage of time:

For propagation loss, **q** = *location* percentage for which the propagation loss **is not** *exceeded*	$\Rightarrow$ *For signal strength,* **q** = *location* percentage for which the received signal level **is** *exceeded*

Point-to-area models derive the variation of the signal strength over a square pixel, often used as part of the calculation of the coverage area.

Another point to note is that some models can be used on their own (in scenarios for which they are suitable), but others just describe components that must be combined with others. For example, while P.452 is a complete propagation model, P.526 only describes how to calculate the diffraction loss and therefore would have to be combined with another model to calculate the total path loss.

Some Recommendations give additional information to the propagation model or models. For example, P.530 covers a wide range of topics relating to planning fixed links in addition to the rain and multipath fade models. It is not always possible to implement all of this information as some is in the form of general guidelines rather than an algorithm that can be implemented in software.

During interference analysis, in many scenarios it is useful to have different propagation model(s) for wanted links and interfering links. An example would be for the fixed service, where point-to-point links might use P.525 for the core loss together with P.530 as the rain fade model to calculate the wanted signal but use P.452 for interfering signal(s).

The polarisation used by a radio wave (as described in Section 5.4) can have an impact on the propagation loss. For example:

- For P.452 (see Section 4.3.4) the path loss is slightly greater for vertical polarisation than for horizontal. This difference is typically less than 1 dB, decreasing as the frequency increases
- For P.530 (see Section 4.33) the rain fade is slightly lower for vertical polarization than horizontal
- The polarisation of a radio wave can rotate as it propagates. An example of this would be low-frequency Earth to space paths, which makes it attractive to use circular polarisation (Haslett, 2008)
- The radio wave can change its polarisation, particularly in locations where there could be reflections off clutter.

Note that there are other aspects of propagation that could be modelled as well as loss, in particular delay due to propagation effects. An example of where this is important would be the management of time synchronisation within single frequency networks (SFN) as described in Section 7.3.

4.2   The Propagation Environment

When radio waves propagate in a vacuum, the so-called free space propagation, the only factors to consider are the frequency and distance. For all other cases it is necessary to consider the environment that the radio wave propagates through, the gases, possibly rain, snow, sleet or sand, any buildings, vegetation, hills, waterbodies and so on. The meteorological and geoclimatic data, together with terrain, surface and land use databases, are resources that can be used by propagation models to improve the accuracy of the prediction of loss between TX and RX antennas.

When undertaking generic interference analysis studies, that is, those where the conclusions are considered applicable to a wide range of locations, it is generally best to select specific values that are appropriate for the scenario under consideration and then quote what those values were together with the results. If the analysis is targeted at a specific site, then it is better to take the values from a database for that location, again documenting what was done and why.

4.2.1   Effective Earth Radius

In a vacuum, radio waves will travel from transmitter to receiver in a straight line. In the Earth's atmosphere radio waves travel through various gases, the density of which varies by height. Therefore it is necessary to consider refraction, the scale of which is determined by the radio refractive index, n. This is a dimensionless number very close to 1, and it is more usual to work with the refractivity, N:

$$N = (n-1) \times 10^6 \tag{4.1}$$

The average change in the refractivity N over the lowest 1 km of the atmosphere, the ΔN parameter, has an impact on how far radio waves travel before reaching the horizon on a **smooth Earth** (i.e. the Earth modelled as a sphere without terrain). In Figure 4.1:

- On the left, the radio path is shown on the physical Earth, a sphere with radius R_e. The radio waves propagate a distance d_r on a curve that reaches beyond the straight line horizon
- The right shows how this curved path can be modelled as a straight line on a larger Earth, a sphere with radius a_e, called the **effective Earth radius**. The radio propagates the same distance d_r to reach the straight line horizon.

In the figure O represents the origin or centre of the physical Earth, while on the right O' is the centre of the larger sphere used in many terrestrial propagation models. In particular, the Earth's curvature must be taken into account when calculating the height of an obstacle (such as terrain) on a path. By using the effective Earth radius, the radio wave can be treated as though travelling in a straight line.

The ratio between the effective Earth radius a_e and physical Earth radius R_e is the parameter k as in:

$$a_e = k.R_e \tag{4.2}$$

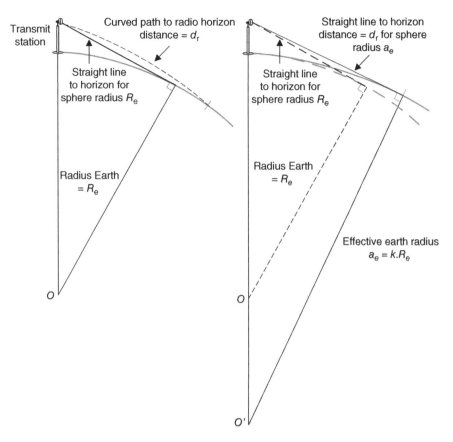

Figure 4.1   Effective Earth radius

Table 4.1   Example calculations of effective Earth radius

ΔN	30.0	45.0	60.0
k_{50}	1.24	1.40	1.62
a_e	7884.7	8940.7	10,323.3

The value of k is not constant but varies by location and also percentage of time. The median value, k_{50}, is given in Rec. ITU-R P.452 (ITU-R, 2013d) as

$$k_{50} = \frac{157}{157 - \Delta N} \tag{4.3}$$

Example values of k_{50} and a_e for various values of ΔN are given in Table 4.1.

While k varies depending upon location and percentage of time, for temperate climates ΔN is typically close to 39, which corresponds to a $k = 1.33$. Thus a $k = 4/3$ is widely used approximation for the median effective Earth radius factor.

At small percentages of time, there can be atmospheric layering effects, which can lead to reflection layers or ducts, for which the effective Earth radius concept does not apply.

4.2.2   Geoclimatic and Meteorological Parameters

The previous section described how the ΔN alters the geometry of the radio path, increasing or decreasing the refractivity. It is one of a number of geoclimatic and meteorological parameters that are used by the propagation models, some of which are available in global databases, including:

- ΔN = the change in radio refractivity over the lowest 1 km of the atmosphere (as described in the previous section)
- N_0 = average value of atmospheric refractivity extrapolated to sea level
- Temperature, typically in °C or K, which varies by (latitude, longitude) and height
- Pressure, typically in hectopascal (hPA = millibar)
- Water vapour density, typically in g/m^3
- Rain rate, typically given as the rainfall rate in mm/h exceeded for 0.01% of the time.

Many of these are specified in ITU-R Recommendations, such as:

- Rec. ITU-R P.453-10: The radio refractive index: its formula and refractivity data (ITU-R, 2012d)
- Rec. ITU-R P.835-5: Reference standard atmospheres (ITU-R, 2013i)
- Rec. ITU-R P.836-5: Water vapour: surface density and total columnar content (ITU-R, 2013j)
- Rec. ITU-R P.837-6: Characteristics of precipitation for propagation modelling (ITU-R, 2012f). This references a set of electronic data sets with instructions on how to interpolate between data points
- Rec. ITU-R P.838-3: Specific attenuation model for rain for use in prediction methods (ITU-R, 2005d)
- Rec. ITU-R P.1510: Annual mean surface temperature (ITU-R, 2001b).

Another, older, approach to define rainfall is via the rain climatic zone, which in Rec. ITU-R SM.1448 is one of {A, B}, {C, D, E}, {F, G, H, J, K}, {L, M} or {N, P, Q}, which maps onto the rainfall rate in mm/h $R(p)$, which is exceeded on average for $p\%$ of a year.

4.2.3   Radio Climatic Zones

Radio waves propagate further over water than land, and so in some cases, particularly terrestrial propagation, key inputs would be the zone or zones as in Rec. ITU-R SM.1448 (ITU-R, 2000b):

- Zone A1: coastal land and shore areas, that is, land adjacent to the sea up to an altitude of 100 m relative to mean sea or water level but limited to a distance of 50 km from the nearest sea area

- Zone A2: all land, other than coastal and shore areas defined as 'coastal land'
- Zone B: 'cold' seas, oceans and large bodies of inland water situated at latitudes above 30°, with the exception of the Mediterranean Sea and the Black Sea
- Zone C: 'warm' seas, oceans and large bodies of inland water situated at latitudes below 30°, as well as the Mediterranean Sea and the Black Sea.

Sometimes Zones B and C are combined together to be simply 'sea', and propagation models such as Rec. ITU-R P.452 are able to handle paths that cross multiple zones.

These zones can be extracted either from using the databases in the following sections (such as terrain plus land use classification) or are available directly in the ITU Digitized World Map (IDWM) available from the ITU's Radiocommunications Bureau.

Example 4.1

An example of how the radio climatic zone changes the interference zone predicted using P.1812 for 10% of the time is given in Figure 4.2. In particular, the calculation without the radio climatic zones assumed the path was always over land and did not take account of how radio waves propagate further over water bodies such as seas.

The 12.5 kHz private mobile radio (PMR) system is transmitting at 12 dBW on 420 MHz with base station located at (50.334453°N, −4.635703°E) and antenna 10 m above smooth Earth. Grid lines are 1° apart in latitude and longitude, and the map projection is Mercator.

The following resources are available:

Resource 4.2 Visualyse Professional simulation file 'LM coverage no surface.sim' was used to generate the coverage without surface data or zones.

Resource 4.3 Visualyse Professional simulation file 'LM coverage no surface zones.sim' was used to generate the coverage using zones but no surface data.

Note that terrain or surface data is not included in these simulation files.

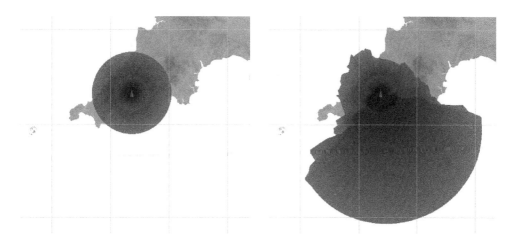

Figure 4.2 Interference zone with (right) and without (left) radio climatic zone data

4.2.4   Terrain and Surface Databases

Section 3.8 described models of the Earth as either a sphere or oblate spheroid, but the reality is more complex with terrain features such as hills and valleys and, on top of these, buildings and vegetation. There are two types of database that define these features:

1. Terrain databases: these include just phenomena such as hills, valleys, etc.
2. Surface databases: these include buildings and vegetation on top of the terrain data.

   These databases are typically provided in a raster grid format, with a spot height at each point. Heights between grid points can be calculated using the bilinear interpolation in Equations 3.101, 3.102 and 3.103 with (azimuth, elevation) replaced by (latitude, longitude) or other units such as (northings, eastings). Other approaches to interpolation can be used to calculate the path profile, such as taking the nearest terrain spot height, but this can lead to inaccuracies.
   An example of the raster grid of spot heights is shown in Figure 4.3, together with the line from TX to RX station, the so-called path profile.
   The path profile is generated along the great circle between the TX and RX stations, though for very short distances, it can be acceptable to use flat Earth geometry. The path profile points are usually extracted at equally spaced distances between the two stations, with the distance between profile points related to the resolution of the terrain or surface database. It is, in theory, possible to have path profiles with non-equal spacing, for example, for long paths to have high resolution between a station and its horizon and lower resolution over the longer distance. However, in most cases equal spacing is used in software tools, and this is reflected in recent updates to the Recommendations.

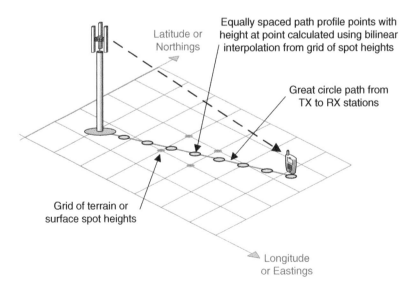

Figure 4.3   Extracting path profile from grid of spot heights

It is generally assumed that the spot's (latitude, longitude) point defines the precise location of that height, but there are cases where it is the average height across a pixel, and the position specifies either the centre of the pixel or its (bottom, left) corner.

There can also be inaccuracies in the generation of these data sets, in particular, in the height axis where errors can be in the order of magnitude of metres. Note that different geoids can have different height references as well as (latitude, longitude) offsets.

As link lengths are not likely to be an integer multiple of the requested path profile resolution, it is necessary for software tools to round up or down the number of samples required when generating each path profile: it is generally considered better practice to round up so that the distance between samples is never greater than the requested resolution. In addition, it is usually necessary to have at least three samples: the TX and RX stations and the point half way in between.

A useful rule of thumb is to have a resolution of the path profile approximately equal to the resolution of the underlying database, whether terrain or surface, so that detail is not lost. However with very high-resolution databases, there are computational overheads involved that can become significant. There is some research that suggests that some reduction of the path profile sample resolution can be achieved without significant degradation in the standard deviation of difference between prediction and measurement, but the results are path dependent.

This path profile can have a dramatic impact on the received signal strength and is a key input to some of the main terrestrial propagation models (e.g. P.452, P.1546, P.1812 and P.2001).

Example 4.2
An example path profile is given in Figure 4.4.

The path profile extracted will depend upon the type of database, that is, whether it's a terrain or surface database as in Figure 4.5.

Increasingly, very high-resolution surface databases are becoming available, thanks to remote sensing techniques. For example, aircraft can be flown over urban areas to generate a surface database with resolution a metre in horizontal directions and errors of a few millimetres in the vertical, capturing each building in detail. This can be used for more accurate predictions of signal propagation in specific urban areas.

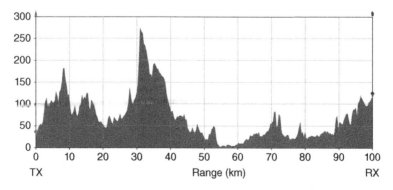

Figure 4.4 Example path profile

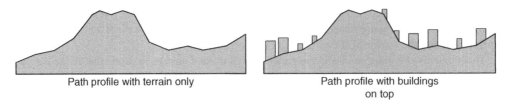

Path profile with terrain only   Path profile with buildings
                                                              on top

Figure 4.5 Difference between terrain and surface path profiles

While the increased detail in the terrain or surface database can lead to more accurate results, it requires the equal accuracy in the other input parameters, in particular the location of the TX and RX stations, including ensuring the heights are selected to be consistent with the database used. Note that surface data is also more likely to become out of date quicker than a simpler terrain database.

A number of freely available global terrain and surface databases are available, in particular:

- Shuttle Radar Topographic Mission (SRTM): a radar was flown on a Space Shuttle mission that scanned the whole Earth between 60°N and 54°S. As it was observing from space, there are occasional holes in the data due to (say) the far side of the mountain being hidden from view or the orbit not covering a location. Where possible, the released data set has tried to fill in these holes. It is a surface database with resolution 30 or 90 m, so it is not always possible to identify specific buildings but rather city blocks, and there is some noise, in particular in the vertical axis (NASA, n.d.)
- Advanced Spaceborne Thermal Emission and Reflection Radiometer (ASTER): this is similar in concept to the SRTM mission except it covers a wider range of latitudes, from 83°N to 83°S. Again there are anomalies and slightly greater noise than SRTM though the resolution is the same at 30 m (NASA JPL, n.d.).

Both these data sets are referenced to the WGS84 geoid. A problem with both is that they define the surface height, but there is no way of identifying what the underlying terrain height is. This creates difficulties if stations are specified in terms of their heights above terrain, and therefore a certain amount of judgement is required in their use. Ideally both terrain and surface data would be provided for the same areas.

Another complication is the relationship with land use data, as described in the next section, and the need to avoid double counting.

Example 4.3
An example of how the SRTM 90 m surface database changes the PMR interference zone from Example 4.1 is given in Figure 4.6. In both cases radio climatic data was used, for example, to identify percentage of path over sea.

The following resource is available:

Resource 4.4 Visualyse Professional simulation file 'LM coverage with surface and zones. sim' was used to generate the coverage with surface data and zones. The associated terrain file 'Resource 4-4 terrain.gen' is also available.

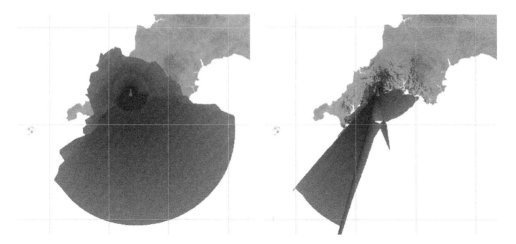

Figure 4.6 Interference zone with (right) and without (left) surface data

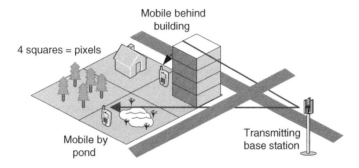

Figure 4.7 Scenario with two mobile signals affected by clutter

4.2.5 Land Use Databases

Consider the scenario in Figure 4.7. This shows a mobile system where there is communication to two mobiles the same distance from the base station. One mobile is behind a large building, the other by an open area where there is a pond. In this situation the received signals will be very different due to the local clutter.

One way to model this would be via the surface database described earlier, but this has a number of limitations:

- It is typically expensive (both to purchase the data and computationally) to model a large area in sufficient detail to capture buildings accurately
- Such data does not take account of soft objects, in particular vegetation such as trees
- Additional information and models would be required to identify how the signal strength would vary across a pixel such as the four shown in Figure 4.7.

Therefore, an alternative approach is to use what is called a land use database, in which each pixel or square has associated with it a code that defines the characteristics of that square. Examples of these codes would be (Ofcom, 2012a):

- Dense Urban
- Urban
- Industry
- Suburban
- Village
- Parks/Recreation
- Open
- Open in Urban
- Forest
- Water.

Another example set of land use codes is given in Rec. ITU-R P.1058: Digital topographic databases for propagation studies (ITU-R, 1999b).

Associated with each of these codes would be a set of parameters, such as typical clutter height, that could then be used in the propagation model to calculate:

1. Clutter loss, that is, how the received signal is reduced due to obstructions near the station (receiver and/or transmitter as appropriate)
2. The standard deviation of the variation in signal across the pixel.

The land use code is applied to all points within a defined pixel, referenced by a location that is either the centre of the pixel or the bottom left. The size of the pixel could be 50 m, 200 m or even 1 km. In general, the higher the resolution the more expensive the data that must be purchased as there aren't the equivalents of the SRTM and ASTER data sets for land use data.

One issue is that the pixel could contain a variety of types of land use, and so allocating just the one is a simplification, particularly when the pixel is large. For example, in Figure 4.7 a larger pixel containing all four squares would include examples of Dense Urban, Suburban, Forest and Parks/Recreation. Conversely, if the pixel is small, the received signal strength could be influenced by surrounding land use codes. For example, a pixel could be defined as Parks/Recreation but be surrounded (further out) by high buildings defined as Dense Urban.

It is important to be aware of the interaction with the terrain or surface database (if used). A surface database that contains details of buildings should not be used with land use codes such as Urban as that could lead to double counting of the terminal clutter loss. Note also how the propagation model in P.1812, described in Section 4.3.6, adjusts the path profile to include the height of the clutter using the land use database, as shown in Figure 4.28.

4.2.6   Signal Variation and Fast Fading

When one end of a terrestrial path is moving, the signal received will vary depending upon a range of factors, categorised in Rec. ITU-R P.1546 (ITU-R, 2013l) into:

- **Multipath variations** that occur over scales of the order of the wavelength due to reflections from the ground and buildings leading to phasor addition

- **Local ground cover variations** that occur due to buildings and vegetation over distances similar in size to these objects, typically much larger than that for multipath
- **Path variations** that occur due to changes in the geometry of the entire propagation path, in particular the presence of terrain. For all except very short paths, these variations are much larger than for local ground cover variations.

Multipath fading can occur over very short distances, and for moving stations, this can lead to very rapid signal variation or **fast fading**. In an urban environment, reflections off buildings result in multiple routes from TX antenna to RX antenna. As these multipath signals combine, they can be in or out of phase depending upon the distances travelled, and this can lead to fading or enhancement.

For a radio system with wavelength λ, a difference in distance Δd between two such paths results in a difference in the combined signal voltage (hence it is necessary to use $20\log_{10}$) in decibels of

$$\Delta E = 20\log_{10}\left[\left|1 + \cos\left(2\pi.\frac{\Delta d}{\lambda}\right)\right|\right] \tag{4.4}$$

Small changes in either frequency or distance can lead to significant differences in voltage, which means for a moving station rapid variations in received signal in frequency or time.

Example 4.4

Figure 4.8 shows a mobile that receives a signal from a base station but also reflected off a building 30 m offset from the direct radio path. For a given direct signal distance d_1 and building with offset x, the reflected signal will travel a distance d_2 given by

$$d_2 = 2\sqrt{x^2 + \left(\frac{d_1}{2}\right)^2} \tag{4.5}$$

The difference in distance between the two paths is then

$$\Delta d = d_2 - d_1 \tag{4.6}$$

This difference in distance will result in a time delay between the two paths, which can be calculated using Equation 3.173, and is a key factor in the design of OFDM systems.

Figure 4.9 shows the impact of multipath on the signal received as the frequency was varied from 700 to 750 MHz for the two cases of distances of 100 and 101 m. It can be seen that the combined signal varies in frequency with enhancements up to 6 dB but also fades down to effectively zero signal. The impact of moving 1 m makes a 6.05 MHz difference in the frequency profile and for the case of $f = 722$ MHz a 13 dB change in signal strength.

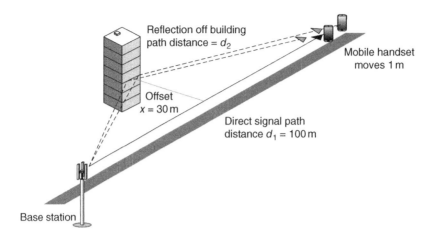

Figure 4.8   Multipath example with reflection off building

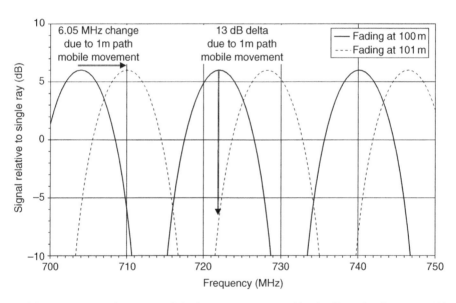

Figure 4.9   Variation in signal strength in frequency due to multipath effects for distance = 100 and 101 m

Fading can be categorised as either fast or slow based upon how rapidly the received signal varies in the time or frequency domains using the terminology in Table 4.2.

For a comparison with multipath fading, consider an example of local ground cover variation, in particular clutter loss. Figure 4.10 shows how the loss for an obstruction of height 20 m varies in frequency between $f = 700$ and 750 MHz, for separation distances of 100 or

Table 4.2 Types of fading

Variation	Slow	Fast
Time domain	Flat slow fading	Flat fast fading
Frequency domain	Frequency selective slow fading	Frequency selective fast fading

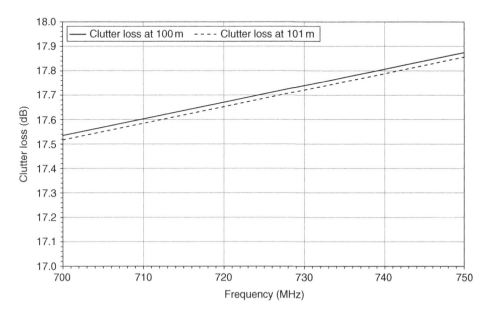

Figure 4.10 Variation in clutter loss by frequency and distance

101 m, using the model in Rec. ITU-R P.452. It can be seen that there is significantly less variation, so this shadowing would represent **slow fading**, whereas multipath would represent fast fading.

Criteria to identify whether fading is slow or fast in the time or frequency domain would be to compare the speed of variation against a radio system's symbol duration or bandwidth, respectively.

The complexity of urban environments makes it extremely challenging to model all possible geometries and hence calculate directly the multipath fading or enhancement. Instead, these multipath effects are typically modelled statistically, in particular using the Rayleigh or Rician distributions. If all the multipath signals have roughly the same strength, then the fading will have Rayleigh distribution. If one signal dominates (e.g. if it is line-of-sight (LoS)) then the fading distribution is better modelled via a Rician distribution (Haslett, 2008).

As noted previously, multipath fading can be highly frequency specific, which means that in a multichannel system, such as OFDM, only some subcarriers are highly attenuated. Fading of subcarriers can be managed by use of coding and processing each channel's *BER*.

Fast fading is typically indirectly a factor in interference analysis. However, inclusion of margin for fast fading in the wanted link budget is likely to result in the need for either higher TX powers or greater density of transmitters to provide the required QoS over the coverage area. This is likely to lead to higher levels of interference than if these fading effects were not included.

4.3 Terrestrial Propagation Models

This section describes the main terrestrial propagation models. Note that a comparison of the main models is given in Section 4.3.12.

4.3.1 P.525: Free Space Path Loss

This model is defined in the following reference:

• Recommendation ITU-R P.525-2: Calculation of free-space attenuation (ITU-R, 1994b).

The characteristics of the free space path loss model in P.525 are given in Table 4.3.

Free space path loss can be calculated using Equation 3.13 as derived in Section 3.3, namely:

$$L_{fs} = 32.45 + 20\log_{10}d_{km} + 20\log_{10}f_{MHz}$$

The Recommendation gives a number of other formulae, such as for point-to-area models and radars, though these are considered less helpful, and typically alternatives are used (e.g. P.1546 or P.1812 in Sections 4.3.5 and 4.3.6, respectively).

Table 4.3 Characteristics of Rec. ITU-R P.525 propagation model

Path types	Any
Frequency ranges	Any
Distances	Any
Antenna heights	Any
Percentages of time	n/a
Point-to-area model	No
Can use terrain data	No
Includes clutter model	No
Includes the core loss	Yes

Example 4.5

Figure 4.11 shows how the free space path loss varies for distances between 1 and 100 km and frequencies = {100 MHz, 1 GHz, 10 GHz}.

The model in Rec. ITU-R P.525 assumes no attenuations, obstructions, reflections and intrusions in the radio path or other atmospheric effects, which could lead to the propagation loss

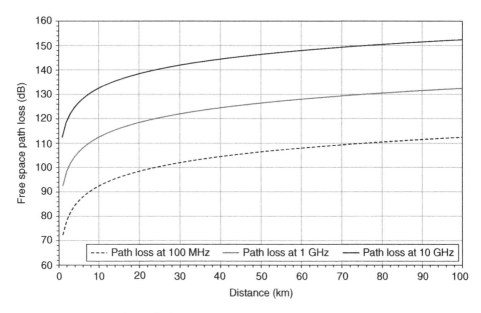

Figure 4.11   Rec. ITU-R P.525: free space path loss

being lower or higher and hence a signal that is either enhanced or faded, respectively. To model some of these factors, P.525 can be combined with other propagation models, such as:

- Rec. ITU-R P.526: diffraction including loss due to obstruction within the path's Fresnel zone, as described in Section 4.3.2
- Rec. ITU-R P.530: fading due to rain or multipath, as described in Section 4.3.3
- Rec. ITU-R P.618: rain fading for satellite links, as described in Section 4.4.2
- Rec. ITU-R P.676: gaseous attenuation, as described in Section 4.4.1.

Alternatively, a more detailed propagation model could be used, such as one of those defined in Recs. ITU-R P.452, P.1546, P.1812 or P.2001 or those designed for specific scenarios, such as the Rec. ITU-R P.528 for aeronautical paths described in Section 4.5.

Using P.525 as the propagation model can be considered a best case for wanted signals or a worst case for interfering signals.

4.3.2   P.526: Diffraction

This model is defined in the following reference:

- Recommendation ITU-R P.526-13: Propagation by diffraction (ITU-R, 2013e).

The characteristics of the diffraction loss model in Annex 1, Section 4.5, of Rec. ITU-R P.526 are given in Table 4.4.

As Rec. ITU-R P.526 does not include a core loss, it must be combined with another model such as the free space path loss model in Rec. ITU-R P.525.

Table 4.4 Characteristics of Rec. ITU-R P.526 propagation model

Path types	Terrestrial
Frequency ranges	No theoretical limit
Distances	Any
Antenna heights	Any
Percentages of time	Not directly, though the effective Earth radius is an input that can have an associated percentage of time
Point-to-area model	No
Can use terrain data	Yes
Includes clutter model	No
Includes the core loss	No

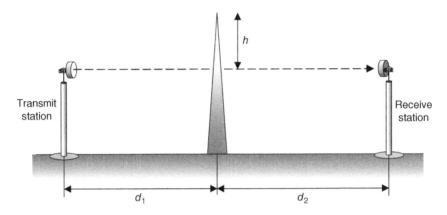

Figure 4.12 Knife-edge diffraction geometry

A simplified model of diffraction over a single knife-edge obstacle considers the geometry in Figure 4.12.

In this case the path loss between the TX and RX stations will be greater than free space path loss by an amount described as the diffraction loss. An important factor in the diffraction loss calculation is called the Fresnel parameter, which for this configuration using self-consistent units for (h, d_1, d_2) is (Haslett, 2008)

$$v = h\sqrt{\frac{2}{\lambda}\left(\frac{1}{d_1} + \frac{1}{d_2}\right)} \tag{4.7}$$

The diffraction loss can be calculated from this parameter. The full calculation is described in Rec. ITU-R P.526, but for most cases the loss is approximately

$$\text{if } v > -0.7 \quad L_d = 6.9 + 20\log\left(\sqrt{(v-0.1)^2 + 1} + v - 0.1\right) \tag{4.8}$$

$$\text{if } v \le -0.7 \quad L_d = 0 \tag{4.9}$$

Example 4.6

With an obstacle height of 50 m above the line between transmit and receive stations, located at 2 km from the transmit station and 1 km from the receive, the diffraction loss at 700 MHz would be

$$v = 50\sqrt{\frac{2}{0.43}\left(\frac{1}{2000} + \frac{1}{1000}\right)} = 4.18$$

$$L_d = 6.9 + 20\log\left(\sqrt{(4.18 - 0.1)^2 + 1} + 4.18 - 0.1\right) = 25.3\,\text{dB}$$

In practice the geometry isn't as simple as in Figure 4.12 as the Earth is curved (modelled using the effective Earth radius from Section 4.2.1), and if a terrain database were to be used, there could be multiple edges. So even if the analysis is undertaken without a terrain database, the so-called smooth Earth, there can still be diffraction if the RX station is sufficiently far from the TX station that it is beyond the radio horizon. Rather than the diffraction point being a terrain obstacle, it is the Earth itself, and the radio path is something like that in Figure 4.13.

The radio path for beyond the radio horizon can be modelled as a great circle between the TX and RX stations that follows:

- Straight line from TX station to horizon point in the direction of the RX station with distance $= d_1$
- Curve around the Earth with effective radius a_e and a distance $= d_2$
- Straight line to the RX station from the horizon point in the direction of the TX station with distance $= d_3$.

This implies that the total propagation loss between the two is a combination of free space path loss and diffraction:

$$L_p = L_{fs}(d) + L_d \tag{4.10}$$

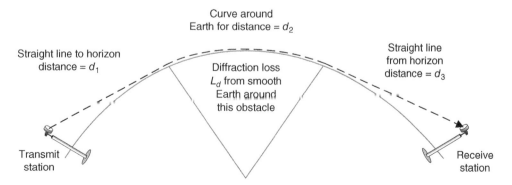

Figure 4.13   Radio path on smooth Earth

where

$$d = d_1 + d_2 + d_3 \tag{4.11}$$

In other words the free space path loss component of the total propagation loss is calculated on the total distance between the TX and RX stations, even if there is no LoS between the two. This differs from the behaviour for Earth to/from space paths where in most circumstances if there is no LoS, the signal can be assumed to be lost.

When there is a terrain database, there can be multiple edges on the path between the TX and RX stations, and it is necessary to extend the calculation in Equations 4.7 and 4.8. A number of methods have been used in various updates to P.526 including the Epstein–Peterson, Deygout and Bullington methods. The most recent version of Rec. ITU-R P.526 is based upon Bullington, and so that is the version assumed in this book.

The diffraction model described in Rec. ITU-R P.526 is included in a number of other propagation models, such as Rec. ITU-R P.452, P.1812 and P.2001, which are discussed later. Note that these do not have a fixed effective Earth radius but vary it according to location and required percentage of time.

There can be a degradation in the radio signal even if there is no obstacle directly in the line between the TX and RX stations if it is within what is called the first **Fresnel zone**. This is given by a radius from the radio path as shown in Figure 4.14, and it can be calculated using

$$R = \sqrt{\frac{\lambda d_1 d_2}{d_1 + d_2}} \tag{4.12}$$

To avoid loss it is usually necessary for there to be no obstacles within the first Fresnel zone (or if a single obstacle, to within a fraction up to 0.6 of its radius): this can have an impact on issues such as fading discussed in the following section.

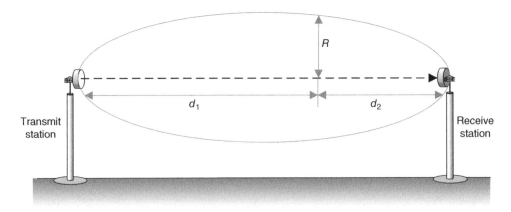

Figure 4.14   The Fresnel zone

Example 4.7

Figure 4.15 shows how the diffraction loss due to smooth Earth varies for distances between 1 and 100 km and frequencies = {100 MHz, 1 GHz, 10 GHz}. The transmit and receive stations were at a height of 10 m.

As the Rec. ITU-R P.526 diffraction loss does not include a core path loss, it is necessary to include one. In this case free space path loss was used to calculate the total path loss shown in Figure 4.16. In addition, it could be helpful to include further losses, such as gaseous attenuations in Rec. ITU-R P.676, described in Section 4.4.1.

Rec. ITU-R P.526 is a straightforward but rather basic model. It can be used for both wanted and interfering paths for a wide range of frequencies and heights but does not directly include a time element.

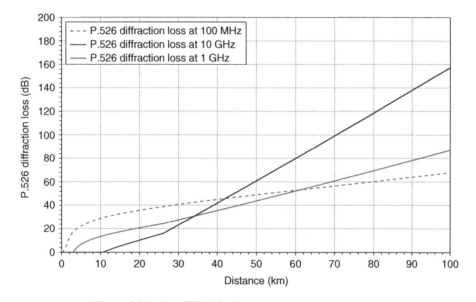

Figure 4.15   Rec. ITU-R P.526: smooth Earth diffraction loss

4.3.3   *P.530: Multipath and Rain Fade*

This model is defined in the following reference:

- Recommendation ITU-R P.530-15: Propagation data and prediction methods required for the design of terrestrial line-of-sight systems (ITU-R, 2013f).

The characteristics of the fade loss model in Rec. ITU-R P.530 are given in Table 4.5.

This Recommendation focusses on a specific scenario, namely, predicting how the wanted signal of a fixed point to point link will vary in time due to propagation phenomena. It is the result of many years of measurement and analysis and is the basis for most fixed link planning. The model is designed to be used with Rec. ITU-R P.525 as the core propagation loss and then gives the difference compared to that including the following propagation elements:

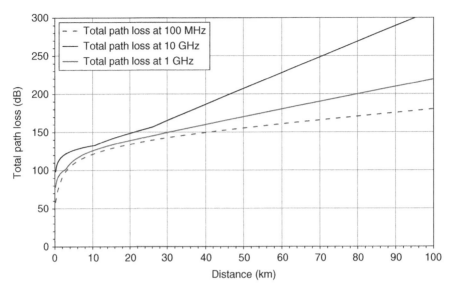

Figure 4.16   Rec. ITU-R P.526: smooth Earth diffraction plus free space path loss

Table 4.5   Characteristics of Rec. ITU-R P.530 propagation model

Path types	Terrestrial
Frequency ranges	Not specified, though text covers frequencies 2–100 GHz
Distances	The rain model is valid up to 60 km
Antenna heights	Not specified, though in practice should be consistent with feasible tower height above terrain
Percentages of time	Multipath: 'all'
	Rain fade: 0.001–1%
Point-to-area model	No
Can use terrain data	Yes (for path clearance)
Includes clutter model	No
Includes the core loss	No

- Diffraction fading due to obstruction of the path by terrain obstacles (see description of Fresnel zone in Section 4.3.2)
- Attenuation due to atmospheric gases, referencing the model in Rec. ITU-R P.676 described in Section 4.4.1
- Fading due to atmospheric multipath associated with abnormal refractive layers or from reflection from surfaces including bodies of water such as between islands (e.g. see boxed text after Section 7.1.5)
- Attenuation due to precipitation including rain, sleet, snow and wet snow.

The Recommendation discusses a wide range of propagation phenomena, but the two main sub-models are the multipath fade/enhancement model and the rain model, as shown in

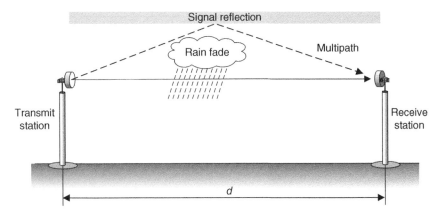

Figure 4.17   Rec. ITU-R P.530: multipath and rain fade models

Figure 4.17. The rain model can also include other precipitations including snow and sleet. The prediction methodology can be applied to the average annual or worst month statistics.

Example 4.8
The Rec. ITU-R P.530 multipath and rain fade losses were calculated for a set of links with the following characteristics:

Transmit and receive antenna heights	20 m
Path length	5 and 10 km
Frequencies	3.6, 18, 36 GHz

The resulting plots are shown in Figures 4.18, 4.19 and 4.20.
Note the following:

- The multipath model has both fade and enhancement components: to show the enhancement it was necessary to plot $100 - p\%$ on the x-axis
- For very small percentages of time, the multipath fade depth becomes linear to log(percentage of time)
- The rain fade is only given between 0.001% and 1% of the time
- For lower frequencies, rain fading is much less than multipath loss
- For higher frequencies, rain fading is more than multipath losses.

The cross-over between where multipath or rain fading is greatest depends upon path length, but typically multipath dominates at frequencies under 10 GHz and rain over 15 GHz. Rain fade for the 10 km fixed link becomes excessively large for the 18 and 36 GHz cases, making these higher frequencies only suitable for short links.

In cases where it is necessary to consider both rain fade and multipath, it is necessary to combine the two models. One way to do that is to identify the fade depth A such that

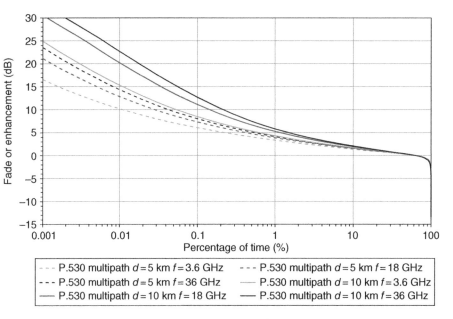

Figure 4.18   Rec. ITU-R P.530: multipath fading, for example, links

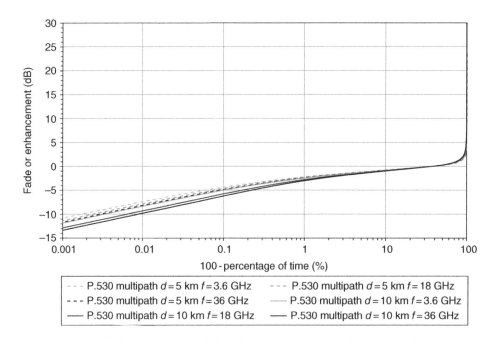

Figure 4.19   Rec. ITU-R P.530: multipath enhancements, for example, links

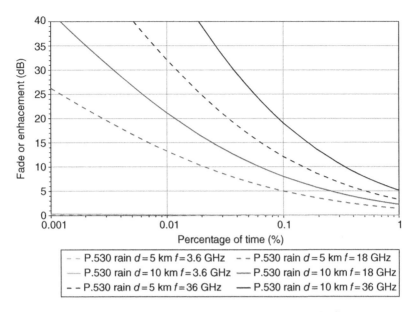

Figure 4.20   Rec. ITU-R P.530: rain fade, for example, links

$$A_{530, fade}(p_1) = A_{530, multi}(p_2) \tag{4.13}$$

where

$$p_1 + p_2 = p \tag{4.14}$$

where p is the required percentage of time. This is likely to require iteration.

Note that there is an American specific fade model with similar characteristics to Rec. ITU-R P.530 described in document TIA Bulletin 10F (Telecommunications Industry Association, 1994). This model is recognised in the FCC rules associated with fixed links.

4.3.4   P.452: Interference Prediction

This model is defined in the following reference:

- Recommendation ITU-R P.452-15: Prediction procedure for the evaluation of interference between stations on the surface of the Earth at frequencies above about 0.1 GHz (ITU-R, 2013d).

The characteristics of the propagation model in Rec. ITU-R P.452 are given in Table 4.6.

This model is a well-established and trusted model for interfering paths used in spectrum management, frequency assignment and interference analysis. It covers a range of phenomena as in Figure 4.21 and can be used with or without a terrain database, in which case the model is smooth Earth. Without a terrain database it is close to the 'worst case' and hence can be used for generic studies for regulatory analysis. With terrain the results are site specific and so can be

Table 4.6   Characteristics of Rec. ITU-R P.452 propagation model

Path types	Terrestrial
Frequency ranges	Has been tested for frequencies between 100 MHz and 50 GHz
Distances	<10,000 km
Antenna heights	Not considered appropriate for aircraft but any reasonable tower height should be acceptable
Percentages of time	$0.001 \leq p \leq 50\%$
Point-to-area model	No
Can use terrain data	Yes
Includes clutter model	Yes
Includes the core loss	Yes

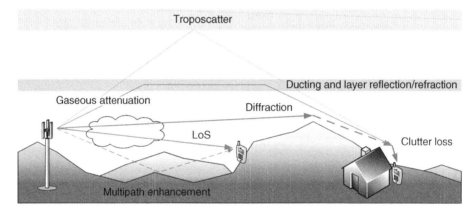

Figure 4.21   Elements in Rec. ITU-R P.452: propagation model

used for coordination and frequency assignment. It could also be used for wanted paths, for example, by setting the percentage of time to 50%. As with Rec. ITU-R P.530 there is the concept of worst month or average annual statistics.

Rec. ITU-R P.452 builds upon the work of other Recommendations, in particular the diffraction model in Rec. ITU-R P.526 described in Section 4.3.2 and the gaseous attenuation model in Rec. ITU-R P.676 described in Section 4.4.1, incorporating versions (typically earlier) of them in its text. The model requires a terrain path profile (as described in Section 4.2.4) and also a classification of each segment of the path as one of {Inland, coastal land, sea} as described in Section 4.2.3. It uses a number of radio-meteorological parameters such as the ΔN and N_0 as described in Section 4.2.2.

There are separate sub-calculations of the loss according to the three potential radio paths due to diffraction, ducting/layer-reflection and troposcatter. At the end of the calculation is a blending mechanism that combines each of the sub-models together to produce a total loss taking into account:

- Free space path loss
- Multipath enhancement

- Diffraction loss
- Gaseous attenuation
- Ducting and layer reflection/refraction
- Troposcatter
- Clutter loss.

Note that the free space path loss component uses the length of the path profile rather than the straight line vector distance between the TX and RX antennas. If there is a large difference in antenna heights, then this could lead to Rec. ITU-R P.452 predicting a lower path loss than would actually be measured.

The clutter loss takes account of:

- d_k = distance in kilometres from clutter to the antenna
- h = antenna height in metres above local ground level
- h_a = clutter height in metres above local ground level.

These parameters are often extracted from a look-up table based upon a land-use code. The clutter loss is then

$$A_h = 10.25F_{fc}e^{-d_k}\left\{1 - \tanh\left[6\left(\frac{h}{h_a} - 0.625\right)\right]\right\} - 0.33 \tag{4.15}$$

where

$$F_{fc} = 0.25 + 0.375\{1 + \tanh[7.5(f_{GHz} - 0.5)]\} \tag{4.16}$$

This could be applied at either the TX or RX station, which can be any height: unlike the models in Rec. ITU-R P.1546 or Hata/European Cooperation in Science and Technology (COST) 231, there is no requirement for the path to be from a greater height antenna to lower.

Example 4.9
A satellite Earth station receiving at a frequency = 3.6 GHz and with antenna height = 5 m is protected from interference by site shielding: an obstruction of height 20 m at a distance of 100 m. In this case the clutter loss would be 18.0 dB.

Example 4.10
Figure 4.22 shows how the Rec. ITU-R P.452 loss varies for distances between 1 and 100 km, frequencies = {100 MHz, 1 GHz, 10 GHz} and percentages of time = 50 and 1% assuming a smooth Earth. The transmit and receive stations were at a height of 10 m. Figure 4.23 shows how the results change when terrain data is introduced for a path starting at (51°N, 0°E) and heading north. In both cases the $\Delta N = 45$ and $N_0 = 325$, and the path profile for this link is given in Figure 4.4.

It can be seen that:

- Propagation loss goes up with frequency but down with percentage of time
- In some cases for smooth Earth, the loss can go *down* with distance (e.g. at $f = 1$ GHz, $p = 1\%$ around $d = 30$ km). This is a 'feature' of the blending mechanism in P.452-15 (and also

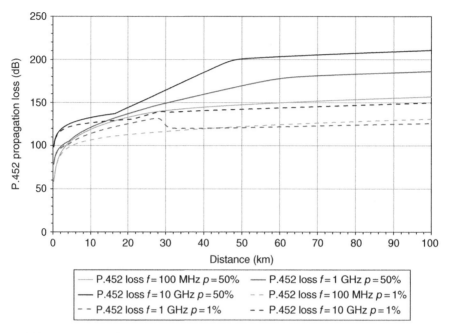

Figure 4.22 P.452: smooth Earth propagation loss for $f = \{100\,\text{MHz}, 1\,\text{GHz}, 10\,\text{GHz}\}$ and $p = \{50\%, 1\%\}$

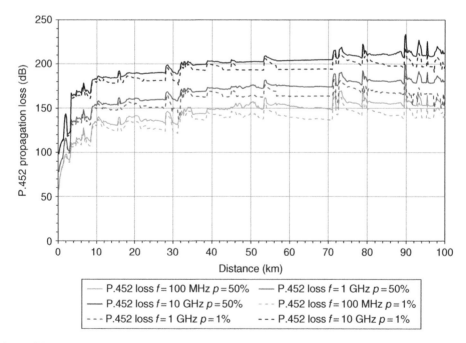

Figure 4.23 P.452: propagation loss including terrain for $f = \{100\,\text{MHz}, 1\,\text{GHz}, 10\,\text{GHz}\}$ and $p = \{50\%, 1\%\}$

P.1812-3), which combines the various components of the model (diffraction, ducting, troposcatter, etc.). If this blending mechanism were to be updated in a later release of these Recommendations, it could remove this feature – for example, if it were to use the one in P.2001

- There is a very large difference between the propagation loss with terrain and smooth Earth (in this case up to 70 dB).

The Longley–Rice propagation model is a US-specific equivalent of P.452 that also derives the point-to-point propagation loss for irregular terrain. It includes additional factors, such as a term to define the confidence in the results, and a point-to-area component, similar to P.1546.

4.3.5 P.1546: Point-to-Area Prediction

This model is defined in the following reference:

- Recommendation ITU-R P.1546-5: Method for point to area predictions for terrestrial services in the frequency range 30 MHz to 3 000 MHz (ITU-R, 2013l).

The characteristics of the propagation model in Rec. ITU-R P.1546 are given in Table 4.7.

This is a well-established and trusted model designed to predict coverage of point-to-area services, in particular for high TX to low RX systems such as broadcasting. When used for planning the percentage of time is typically set at 50%, while for interference studies lower percentages are used such as 1%. It covers a range of phenomena as shown in Figure 4.24 and can be used with or without a terrain database. By setting the percentage of locations $q = 50\%$, it can be used as a point-to-point model.

P.1546 is based upon Recommendation ITU-R P.370, which is now suppressed but gave curves of propagation loss for various distances, frequencies, heights and climatic zones. These curves have been extended within Rec. ITU-R P.1546 to include the following components:

Table 4.7 Characteristics of Rec. ITU-R P.1546 propagation model

Path types	Terrestrial
Frequency ranges	$30\,\text{MHz} \leq f \leq 3\,\text{GHz}$
Distances	$\leq 1000\,\text{km}$
Antenna heights	$<3000\,\text{m}$
Percentages of time	$1\% \leq p \leq 50\%$
Point-to-area model	Yes: $1\% \leq q \leq 99\%$
Can use terrain data	Yes
Includes clutter model	Yes
Includes the core loss	Yes

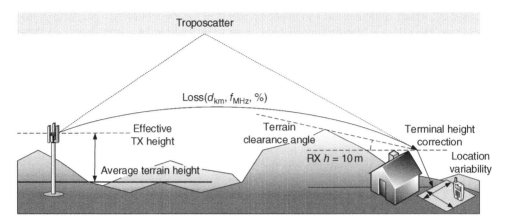

Figure 4.24 Elements in Rec. ITU-R P.1546

- Core propagation loss, derived from the field strength versus distance curves for various frequencies, percentages of time and heights. Note that the reference receiver assumed for these curves is at a height of 10 m and the field strength assumes a TX ERP = 1 kW
- Adjustments to calculate loss for other percentages of time
- Adjustments to take account of TX height, using the average height around the transmitter (where terrain is available) for distances between 3 and 15 km in the direction of the receiver
- Adjustments to take account of the clearance angle of the radio path at the receiver (when terrain is available)
- Adjustment for the height of the receiver if it's below the 10 m reference height. This has a similar role to the Rec. ITU-R P.452 clutter loss though is a slightly more detailed model
- Adjustment for the location variation of signal across the RX pixel
- Adjustment to take account of limit in loss due to tropospheric propagation.

The development of Rec. ITU-R P.1546 can be seen as an increasing number of adjustments to the model based on field strength curves, and one of the motivations for Rec. ITU-R P.1812 was to develop a point-to-area propagation model based upon physical modelling.

Note that the model is not symmetric so that the loss from Station A to Station B might not be the same as the loss in the reverse direction, unlike (say) Rec. ITU-R P.452, P.1812 and P.2001. This relates to Rec. ITU-R P.1546's history as being designed for scenarios involving high TX to lower RX height and caution should be used when applying Rec. ITU-R P.1546 for other cases.

The model is not ideal for paths under about 3 km. For short-range paths the variation due to environmental factors (such as buildings) becomes increasingly dominant, and so basing prediction on averaged field strength curves becomes less accurate. For very short paths the model converges with free space path loss.

One of the key elements of Rec. ITU-R P.1546 is that it is designed as a point-to-area model, able to calculate the path loss not exceeded for $q\%$ of locations within a pixel. The model assumes a log-normal location variation in the signal across the pixel, with mean zero and standard deviation for a 500×500 m pixel:

$$\sigma = K + 1.3 \log(f_{MHz}) \qquad (4.17)$$

where

$K = 1.2$, for receivers with antennas below clutter height in urban or suburban environments for mobile systems with omnidirectional antennas at car-roof height
$K = 1.0$, for receivers with rooftop antennas near the clutter height
$K = 0.5$, for receivers in rural areas.

Example 4.11

An analogue TV signal is being transmitted on UHF channel 38 in the United Kingdom (where the channels are 8 MHz wide starting at 306 MHz): what is the location variability standard deviation for a rooftop receiver? UHF channel 38 corresponds to a frequency = 610 MHz, so the standard deviation of the location variability = 4.6 dB.

In specific scenarios, Rec. ITU-R P.1546 recommends use of given standard deviations, and for all digital scenarios this is 5.5 dB. This value is often used in scenarios ranging from broadcasting to mobile and for pixel sizes from 1 km to 50 m, though in some cases this could be due to lack of alternatives. In a study for Ofcom of sharing between LTE and DVB at 800 MHz (Aegis Systems Ltd, 2011), measurement data suggested location variability should be 1 dB for ranges under 100 m and 5.5 dB for ranges over 1 km, with interpolation between these values.

Rec. ITU-R P.1546 notes that '*q can vary between 1 and 99. This Recommendation is not valid for percentage locations less than 1% or greater than 99%*'. There is no information about how to handle values of q outside this range (e.g. during Monte Carlo analysis), but one possible approach would be clip the percentage using

$$q' = max(min(q, 99), 1) \qquad (4.18)$$

Example 4.12

For a digital system the location variability standard deviation is 5.5 dB. The loss for the range of percentage of locations $q = [1\%, 99\%]$ is shown in Figure 4.25.

It is unlikely that this is the actual distribution of location variability in all cases. For example, in suburban areas the distribution could be clustered around two peaks in the histogram relating to whether a point in the pixel has LoS or not with the transmitter. However unless there are alternatives, this distribution should be used as the best available. Note that there is no location variability for sea pixels.

Example 4.13

Figure 4.26 shows how the Rec. ITU-R P.1546 loss due to smooth Earth varies for distances between 1 and 100 km, frequencies = {100 MHz, 1 GHz} and percentages of time $p = 50$ and 1%. The transmit and receive stations were at a height of 10 m, and the percentage of locations was set to $q = 50\%$.

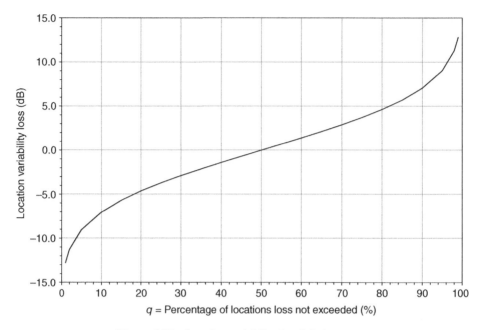

Figure 4.25   Location variability for digital systems

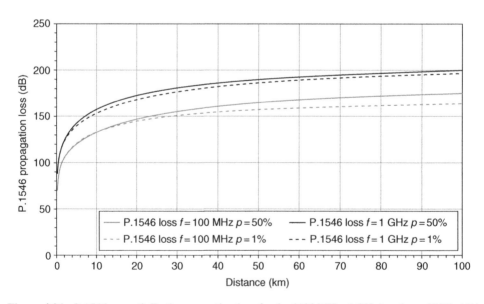

Figure 4.26   P.1546: smooth Earth propagation loss for $f = \{100\,\text{MHz}, 1\,\text{GHz}\}$ and $p = \{50\%, 1\%\}$

4.3.6   P.1812: Point-to-Area Prediction

This model is defined in the following reference:

- Recommendation ITU-R P.1812-3 A path-specific propagation prediction method for point-to-area terrestrial services in the VHF and UHF bands (ITU-R, 2013m).

The characteristics of the propagation model in P.1812 are given in Table 4.8.

This propagation model, more recent than either P.1546 or P.452, was designed to be a generic point-to-area model for systems in VHF and UHF bands. P.1812 can be used for deriving the coverage of a wanted signal or as an interfering signal for down to $p = 1\%$ of the time. By setting the percentage of locations $q = 50\%$, it can be used as a point-to-point model.

P.1812 covers a range of phenomena as shown in Figure 4.27 and can be used with or without a terrain database. By comparing with Figures 4.21 and 4.24, it can be seen that the model has elements of both P.452 and P.1546, and that reflects its development, which was as a 'low-frequency P.452' to be used for point-to-area applications.

Therefore the following components were based upon P.452:

Table 4.8   Characteristics of P.1812 propagation model

Path types	Terrestrial
Frequency ranges	$30\,\mathrm{MHz} \le f \le 3\,\mathrm{GHz}$
Distances	$0.25\,\mathrm{km} \le d \le 3000\,\mathrm{km}$
Antenna heights	$<3000\,\mathrm{m}$
Percentages of time	$1\% \le p \le 50\%$
Point-to-area model	Yes: $1\% \le q \le 99\%$
Can use terrain data	Yes
Includes clutter model	Yes
Includes the core loss	Yes

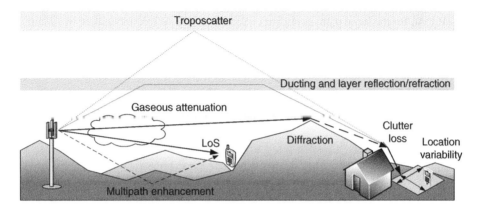

Figure 4.27   Elements in Rec. ITU-R P.1812

- Free space path loss
- Multipath enhancement
- Diffraction loss
- Gaseous attenuation
- Ducting and layer reflection/refraction
- Troposcatter
- Clutter model.

However the following is from P.1546:

- Location variability, with an extension to taper this effect as the antenna height exceeds the clutter height reaching zero when the antenna is 10m above the clutter height.

An implication of this approach to location variability is it will always be zero where there is no land use data available so that the clutter height defaults to zero: this is not the case for P.1546, which does not include the tapering effect.

Note there is a typographic error in P.1812-3's location variability equation (66) in that the frequency should be in MHz as in Equation 4.17. Such editorials are rare but can be present in any document including ITU-R Recommendations, and the reader should, wherever possible, check the equations and software implementations.

A difference with P.452 is in how P.1812 generates the path profile, in that the height is adjusted to be terrain + height of clutter from a land use database as shown in Figure 4.28. Note that care must be taken when using P.1812 with a surface database to avoid double counting the height of buildings.

The range of valid frequency is lower than that for P.452, going down to 30 MHz rather than stopping at 100 MHz. However this led to a restriction on the minimum percentage of time to 1%, the same as P.1546.

P.1812 has many attractive qualities: it uses an approach to propagation modelling that is consistent with that used by P.452, ensuring there are no discontinuities between point-to-point models and the point-to-area median loss. It takes account of the full path profile, as in P.452, rather than just horizon angles at the TX and RX stations as in P.1546.

There can be significant differences in the propagation loss predicted by P.1812 compared to that by P.1546 for the same path, as shown in Section 4.3.12. Work undertaken by Ofcom has

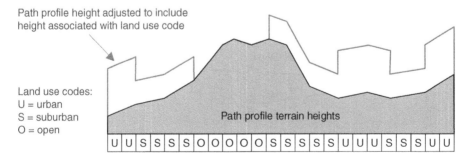

Figure 4.28   Path profile in P.1812 adjusted upwards to include clutter height

identified that P.1812 has a lower standard deviation of error compared to measurement than P.1546 (ITU-R, 2008a). However P.1812 is not as widely used as P.1546, which has decades of use planning broadcast and land mobile systems, including references within the Regional Radiocommunication Conferences (RRCs), such as that in Geneva in 2006 (ITU-R, 2006a).

Example 4.14
Figure 4.29 shows how the P.1812 loss due to smooth Earth varies for distances between 1 and 100 km, frequencies = {100 MHz, 1 GHz, 10 GHz} and percentages of time $p = 50$ and 1%. The transmit and receive stations were at a height of 10 m, and the percentage of locations was set to $q = 50\%$.
   Note that:

1. The model gives results extremely close to those from P.452
2. The model generates reasonable values for $f = 10$ GHz even though the Recommendation gives the maximum frequency as 3 GHz (as it is based upon P.452, which has a much higher upper limit).

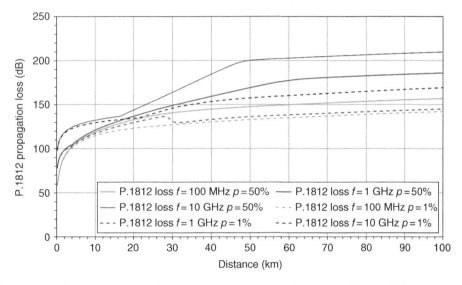

Figure 4.29   P.1812: smooth Earth propagation loss for $f = \{100\,\text{MHz}, 1\,\text{GHz}, 10\,\text{GHz}\}$ and $p = \{50\%, 1\%\}$

4.3.7   P.2001: Wide-Range Propagation Model

This model is defined in the following reference:

• Recommendation ITU-R P.2001-1: A general purpose wide-range terrestrial propagation model in the frequency range 30 MHz to 50 GHz (ITU-R, 2013n).

The characteristics of the propagation model in P.2001 are given in Table 4.9.

Table 4.9   Characteristics of P.2001 propagation model

Path types	Terrestrial
Frequency ranges	30 MHz $\leq f \leq$ 50 GHz
Distances	$\leq$3000 km
Antenna heights	Assumed to be <3000 m
Percentages of time	0.00001% $\leq p \leq$ 99.99999%
Point-to-area model	No
Can use terrain data	Yes
Includes clutter model	No
Includes the core loss	Yes

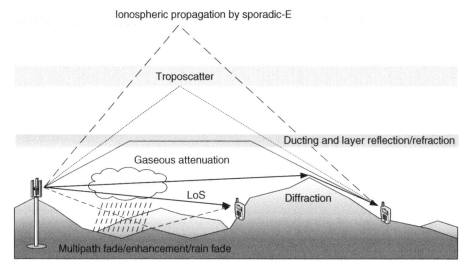

Figure 4.30   Elements in Rec. ITU-R P.2001

The various components of P.2001 are shown in Figure 4.30 and include many sub-models also used by P.452 such as:

- Core path loss
- Multipath enhancement and fading
- Rain fading
- Diffraction loss
- Gaseous attenuation
- Ducting and layer reflection/refraction
- Troposcatter
- Ionospheric propagation by sporadic-E, which may be significant for long paths and low frequencies (not included in P.452).

P.452 and P.1812 give propagation models that can be used for point to point and point to area propagation analysis for most of the frequency ranges used by radiocommunication systems, taking into account terrain, geoclimatic parameters and clutter. However both are limited in the time domain to a maximum $p = 50\%$, which results in two limitations:

1. They can't model the reduction in the wanted signal due to fading, that is, to calculate the signal met for (say) 99% of the time
2. Monte Carlo analysis (as described in Section 6.9) involves the selection of a percentage of time uniformly in the range [0, 100]. Therefore using P.452 or P.1812 (which are limited to 50% of the time) in Monte Carlo analysis will result in (at best) biased statistics.

For that reason, ITU-R Study Group 3 worked to extend these models to develop a wide-range propagation model (WRPM) that is able to calculate both enhanced and faded signal levels, so it could be used in wanted calculations and Monte Carlo modelling.

The starting point was the models in P.452 and was then extended to include fading (multi-path and rain fade as per Section 4.3.3) plus an ionospheric model. The result is a multipurpose propagation model that could be used as the basis of a very wide range of studies.

The model in P.2001 is more recent than those in P.452 or P.1546 but has great potential, though there are a few (existing) limitations:

1. There isn't a clutter model, though there appears to be no reason why the model used by P.452 and P.1812 could not be added to the P.2001 loss
2. It doesn't include a location variability term, though there appears to be no reason why the model used in P.1546 and P.1812 couldn't also be added to the P.2001 loss.

Whereas P.452 is currently on revision 15, P.2001 had at the time of writing only just had its first revision so is still being assessed within the community.

Example 4.15

Figure 4.31 shows how the P.2001 loss due to smooth Earth varies for distances between 1 and 100 km, frequencies = {800 MHz, 3 600 MHz} and percentages of time = {1%, 50%, 99%}. The transmit and receive stations were at a height of 10 m. Note how:

• While the 1% of time curves are similar to those for P.452 and P.1812, there is no sign of the blending feature in those models at a distance of 30 km
• The 99% of curves were faded by up to 17 dB from the median loss, a significant reduction in potential coverage and/or service.

As with P.1812 et al. there are notes in the Recommendation warning about how short paths could result in significant differences from measurement due to variations in local clutter. An approach to reduce this difference is to use high-resolution surface databases, which should give reasonably accurate results even for short paths. However, this leads to highly site-specific results: for generic studies it is useful to have a median loss for defined environments, as in the Hata/COST 231 model.

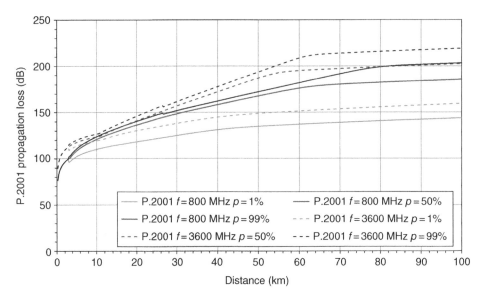

Figure 4.31   P.2001: smooth Earth propagation loss for $f = \{800\,\text{MHz},\ 3\,600\,\text{MHz}\}$ and $p = \{1\%,$ 50%, 99%\}

4.3.8   Hata/COST 231 Median Loss Model

This model has been developed in stages and hence has multiple references. The Hata model was derived from measurement data generated by Okumara in 1966 as documented in:

• Empirical Formula for Propagation Loss in Land Mobile Radio Services by Masaharu Hata (Hata, 1980).

This was defined for the frequency range 150–1 500 MHz and was then extended by the COST Action 231 project to 2 000 MHz:

• Final Report of the COST Action 231 (COST Action 231, n.d.).

Finally this was developed further up to 3 GHz in:

• ERC Report 68: Monte-Carlo Radio Simulation Methodology for the use in sharing and compatibility studies between different radio services or systems (CEPT ERC, 2002).

The model was developed by curve matching to measurements of path loss in urban, suburban and open/rural conditions at a range of frequencies and station heights. It is therefore useful for generic studies where the interest is in the average path loss by distance rather than a site-specific loss. While ITU-R Rec. P.452 can be used with smooth Earth to give a generic loss, it tends to be 'worst case' in that it implies there is no terrain or other obstructions along the path.

Table 4.10   Characteristics of Hata/COST 231 propagation model

Path types	Terrestrial
Frequency ranges	30 MHz $\leq f \leq$ 3 000 MHz
Distances	$\leq$20 – 100 km
Antenna heights	1 m $\leq h \leq$ about 200 m
Percentages of time	Median loss 50% of time
Point-to-area model	No
Can use terrain data	No
Includes clutter model	No
Includes the core loss	Yes

However, in particular for short paths, this is unlikely to be the case, and the Hata/COST 231 model takes that into account. Note that it only gives the median loss as there is no associated percentage of time.

The characteristics of the propagation model in Hata/COST 231 are given in Table 4.10 based on the extended version in CEPT ERC Report 68.

The core Hata model is given as a median loss for urban environments with adjustments for other environments, as in the following equations:

$$L_{50}(urban) =$$

$$69.55 + 26.16\log_{10}(f_{MHz}) - 13.82\log_{10}(h_{tx}) - a(h_{rx}) + [44.9 - 6.55\log_{10}(h_{tx})]\log_{10}(d_{km})$$

$$(4.19)$$

where

f_{MHz} = frequency in MHz
d_{km} = distance between TX and RX in km
h_{tx} = effective height of TX station in metres
h_{rx} = effective height of RX station in metres.

The $a(h_{rx})$ term depends upon city size and frequency so that for small- and medium-sized cities, it is

$$a(h_{rx}) = [1.1\log_{10}(f_{MHz}) - 0.7]h_{rx} - [1.56\log_{10}(f_{MHz}) - 0.8] \qquad (4.20)$$

For a large city and frequencies below 300 MHz, it is

$$a(h_{rx}) = 8.29[\log_{10}(1.54h_{rx})]^2 - 1.1 \qquad (4.21)$$

For a large city and frequencies above 300 MHz, it is

$$a(h_{rx}) = 3.2[\log_{10}(11.75h_{rx})]^2 - 4.97 \qquad (4.22)$$

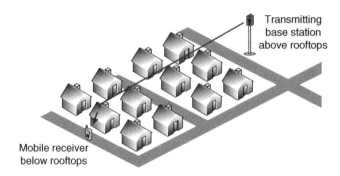

Figure 4.32   Rooftop level with base station above and mobile below

The loss for suburban environments is calculated based upon the loss for urban environments as in

$$L_{50}(suburban) = L_{50}(urban) - 2\left[\log_{10}\left(\frac{f_{\text{MHz}}}{28}\right)\right]^2 - 5.4 \tag{4.23}$$

The loss for rural/open environments is then

$$L_{50}(open) = L_{50}(urban) - 4.78\left[\log_{10}(f_{\text{MHz}})\right]^2 + 18.33\log_{10}(f_{\text{MHz}}) - 40.98 \tag{4.24}$$

The extended versions build upon these equations and are available in the references given.

In the propagation model earlier, the two stations have been described as TX and RX, but that is based upon the assumption that this is a high to low path. The model in ERC Report 68 takes care of potential low to high paths by selecting the TX antenna height to be the larger of the two and the RX to be the smaller. It also expands the model by including a log-normal variation term with standard deviation that varies depending upon whether the loss is above rooftops or below. The concept of a radio path above or below the rooftops is a useful one for understanding propagation in urban and suburban environments, and an example is shown in Figure 4.32.

The models in Hata/COST 231 and the extensions in CEPT ERC Report 68 are widely used for generic sharing studies, particularly for scenarios involving mobile systems where it is sufficient to only have the median 50% of time loss.

Example 4.16
Figure 4.33 shows how the Hata/COST 231 propagation loss varies for distances up to 40 km, frequencies = {800 MHz, 1 800 MHz} and environments = {urban, suburban, open}. The transmit station's height was 20 m and the receive 2 m.

4.3.9   Appendix 7

This model is defined in both the Radio Regulations and an ITU-R Recommendation, in particular:

- Radio Regulations, Appendix 7: Methods for the determination of the coordination area around an earth station in frequency bands between 100 MHz and 105 GHz (ITU, 2012a)

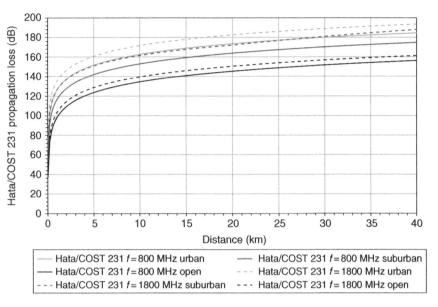

Figure 4.33 Hata/COST 231 Earth propagation loss for $f = \{800\,\text{MHz}, 1\,800\,\text{MHz}\}$ and environments = {urban, suburban, open}

Table 4.11 Characteristics of Appendix 7 propagation model

Path types	Terrestrial
Frequency ranges	$100\,\text{MHz} \le f \le 105\,\text{GHz}$
Distances	See description of model
Antenna heights	Not given
Percentages of time	$0.001\% \le p \le 50\%$
Point-to-area model	No
Can use terrain data	Limited
Includes clutter model	No
Includes the core loss	Yes

- Rec. ITU-R SM.1448: Determination of the coordination area around an Earth station in the frequency bands between 100 MHz and 105 GHz (ITU-R, 2000b).

There is an almost word for word match between the two, as the Recommendation text is included directly in the Radio Regulations. The characteristics of this propagation model are given in Table 4.11.

The objective of this propagation model is to support the coordination of satellite Earth stations and in particular the calculation of the coordination contour. This is the area around the Earth station where if there are any other co-frequency stations, in particular those in the fixed service or fixed-satellite service, then it is necessary to undertake detailed coordination. The contour is the line around the Earth station that identifies the edge of the area or zone within which coordination is required. If the contour intersects another country or countries, then it implies that international coordination is necessary.

Example 4.17

Figure 4.34 shows an example coordination contour around a 2.4 m diameter antenna Earth station located at (50.334453°N, −4.635703°E) transmitting in C-band with power 10 dBW and 30 MHz carrier towards a GSO satellite at 30°E.

The contour is calculated using pessimistic assumptions on the grounds that it is better that cases that would not suffer harmful interference are double checked rather than cases where there could be harmful interference are missed. Therefore, for example, fixed service point-to-point links are assumed to be directly aligned with the Earth station even though this is very unlikely. Detailed analysis would take account of the actual pointing directions, as described in Section 7.4.

The contour is generated using typical parameters for the other service and a propagation model that has two components with geometry similar to that in Figure 4.35:

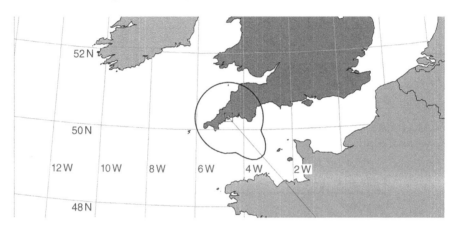

Figure 4.34 Example of Appendix 7: coordination contour. *Screenshot credit: Visualyse Coordinate*

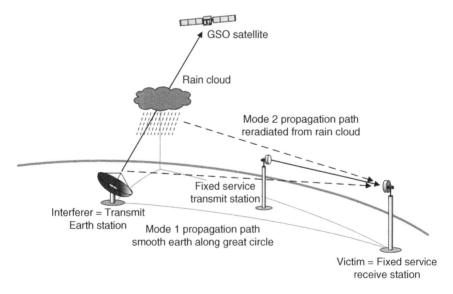

Figure 4.35 Appendix 7, Mode 1 and Mode 2, propagation models for case when interferer is Earth station and victim fixed service

- Mode 1: great circle smooth Earth propagation, similar to the model in P.452 with tropospheric scatter, ducting, layer reflection/refraction and gaseous absorption elements
- Mode 2: hydrometeor scatter, effectively a rain cloud reflecting or reradiating radio energy.

The contour is specified as a set of distances for each azimuth from the Earth station, and there is a minimum distance that is frequency dependent but is typically around 100 km. There is also a maximum distance that depends upon the climatic zone ranging from 375 km (non-coastal land) to 1200 km (warm sea).

For Mode 2 the rain cell is assumed to be at the 'worst' place possible, that is, directly on the line between the Earth station and its satellite, while the FS station is assumed to be pointing directly at the rain cloud. In most cases the ES will not be pointing vertically upwards, and hence the Mode 2 contour is usually slightly offset from its location.

The Mode 1 contour often has a bulge in the direction of the satellite, which is due to the increased gain of the Earth station antenna. The contour also takes account of the various radio climatic zones as described in Section 4.2.3. The overall contour is made up of the larger of the Mode 1 and Mode 2 distances for each azimuth.

It is also possible to add what are called **auxiliary contours**. These are the contours taking into account additional attenuations, for example, an extra 5 or 10 dB of antenna discrimination at the fixed service station or site shielding.

Example 4.18

Figure 4.36 shows the Mode 1 contour (solid line) and Mode 2 contour (dashed line) for the transmit Earth station in Example 4.17. Figure 4.37 shows the reverse band contour (i.e. between two Earth stations) with the characteristic diamond shape that reflects the geometry involved.

The propagation model is by design conservative and hence baselined upon smooth Earth. However it can take some account of terrain by including the horizon elevation angle around

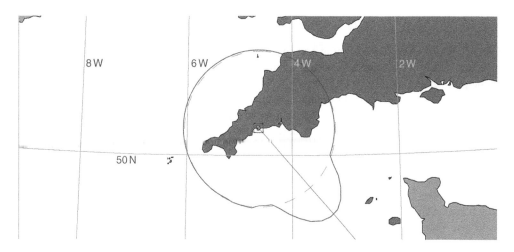

Figure 4.36   Example of Appendix 7: Mode 1 and Mode 2 contours

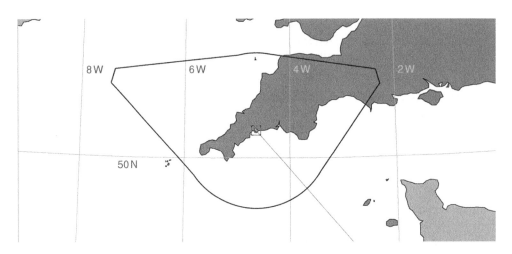

Figure 4.37   Example of Appendix 7: reverse band contour

the ES. This can alter the shape and size of the contour, depending upon the location, though less than if the whole path profile was included.

Example 4.19
Figure 4.38 shows the coordination contour of a receive Earth station at the same location as in Example 4.17, with and without a horizon elevation angle extracted using the SRTM terrain database. One benefit of using terrain in this example is it removes the requirement to coordinate with Belgium. In both cases the contour just touches the Spanish northern coastline, and coordination could be avoided by arguing that it is appropriate in this case to use the 5 dB or greater auxiliary contours (as shown) on the grounds that fixed link stations located on a coastline rarely point directly out to sea.

4.3.10   Generic Models

One of the difficulties in designing and selecting propagation models is to balance accuracy, selecting which of the range of phenomena that can affect path loss to include, with genericity. Some propagation models, such as P.1812, raised concerns about short paths, noting that the signal strength at a specific distance from a transmitter can vary significantly due to differences in nearby structures. Examples of this are shown in Figure 4.39, where a mobile network's base station is below the clutter level in one street: in this case communication with mobiles in that street will be much better than those in adjacent roads, but to a degree depending upon building heights.

While there are models to analyse specific scenarios given, for example, a surface database that can resolve individual buildings, the results become highly dependent upon assumptions, which increasingly become complicated. What is often required is a model that can be specified by a few parameters, which can be agreed by parties undertaking interference analysis. Examples of these include generic models based upon dual or triple slopes.

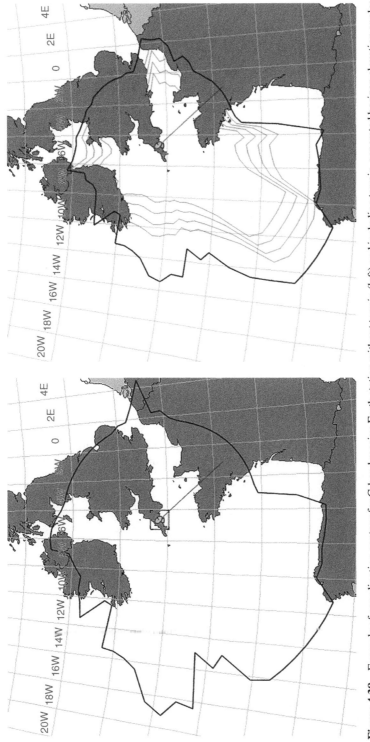

Figure 4.38 Example of coordination contour for C-band receive Earth station without terrain (left) and including terrain-generated horizon elevation angles with 5 dB auxiliary contours (right)

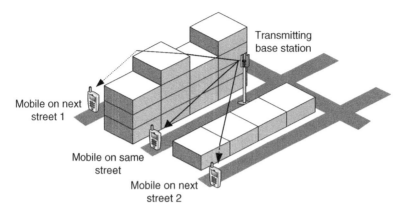

Figure 4.39 Variation between streets in urban environments

Propagation models based upon dual slopes can be found in these sources:

- Rec. ITU-R P.1238-7: Propagation data and prediction methods for the planning of indoor radiocommunication systems and radio local area networks in the frequency range 900 MHz to 100 GHz (ITU-R, 2012g)
- Rec. ITU-R P.1411-7: Propagation data and prediction methods for the planning of short-range outdoor radiocommunication systems and radio local area networks in the frequency range 300 MHz to 100 GHz (ITU-R, 2013k)
- Rec. ITU-R P.1791: Propagation prediction methods for assessment of the impact of ultra-wideband devices (ITU-R, 2007e).

These models use free space path loss up to a break point, such as the nearest building, and then the loss increases more quickly afterwards. Mathematically they are defined via the following equations:

$$L = L_1(d_{km}, f_{MHz}) \quad \text{if } d_{km} \leq d_1 \tag{4.25}$$

$$L = L_2(d_{km}, f_{MHz}) \quad \text{if } d_{km} > d_1 \tag{4.26}$$

where

$$L_1(d_{km}) = 32.45 + 20 \log_{10} d_{km} + 20 \log_{10} f_{MHz} \tag{4.27}$$

$$L_2(d_{km}) = L_1(d_1) + N_1 10 \log_{10}\left(\frac{d_{km}}{d_1}\right) \tag{4.28}$$

This model is parameterised by distance d_1 in km and slope after breakpoint of N_1. It can easily be extended to have two or more breakpoints and have additional factors included. The distance of the breakpoint can vary depending upon scenario, with example values in Table 4.12. For the in-building scenario the first break point is the nearest wall, in this case 5 m from the transmitter.

Table 4.12 Example dual slope parameters

Scenario	d_1 (km)	N_1
In building	0.005	4
Urban	0.05	4
Suburban	0.2	3

Table 4.13 Characteristics of dual slope propagation model

Path types	Terrestrial
Frequency ranges	Depends upon parameters selected
Distances	Depends upon parameters selected
Antenna heights	Depends upon parameters selected
Percentages of time	Typically median 50% of time
Point-to-area model	Depends upon parameters selected
Can use terrain data	No
Includes clutter model	No
Includes the core loss	Yes

Each set of parameters would have a specific range of validity, so the characteristics in Table 4.13 will depend upon the scenario involved.

A number of possible values of slope parameter N are given in Rec. ITU-R P.1238, though these are for a single slope model rather than dual slope.

The dual (or triple) slope models typically give the median loss as a fixed value for a given distance. In practice it could vary:

- By azimuth, depending upon local clutter
- In time, depending upon motion of people and vehicles, leading to obstructions and multipath losses or enhancements.

One way to model these is by adding onto the dual slope loss an additional factor, such as a log-normal variation, typically with mean zero and given standard deviation.

Example 4.20

Figure 4.40 shows path loss by distance for a dual slope model with or without random normal fade/enhancement component as given in Table 4.14.

This example shows one problem with using variation with a dual slope model: the result can be a path loss less than free space. There are three approaches to this situation:

1. Accept that there could be enhancements, for example, in urban canyons propagation can be significantly better than free space due to multiple reflections
2. Only include the variation term after the first break point
3. Cap the path loss to the free space value.

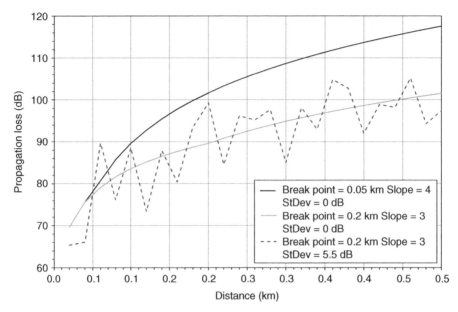

Figure 4.40 Example of dual slope with variation models

Table 4.14 Example parameters for dual slope with variation models

Scenario	d_1 (km)	N_1	σ
Urban	0.05	4	0.0
Suburban	0.2	3	0.0
Suburban with variation	0.2	4	5.5

This model can be further extended by having a generic clutter model, whereby a database is queried and then a look-up table used to convert a land use code to equivalent clutter loss. This could be combined with a dual slope or even a model such as Rec. ITU-R P.2001.

Finally there could be a term to model the loss when a radio wave enters or exits a building. This can be achieved by a range of approaches, including using a fixed value or a [min, max] range of loss with the figure to use selected by a random number as described in Section 4.6.

4.3.11   Other Propagation Models

A number of other propagation models could be considered depending upon scenario. For example, detailed modelling of short-range scenarios can use ray tracing. These models analyse how radio waves reflect off walls, ceilings and floors and how they diffract around obstacles. The result is a 3D model of radio wave propagation within buildings and in urban areas. Such models are computationally intense but give more accurate results for highly specific scenarios.

An alternative to ray tracing would be the Walfisch and Bertoni model (Rappaport, 1996) and that in Rec. ITU-R P.1411. These analyse how the path relates to buildings in an urban area and determines, for example, whether the path is LoS, urban canyon or diffracted into an adjacent street. Again, this is more widely used for planning radio systems than interference analysis. There are also a number of alternatives to the Hata model that could be considered, such as Lee (1993) and Egli (1957).

At low frequencies it is necessary to consider sky-wave and ground-wave propagation for which the references are:

- Rec. ITU-R P.368-9: Ground-wave propagation curves for frequencies between 10 kHz and 30 MHz (ITU-R, 2007c). This Recommendation consists of a series of curves of field strength versus distance for a range of frequencies and environments (land, sea, etc.). The GRWAVE software tool, which implements these Recommendations is available from the ITU-R
- Rec. ITU-R P.1147-4: Prediction of sky-wave field strength at frequencies between about 150 and 1 700 kHz (ITU-R, 2007d).

Radar systems have additional requirements on propagation models, namely, to predict reflections from clutter such as waves, terrain, etc. This represents noise that can reduce the likelihood of detection as described in Section 7.7.

4.3.12   Comparing Terrestrial Propagation Models

It is useful to have a feel for how the propagation models compare to each other and behave in reaction to (say) terrain data. However there is a very wide range of possible plots that could be displayed, depending upon frequencies, heights, path profiles, percentages of time and percentages of locations.

The following examples take two approaches:

- Comparing how the propagation loss varies by distance
- Showing the predicted coverage of a broadcast digital TV service.

Example 4.21
The following plots were generated for an 800 MHz link with transmit height of 15 m and receive height of 2 m for paths between 0 and 25–50 km. In all cases the transmitter was at (lat, long) = (51.0, 0.0) and the receiver pointing due north, with the path set to be land zone A1. Other propagation parameters were

$\Delta N = 45$
$N_0 = 325$
Temperature $T = 15\,°C$
Water vapour density $= 3\ g/m^3$
Pressure $= 1013\ hPa$

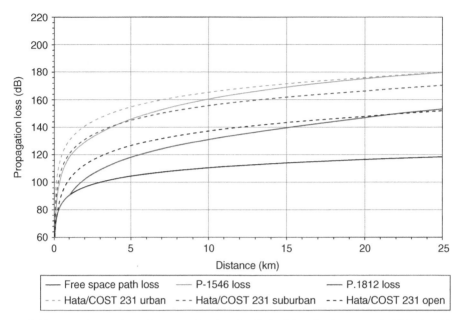

Figure 4.41 Plot of smooth Earth propagation loss for 50% of time for free space, P.1546, P.1812 and Hata models

In Figure 4.41 it can be seen that:

- The least propagation loss was predicted using free space path loss
- P.1546 predicted greater propagation loss than P.1812, which matched free space for short paths. The largest difference was 14.7 dB
- The highest propagation loss was for the Hata urban model followed by Hata suburban
- Hata open was in general between P.1546 and P.1812.

In Figure 4.42 it can be seen that:

- The lowest propagation loss was predicted using free space path loss
- The highest propagation loss was predicted using P.1546
- P.452, P.1812 and P.2001 predictions were very similar (as expected given the similarities of their construction)
- Longley–Rice was initially similar to P.452/P.1812/P.2001 but for longer paths tended towards P.1546.

In Figure 4.43 it can be seen that while P.1546 remains the highest propagation loss, there are distances for which P.2001 matches free space path loss as the lowest propagation loss. Note that for percentages of time less than 1%, this effect becomes more pronounced, with P.452, P.1812 and P.2001 predicting path losses lower than free space. As before, Longley–Rice is similar to P.452/P.1812/P.2001 for short paths but for longer paths tended towards P.1546.

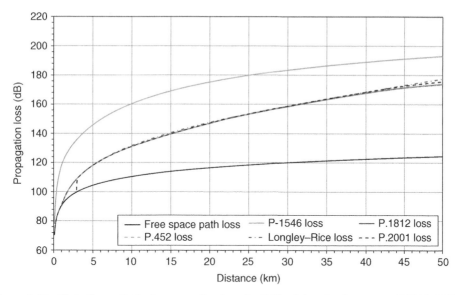

Figure 4.42 Plot of smooth Earth propagation loss for 50% of time for free space, P.1546, P.1812, P.452, Longley–Rice and P.2001 models

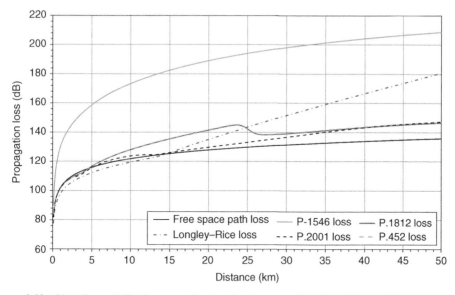

Figure 4.43 Plot of smooth Earth propagation loss for free space, P.1546, P.1812, P.452, Longley–Rice and P.2001 models at 1% of time

Figure 4.44 includes the effects of terrain (in this case SRTM), so the results will be highly dependent upon the path chosen. As before, the free space path was the lowest propagation loss and P.1546 in general the highest, with Longley–Rice somewhere between P.452, P.1812 and P.2001. In particular it can be seen that there are large differences compared to the results without terrain given in Figure 4.42.

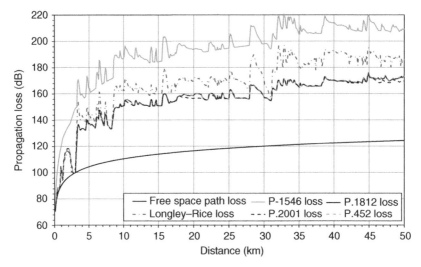

Figure 4.44   Plot of propagation loss with terrain for 50% of time for free space, P.1546, P.1812, P.452, Longley–Rice and P.2001 models

Table 4.15   Digital TV coverage example parameters

Transmitter latitude	51.424167°N
Transmitter longitude	−0.075°E
Transmitter height above terrain	150 m
Transmit frequency	610 MHz
Transmit power	40 dBW
Receive antenna height above terrain	10 m
Receive antenna peak gain including losses	9.15 dBi
Receive noise figure	7 dB
Noise bandwidth	7.6 MHz
Required C/N for 256-QAM DVB-T2	19 dB

Example 4.22
Coverage plots of a digital TV broadcast transmitter in the United Kingdom were generated using a terrain and a land use database using the parameters as in Table 4.15, some of which were taken from an input contribution from the European Broadcasting Union (EBU) to CEPT ECC PTD. The analysis used the same geoclimatic parameters as in Example 4.21 except that full use was made of the IDWM climatic zone data to determine percentage of path over water.

The coverage assuming the propagation model in P.1546 and P.1812 is given in Figures 4.45 and 4.46, respectively. The following can be observed:

• The P.1546 coverage is more circular than that for P.1812, taking less account of terrain. This relates to how it only uses part of the path profile, namely, the average terrain height and horizon elevation angle
• The P.1812 coverage is in general wider corresponding to a lower average loss as shown in Example 4.21. Its shape also is more reflective of the underlying terrain and less circular. P.1812 is also more 'bitty' in that there are more outlying pixels.

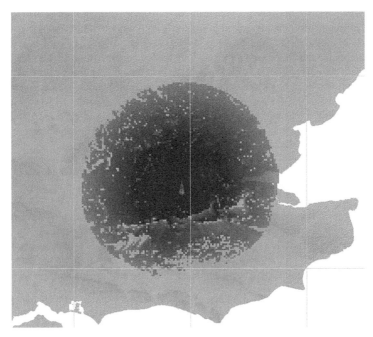

Figure 4.45   Coverage predicted using P.1546, $p = 50\%$, $q = 50\%$

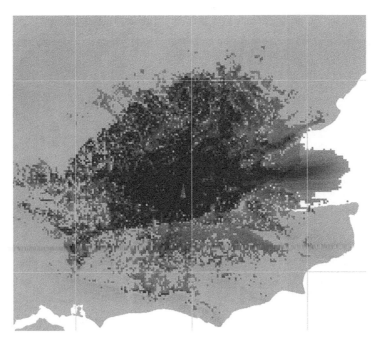

Figure 4.46 Coverage predicted using P.1812, $p = 50\%$, $q = 50\%$. *Screenshot credit: Visualyse Professional. Terrain and land use credit: Ofcom and OS*

Modelling 4G Coverage on Mars

How would you model a 4G network's coverage on Mars? That was the question I had to address when analysing the requirements for communication and navigation services (Pahl, 2006) as part of Project Boreas, a study of a base in the Martian polar regions (Cockell, 2006).

The propagation models described in the previous sections are the result of decades of measurement, and to date there have been no missions to Mars that included analysis of surface radio wave propagation. All missions have required communications from the Martian surface to either orbit or Earth, and this has led to the development by the Jet Propulsion Laboratory (JPL) of a Radio Wave Propagation Handbook for Communication on and Around Mars (Ho et al., 2002). This document identified a number of significant differences with Mars compared to Earth, such as:

- Smaller planetary radius and lower average temperature
- Much reduced ionosphere, an order of magnitude less than Earth's, making it almost transparent to radio waves above 450 MHz. While it could be used for low-frequency ground to ground communication, the bandwidth would be limited
- Minimal troposphere (two orders of magnitude less than Earth) and hence minimal tropospheric effects
- Gaseous attenuation is much lower due to the lower atmospheric density and lack of rain or sleet, so typically less than 1 dB
- Dust storms could be significant, causing at least a 3 dB loss at Ka band.

What would be unchanged compared to terrestrial propagation are terrain-based phenomena such as diffraction and multipath, and hence a starting point could be the diffraction model in P.526. This could be used to predict the coverage of a 4G base station located within (say) the Grand Canyon of Mars, the Valles Marineris, which is up to 4000 km long, 200 km wide and 7 km deep. To create the prediction it would be necessary to have a terrain database, and the one used was generated by the Mars Orbiter Laser Altimeter (MOLA) instrument on the Mars Global Surveyor spacecraft. An example of the coverage along the Valles Marineris is shown in Figure 4.47: it can be seen how the canyon edges are 'lit up' due to their LoS geometry with the base station.

Figure 4.47   4G coverage along the Martian Valles Marineris. *Screenshot credit: Google Earth data ESA/DLR/FU Berlin (G. Neukum)*

While operations within the vicinity of Mars remain at a low density, it is possible to operate non-co-frequency to avoid interference. As the number of missions increases, it will become necessary to undertake interference analysis on Mars, particularly between operations from different space agencies.

The coverage plots for P.452 and P.2001 were found to be similar to those plots for P.1812 and so are not shown. Note that P.452 and P.2001 are point-to-point models but can in some scenarios be used for point-to-area analysis. This is an important issue: while defined as point to point models, there is no reason why they can't be used to analyse interfering (or wanted) signals over an area. What is being modelled in all cases is the signal at the centre of each pixel, and then the location variability term in P.1546 and P.1812 are then used to make them point-to-area models.

Hence with a $q = 50\%$, P.1546 and P.1812 are effectively the same as P.452 and P.2001 in that they give the point-to-point loss for the centre of the relevant pixel. It could also be argued that P.452 and in particular P.2001 could be extended to point to area models by including a log-normal term to model location variation. Care must be taken to ensure underlying assumptions are valid. For example, P.452 is often used for transmitters above the clutter level such as fixed service towers, and hence there would be no location variation (as per the taper in P.1812).

4.4   Earth to Space Propagation Models

This section covers propagation models for Earth to space and space to Earth paths.

4.4.1   P.676: Gaseous Attenuation

This model is defined in the following reference:

- Recommendation ITU-R P.676-10: Attenuation by atmospheric gases (ITU-R, 2013h).

The characteristics of the propagation model in P.676 are given in Table 4.16.

This Recommendation covers both terrestrial and Earth to space paths, but the terrestrial component is usually built into other models, such as in P.452 or P.1812. Therefore this

Table 4.16   Characteristics of P.676 propagation model

Path types	Terrestrial and Earth to space
Frequency ranges	$1\,\mathrm{GHz} \leq f \leq 350\,\mathrm{GHz}$
Distances	n/a
Antenna heights	n/a
Percentages of time	n/a
Point-to-area model	No
Can use terrain data	No
Includes clutter model	No
Includes the core loss	No

section will concentrate on the Earth to space component. There are two possible implementations:

1. Line-by-line calculation of gaseous attenuation: this is a more detailed method that integrates over layers of the atmosphere
2. Approximate estimation of gaseous attenuation: this is a more straightforward approach that calculates the loss as a function of elevation angle and other variables.

In general it is the second, more computationally efficient, method that is used in interference analysis, as it gives very similar results apart from at very low elevation angles (below about 5°). In some scenarios it has been useful to have an accurate gaseous attenuation loss for low elevations, and these cases are covered by:

• Rec. ITU-R SF.1395: Minimum propagation attenuation due to atmospheric gases for use in frequency sharing studies between the fixed-satellite service and the fixed service (ITU-R, 1999c).

The key inputs of Rec. ITU-R P.676 are:

• Frequency (GHz) in the range 1–350 GHz
• Air pressure, available in Rec. ITU-R P.836
• Temperature, available in Rec. ITU-R P.1510.

The model includes both dry air and also attenuation due to water vapour, excluding rain loss, which is described in Rec. ITU-R P.618. It is very useful in modelling Earth to space links and is included in many interference analysis studies. Note however that some studies are undertaken in 'clear air', that is, assuming no attenuations or additional losses including rain. Examples of such clear air analysis include the methodology in Appendix 8 described in Section 7.5.

Example 4.23
The Earth to space path between London and a GSO satellite at 25°E (as in Example 3.33, so the elevation is about 26.5°) is assumed to have the following geoclimatic parameters:

Temperature: 15 °C
Pressure: 1013 hPA
Water vapour density: 9 g/m^3

The losses due to dry air and water vapour between 1 and 50 GHz are shown in Figure 4.48. Note the peak in water vapour loss around 22.3 GHz, which makes those frequency bands, plus those over about 45 GHz, more difficult for space applications.

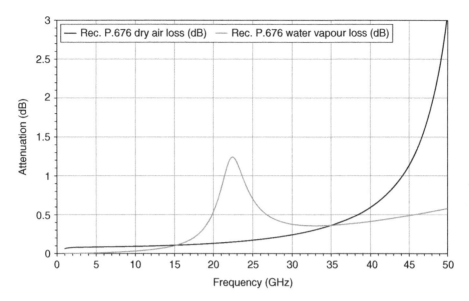

Figure 4.48   Example of gaseous absorption loss from Rec. ITU-R P.676

4.4.2   *P.618: Rain Loss and Noise Rise*

This model is defined in the following reference:

- Recommendation ITU-R P.618-11: Propagation data and prediction methods required for the design of Earth-space telecommunication systems (ITU-R, 2013g).

The characteristics of the propagation model in P.618 are given in Table 4.17.

Table 4.17   Characteristics of P.618 propagation model

Path types	Earth to space
Frequency ranges	$1\,\text{GHz} \leq f \leq 350\,\text{GHz}$
Distances	n/a
Antenna heights	n/a
Percentages of time	$0.001\% \leq p \leq 5\%$
Point-to-area model	No
Can use terrain data	No
Includes clutter model	No
Includes the core loss	No

Example 4.24

The Earth to space path in Example 4.23 was assumed to have a rain rate exceeded for 0.01% of the time of 30.77 mm on the link between the London Earth station and the GSO

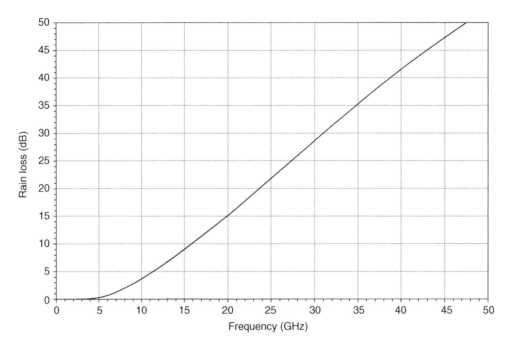

Figure 4.49   Example of P.618 rain loss for 0.01% of the time

satellite at 25°E. The rain loss for 0.01% of the time between 1 and 50 GHz is shown in Figure 4.49.

Note how the loss or fade is minimal below about 6 GHz but then increases with frequency. Reduced rain loss is one of the main motivations for satellite communications to use lower frequencies. The downside is that larger diameter Earth stations can be required to close the link.

The rain has a secondary effect in that it can increase the noise on a satellite link by reradiated energy that has been absorbed by rain. This adds to the total noise an amount that can reach the temperature of the medium depending upon the fade depth, calculated via

$$T_s = T_m \left(1 - 10^{-A/10} \right) \tag{4.29}$$

where

T_s = additional temperature into antenna due to sky (K)
A = path attenuation (dB)
T_m = temperature of the medium (K), typically in the range 250–280 K.

Example 4.25

For the Earth to space path in Example 4.24, the total receiver noise can be seen in Figure 4.50 to increase with frequency due to rain noise until it reaches a value determined by the temperature of the medium and the bandwidth (in this case 30 MHz).

Rec. ITU-R P.618 includes other useful information about the impact of rain on Earth to space paths such as the impact of depolarisation and event lengths.

A key issue for Earth to space paths is the degree of correlation between the rain fades of the wanted and interfering signals. Take the scenario in Figure 4.51: the uplink rain losses are likely to be different unless the two Earth stations are located close together. However, as shown in Figure 4.52, for the downlink it is likely that if the wanted signal is faded, then the interferer will be too (plus there will be noise rise due to rain). The degree of correlation will depend upon whether the radio paths are in similar or different directions and the size of the rain cloud.

The issue of correlation is discussed further in Section 4.8.

Note there is also the Crane rain model developed in the United States, which is defined in the following reference:

- *Electromagnetic Wave Propagation Through Rain* by Robert K. Crane (Crane, 1996).

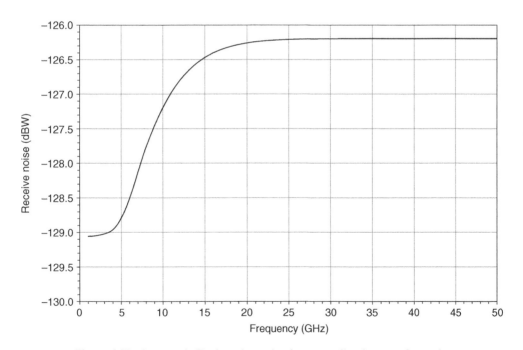

Figure 4.50 Increase in Earth station noise due to reradiated energy from rain

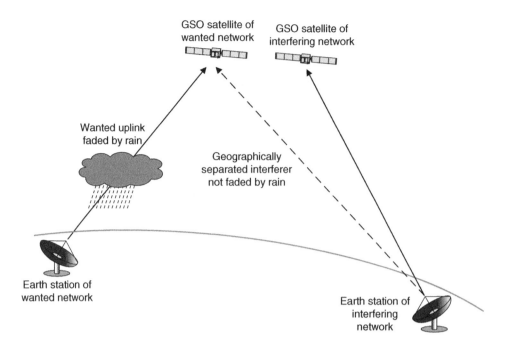

Figure 4.51 Low likelihood of correlated rain fades on uplink for geographically separated wanted and interfering Earth stations

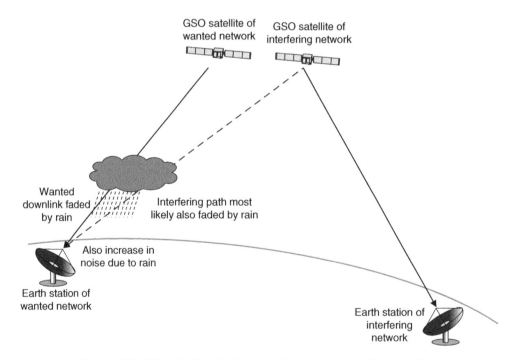

Figure 4.52 High likelihood of correlated rain fades on satellite downlinks

Table 4.18   Characteristics of P.528 propagation model

Path types	Aeronautical
Frequency ranges	$125\,\text{MHz} \leq f \leq 5.5\,\text{GHz}$
Distances	$d \leq 1800\,\text{km}$
Antenna heights	$1.5\,\text{m} \leq h \leq 20{,}000\,\text{m}$
Percentages of time	$1\% \leq p \leq 95\%$
Point-to-area model	No
Can use terrain data	No
Includes clutter model	No
Includes the core loss	Yes

4.5   Aeronautical Propagation Models

An aeronautical propagation model is defined in the following reference:

- Recommendation ITU-R P.528-3: Propagation curves for aeronautical mobile and radionavigation services using the VHF, UHF and SHF bands (ITU-R, 2012e).

The characteristics of the propagation model in P.528 are given in Table 4.18.

This model can be used for air to ground analysis of both wanted and interfering signals. So, for example, the wanted signal might be computed to 95% of the time, that is, faded, to ensure there is sufficient coverage for a safety critical system such as air traffic control. The model includes effects such as diffraction, core path loss and attenuations though the Recommendation notes that *'these curves are based on data obtained mainly for a continental temperate climate'*.

It is a smooth Earth model so that the effects of terrain are not included. For many aviation applications the aircraft is at sufficient a height that terrain can be ignored. For those cases where the aircraft is at a low altitude so that this is not the case, then another model might have to be used. One possibility could be P.1812, which has a maximum height of 3000 m.

Example 4.26
Figure 4.53 shows the propagation loss for a 300 MHz link between a ground station at height = 10 m and aircraft at heights 5,000 and 10,000 m for {5%, 50%, 95%} of the time. Figure 4.54 shows the plots for a 3 GHz link with the same heights and percentages of time.

4.6   Additional Attenuations

In addition to the propagation models described in the sections earlier, there can be further attenuations to the wanted or interfering signal due to:

- Site shielding, where an obstruction is created by design, typically to reduce interfering signals. The site shielding loss could be calculated using the clutter loss model in Rec. ITU-R P.452 as in Section 4.3.4
- Building entry loss when the transmitter is outside and the receiver inside or vice versa

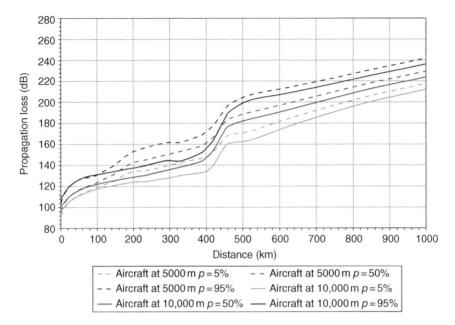

Figure 4.53   Example of P.528 propagation loss for link at $f = 300$ MHz

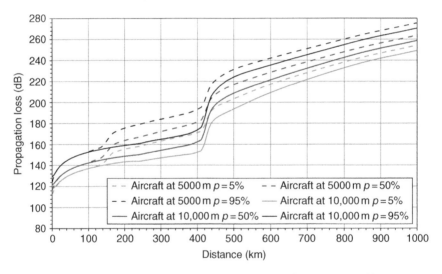

Figure 4.54   Example of P.528 propagation loss for link at $f = 3$ GHz

- Vegetation loss, as in Rec. ITU-R P.833 (ITU-R, 2013s)
- Wall loss within a building
- Ceiling loss within a building
- Person loss due to absorption by the body
- Underground loss, where the transmitter is above ground and the receiver below ground or vice versa.

Table 4.19 Example body/orientation and building entry attenuations

Frequency band (MHz)	Body/orientation loss (dB)	Building entry loss (dB)
800	2.5	13.2
900	2.5	13.7
1 800	2.5	16.5
2 100	2.5	17.0
2 600	2.5	17.9

Some of these attenuations will depend significantly upon the circumstances, as different materials can generate different losses. This makes it difficult to model specific locations unless there is a database of buildings and their construction materials available.

An alternative approach is to use average values, such as (say) that the building entry loss = 10 dB, as assumed in the GRMT project described in Section 7.9. The MASTS algorithm described in Section 7.2 assumed a value of 5 dB for land mobile assignments in the VHF and UHF bands.

The value used could be frequency dependent, as in the methodology used by Ofcom to check the coverage of 4G LTE networks in the United Kingdom (Ofcom, 2012b). The coverage obligation is for indoor signal strength and so includes building entry loss as well as body/orientation loss using the values in Table 4.19.

Information about building entry losses and floor/wall losses can be found in Recs. ITU-R P.1238 and P.1411. For example, P.1411 gives a building entry loss at 5.2 GHz for an office environment of mean 12 dB and standard deviation 5 dB. More detail is given in Rec. ITU-R P.2040 (ITU-R, 2013o).

The values previously mentioned are for terrestrial radio paths, that is, where the main building entry route is via the walls or windows. For Earth to space paths, it is necessary to also consider the number of floors and the elevation angle of the radio path.

This geometry is particularly relevant for the sharing scenario in the 5 GHz band between radio local area network (RLAN) devices and fixed-satellite service (FSS) networks. Satellites can see very large parts of the Earth's surface, and this can lead to aggregation issues. For example, a satellite serving either North America or Europe could see a population of over 500 million. This can lead to the potential for an aggregation of interference by 10 $\log_{10}(500$ million$) = 87$ dB, a highly significant amount. The actual interference would depend upon market penetration, activity factors, etc., but also for indoor RLAN devices the attenuation due to building exit loss. If the building has many stories and the elevation angle is near vertical, this could lead to significant losses.

An example of the methodology for sharing between RLANs and FSS networks can be found in:

- Recommendation ITU-R M.1454: e.i.r.p. density limit and operational restrictions for RLANS or other wireless access transmitters in order to ensure the protection of feeder links of non-geostationary systems in the mobile-satellite service in the frequency band 5 150–5 250 MHz (ITU-R, 2000a).

In that Recommendation, the average building loss was considered to be between 7 and 17 dB.

For Monte Carlo analysis it can be useful to have a statistical model that gives a range of values depending upon the selection of a random number. For example, given a random number p in the range $p = [0, 1]$, the building penetration loss could be calculated as being

$$
\begin{array}{ll}
\text{if } p \le 0.15 & L_B = 5 \\[2mm]
\text{if } 0.15 < p < 0.85 & L_B = 5 + 10\dfrac{(p - 0.15)}{(0.85 - 0.15)} \\[2mm]
\text{if } 0.85 \le p & L_B = 15
\end{array}
\tag{4.30}
$$

This model is saying that for 15% of the time, the indoor station is close to the edge of the building, so the loss is the minimum, in this case 5 dB. For 15% of the time the station is deep within the building, where the loss is at the maximum value of 15 dB. For all other times a value between 5 and 15 dB is selected.

Buildings can be coated with frequency selective surfaces that block specific frequencies while letting others in. An example might be a concert hall that blocks frequencies used by publicly used mobile phones but permits those used by the emergency services.

Underground radio networks can be considered to be either fully isolated from above ground or a high attenuation used such as 20 dB.

4.7 Radio Path Geometry

The propagation model has an impact on other aspects of the link calculation, in particular on the radio path geometry and hence the antenna gains.

Consider the example in Figure 4.55: the radio path is over the top of a building (and not through it), and this will have an impact on the angles that should be used for the gain pattern calculation. Calculating the TX or RX gain using the angles of the path directly from the base

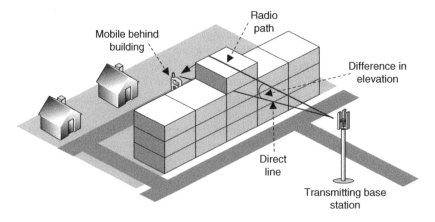

Figure 4.55   Impact of propagation path on antenna gain calculation

station to the mobile would be incorrect, as here the radio waves travel over the top. Hence the radio path calculated during the propagation model should also be used to calculate the antenna gains.

There can also be propagation paths around the sides of buildings, in particular, those that are significantly taller than they are wide. In general, this is a secondary effect to be considered on a case by case basis, such as when planning of networks (e.g. mobile or broadcasting) in very specific dense urban locations where such buildings are known to occur.

4.8   Percentages of Time and Correlation

Many of the propagation models described in the previous sections give the loss for an associated percentage of time p where:

For propagation loss, p = *time percentage for which the propagation loss* **is not** *exceeded*	⇒	*For signal strength,* p = *time percentage for which the received signal level* **is** *exceeded*

These models, therefore, are statistical: they do not give information about specific propagation events, their times and durations but likelihood over a measurement period such as a month or year. While actual measurement campaigns will have a sample duration and frequency, this information is not available in the models.

For this reason it is not possible to use these models to analyse (say) the impact on a radio system of a snow storm approaching from the north-east. This is partly due to modelling complexity but also because the results would be very scenario specific. There is a high degree of variability in weather conditions, and statistics only become sufficiently stable to be used with confidence over very long periods of time.

The complexity is even greater when considering scenarios that involve multiple propagation paths. An example of this can be seen in Figure 4.56, which includes:

- Wanted system: a point-to-point fixed service link with fading (or enhancement) modelled via Rec. ITU-R P.530 and percentage of time $p = p_w$
- Interfering systems: two satellite Earth stations, using the Rec. ITU-R P.452 propagation model with percentages of time p_1 and p_2.

The question is then how should the three percentages of time (p_w, p_1, p_2) be selected during interference analysis?

The issue relates to the correlation of propagation effects between radio paths, and two examples of this were given in Section 4.1.2:

- Satellite uplink in Figure 4.51, likely to be uncorrelated if the wanted and interfering satellite Earth stations are separated by a large distance
- Satellite downlink in Figure 4.52, likely to be correlated in the wanted and interfering paths as both can be expected to suffer the same attenuations, in particular rain, near the victim Earth station.

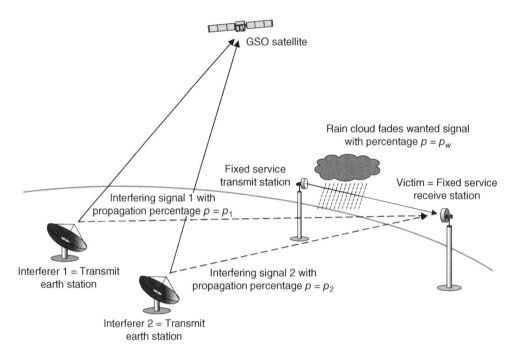

Figure 4.56 Multiple percentages of time for aggregate interference scenario

These represent the extremes between fully correlated and uncorrelated, and for these two cases it is possible to select the propagation model's percentages of time using:

- **Fully correlated**, where the same percentage of time is used for all propagation paths so that in Figure 4.56,

$$p_w = p_1 = p_2 = Random[0,100] \qquad (4.31)$$

- **Fully uncorrelated**, where each propagation path has a separately selected and independent percentage of time so that

$$p_w = Random[0,100] \qquad (4.32)$$

$$p_1 = Random[0,100] \qquad (4.33)$$

$$p_2 = Random[0,100] \qquad (4.34)$$

Note the random number is selected from the full range of percentages, but it might be necessary to cap it to the range required by the propagation model, for example, ITU-R Rec. P.452 is valid for $0.001\% \le p \le 50\%$.

However, while the Earth station example from Section 4.4.2 could be either fully correlated or fully uncorrelated, there are circumstances, in particular for terrestrial scenarios, where in reality there would be a degree of correlation between paths. So for Figure 4.56 it might be that the two interfering paths from the Earth stations would experience similar atmospheric conditions so that rather than being independent, they could simultaneously have a degree of ducting or fading.

Furthermore, fully uncorrelated becomes increasingly unrealistic as the number of interferers increases. The likelihood of one of N paths having a percentage of time under 0.01% increases with N as

$$p(p_i < 0.01\%) = 1 - (1 - 0.0001)^N \tag{4.35}$$

So with 7000 interferers it becomes likely that at each sample, there is one path that is modelled as being enhanced to the level of 0.01% of the time: this is likely to be unrepresentative of actual atmospheric behaviour.

The topic of correlation of propagation paths is difficult to research and model: correlation statistics are dependent on path and climate, and measurements would have to be taken over many years to become statistically significant. Furthermore, there is limited guidance in the ITU-R Recommendations, and what is available is typically restricted to cases where there are very similar geometries between paths.

One technique that has been developed is the use of copulas, which provide a way to define the degree of correlation ρ between two propagation paths, via the method used to select the associated percentages of time (Craig, 2004).

Consider the simplified case with two probabilities (u, v) such as required for the scenario in Figure 4.56. The fully correlated and fully uncorrelated are modelled by one of the following:

$$u = v = Random[0,1] \tag{4.36}$$

$$u = Random[0,1], \quad v = Random[0,1] \tag{4.37}$$

Copulas can be used to connect the two distributions to allow parameterisation of the degree of correlation. Given the individual cumulative distribution functions (CDFs) $F(u)$ and $F(v)$ with joint CDF $F(u, v)$, then there exists a copula $C(u, v)$ such that

$$F_{uv}(u,v) = C(F_u(u), F_v(v)) \tag{4.38}$$

One such is the Clayton copula, which uses a generating function ϕ where

$$\phi(C(u,v)) = \phi(u) + \phi(v) \tag{4.39}$$

With generating function defined by

$$\phi(t) = \frac{1}{\alpha}(t^{-\alpha} - 1) \tag{4.40}$$

it can be shown that the copula is then

$$C(u,v) = (u^{-\alpha} + v^{-\alpha} - 1)^{-1/\alpha} \tag{4.41}$$

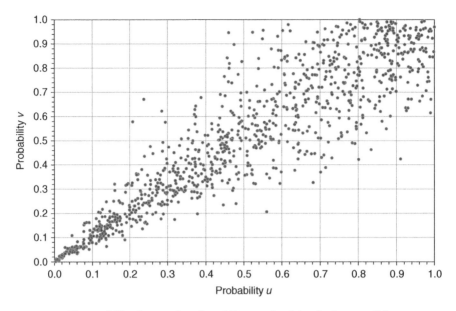

Figure 4.57 Scatter plot of partially correlated (u, v) where $\rho \sim 0.9$

The α parameter defines the degree of correlation between the (u, v) variables. The mapping between degree of correlation term ρ and α can be calculated (Craig, 2004) using

$$\alpha = \left(\frac{\rho}{1-\rho} \right)^{0.8} \tag{4.42}$$

This allows intermediate values of correlation to be defined and sets of partially correlated random numbers (u, v) generated (Nelsen, 1999) as in Figure 4.57.

These two probabilities could then be converted into percentages and used to calculate the interference from the two satellite Earth stations in Figure 4.56. The degree of correlation will depend upon the propagation effect being modelled and distances involved, which could also influence the duration of the effect. For example, ducting could have an impact on links over a wide area for many hours, leading to the likelihood of correlation. On the other hand short-range paths in an urban environment could suffer multipath fading and obstruction losses that change rapidly and are mostly de-correlated.

There can be correlation between propagation effects and other aspects to modelling radio systems. An example would be power control, where radio systems increase their TX power to overcome fading, as discussed in Section 5.5.2. Correlation and percentages of time are important factors to consider when undertaking Monte Carlo style models as described in Section 6.9.

One such Monte Carlo method was used to predict 4G coverage in the United Kingdom (Ofcom, 2012b) based upon an IEEE methodology (IEEE, 2009b). At each pixel the signal strength was calculated from all sectors of the 20 nearest base stations. The variation in signal strength across the pixel was modelled as being log-normal using the methodology in Rec. ITU-R P.1812, with mean zero and with the standard deviation σ dependent on the environment via the land use code for that pixel.

For each Monte Carlo sample, random numbers p were generated:

- $\{p_i\}$ for the ith pixel
- $\{p_{ij}\}$ for the path from the jth base station to the ith pixel.

There was then assumed to be a 0.5 weighting factor between the path and pixel variations so that the total variation Z_{ij} on the path from the jth base station to the ith pixel was

$$Z_{ij} = \frac{1}{\sqrt{2}} Z(p_i, \sigma) + \frac{1}{\sqrt{2}} Z(p_{ij}, \sigma)$$

(4.43)

Here $Z(p, \sigma)$ is the log-normal distribution function with mean zero and standard deviation σ sampled at a probability $= p$.

A pragmatic approach to combine signals from multiple propagation paths was tentatively proposed in Document 3J/146-E (Ofcom, 2006a), which can handle more than two entries. It is based on the following principles:

- For small percentages of time, the aggregate power is typically driven by the highest single entry signal
- For median percentages of time, the aggregate power tends to be driven by the power summation of all signals.

The algorithm requires an interpolation variable, R, that can be calculated using parameters determined by fitting data from measurements in the United Kingdom:

1. For n paths $i = [i...n]$ calculate the signal strength S_i for $p\%$ of time
2. Determine the strongest signal:

$$S_{max} = max\{S_1, S_2, ... S_n\}$$

(4.44)

3. Calculate the power summation of all signals:

$$S_s = 10\log_{10} \sum_{i=1}^{n} 10^{S_i/10}$$

(4.45)

4. Calculate the interpolation variable R:

$$R = 0.1 + 0.213 \log_{10}(1000p)$$

(4.46)

5. Then the aggregate signal is

$$S = S_{max} + R(S_s - S_{max})$$

(4.47)

4.9 Selection of Propagation Model

Given the range of propagation models available, an important part of any interference analysis study is selecting which ones to use. Each of the radio paths, wanted or interfering, could require a separate propagation model or combination of models.

Factors to consider include:

- Specific or generic: is the study relating to specific location(s) or is the aim to produce results that would be generally applicable?
- Frequencies: which models would be valid for the frequencies involved?
- Geometry: is the path terrestrial or between Earth and space or from ground to air?
- Distance: what are the distances involved?
- Environment: does the model need to take account of general environment types such as {urban, suburban, open}?
- Time variability: is it necessary to consider the time variation of signals – for example, if the interference thresholds include one for the short-term?
- Point-to-point or point-to-area? Is it important to model how the signal strength varies across a pixel?
- Databases: what terrain, surface and land use databases are available, and what are their resolutions?
- Data accuracy: to what accuracy are the TX and RX station locations known?
- Regulatory requirements: do the regulations specify which propagation model(s) should be used?

A possible decision tree is shown in Figure 4.58, though it should be used with caution as there are often special cases that override some of the choices. For example, take the impact of distance for terrestrial paths:

- Short range (say, < 50 m): propagation within building could be modelled using a dual slope approach as in Section 4.3.10
- Medium range (say, 50 m < d < 3 km): a number of choices are available:
  - ○ P.452, P.1546, P.1812 or P.2001 with smooth Earth for a generic study, possibly with additional clutter losses
  - ○ P.452, P.1546, P.1812 or P.2001 with a terrain database and possibly land use data for site-specific analysis
  - ○ P.452, P.1812 or P.2001 with a high-resolution surface database for detailed site-specific analysis
  - ○ Hata/COST 231 for a generic study taking into account environment type.
- Long range (say, d > 3 km): any of P.452, P.1546, P.1812 or P.2001 either with databases (terrain, surface, land use) or smooth Earth.

Section 4.3.12 noted that P.452, P.1812 and P.2001 give similar results, unlike, in particular, P.1546. In cases where any of these four models could be used, a decision has to be made as to which is most suitable. Work within SG 3 suggests that the standard deviation error between prediction and measurement of P.1812 is less than that for P.1546. However, the data

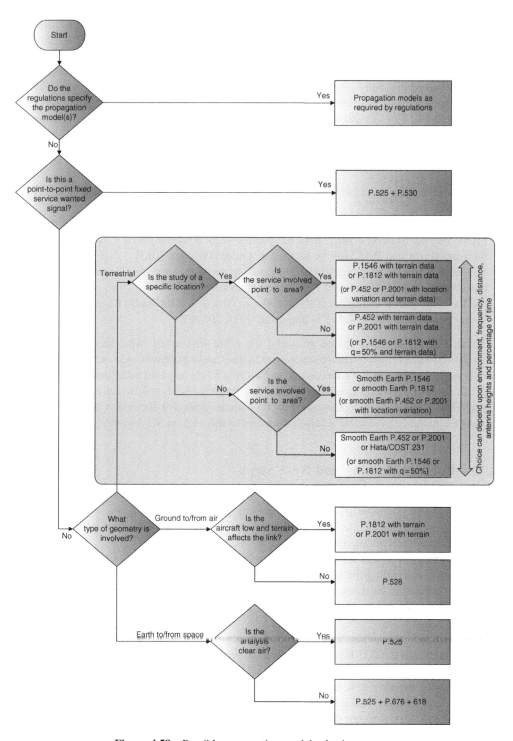

Figure 4.58 Possible propagation model selection process

requirements and accuracy are also greater: if there are uncertainties in the location of transmitter, there could be greater variation in predicted signal strengths using P.1812 than with P.1546.

Sometimes the propagation model that is technically the best might be rejected because some of those involved in a study project are not comfortable or familiar with its use. In particular, if there has been a long period of time in which one model has been used for planning services (such as P.1546 for broadcasting and PMR coverage predictions), then there can be resistance to the use of more recent models (such as P.1812).

National and international regulations can require use of specific propagation models, such as:

- In the ITU-R RRC to replan broadcasting services GE06, use of P.1546 is mandated (ITU-R, 2006a)
- For work within the United States, some FCC rules suggest use of the Longley–Rice model
- Fixed links are planned in the United Kingdom using a combination of P.525 and P.530 for the wanted link and P.452 for the interfering path(s) (Ofcom, 2013)
- The international coordination of satellite Earth stations uses Appendix 7 to calculate the contours and then (typically) P.452 for detailed analysis
- For the coordination trigger for GSO satellites in Article 8 of the Radio Regulations, the use of free space path loss in P.525 is mandated (ITU, 2012a).

Sometimes the question is best answered the other way round: what scenarios are each propagation model valid for? This requires a good understanding of each model, what it is designed for, its strengths and whether it represents the best fit with the scenario under consideration.

If there is any doubt or disagreement with parties, then it is often helpful to consult with relevant experts. This can be done by contacting the national administration or, for work within the ITU-R, a request for guidance by sending a liaison statement to the relevant SG 3 working party.

4.10   Further Reading and Next Steps

This chapter considered the main propagation models including their strengths, weaknesses and applicability. Example plots showed how the propagation loss for these models varies by distance at selected frequencies, and in addition, coverage plots showed how the impact of terrain varies between models.

The propagation model was seen to determine the radio path that should be taken into account when calculating the antenna gains. The issue was raised of the correlation of time-dependent behaviour between different radio paths. A range of approaches were described, including assuming they are fully correlated or uncorrelated and techniques to model partially correlated paths.

More detailed information about each propagation model can be found in the original documents, in particular the ITU-R Recommendations. A wide range of additional reading materials is available, such as the books referenced: Barclay (2003) and Haslett (2008).

This chapter, together with the one on 'Fundamental Concepts', describes how to calculate signal strengths, wanted and interfering, between any two points, whether terrestrial, air or in space. The next step is to consider in more detail the interference calculation.

5

The Interference Calculation

The previous chapters have described the basic link budget, the calculation of the strength of the wanted and interfering signals at the input to the receiver, using various propagation models. A number of additional factors also need to be taken into account, such as the bandwidth and polarisation of both signals, and this chapter describes these and other adjustments that might be required in the interference calculation.

Interference analysis can be undertaken between radio systems operating either co-frequency or non-co-frequency, and for the latter case it is important to be able to calculate how much power transmitted by a system on one frequency would be detected at a receiver on another frequency. In particular this involves the calculation of the mask integration adjustment factor and use of ratios such as the adjacent channel selectivity.

This chapter also looks at how to analyse system behaviour, such as adaptive power and modulation, plus intermodulation products and variation in interference due to deployment models and traffic.

Having calculated the wanted and interfering signals taking these factors into account, it is then possible to determine whether these would be considered acceptable or not by comparing against various thresholds. These thresholds are discussed, in particular in relation to the interference margin and methods to apportion aggregate interference.

The various factors that can influence the interference calculation can be considered options to investigate when attempting to mitigate or reduce interference. The chapter therefore ends with a discussion on interference mitigation techniques.

The following resource is available:

Resource 5.1 The spreadsheet 'Chapter 5 Examples.xlsx' contains the calculations used in the examples of bandwidth adjustment, adjacent channel interference ratio (ACIR), end to end performance, Appendix 7 and SF.1006 thresholds and the derivation of the C-band PFD threshold.

Interference Analysis: Modelling Radio Systems for Spectrum Management, First Edition. John Pahl.
© 2016 John Wiley & Sons, Ltd. Published 2016 by John Wiley & Sons, Ltd.
Companion website: www.wiley.com/go/pahl1015

5.1 Bandwidths and Domains

Section 3.4 introduced the concept of the carrier and how information is encoded onto the radio signal using the modulation. This modulation changes the shape of the power in the frequency domain, which can be modelled via parameters such as the bandwidth. When there are multiple interferers operating on different frequencies, there can also be aggregation effects, and so it can be necessary to consider the channel plan.

Some of the key concepts are shown in Figure 5.1. This shows a simplified model of carriers as rectangular blocks of power in the frequency domain, as is typically used for co-frequency scenarios. The total power of the carrier p_{tx} is the power density ρ times the occupied bandwidth (OBW) B, that is, in absolute:

$$p_{tx} = \rho_{tx}.B \tag{5.1}$$

More advanced analysis can use spectrum masks which are able to model the carrier in the in-band, out-of-band and spurious domains as in Figure 5.2.

The central frequencies $\{f_1, f_2, f_3, \dots\}$ are often selected from the **channel plan** for the service during the assignment process. Channel plans are defined for the broadcast, mobile and fixed services (FS) in regulatory documents including ITU-R Recommendations. For some services (e.g. duplex land mobile and fixed), the channel plan gives a pair of frequencies.

Example 5.1
Fixed service systems using 28 MHz carriers within Europe in the 24.5–26.5 GHz band are recommended to select paired centre frequencies (f, f') from the set:

$$f_n = f_0 - 966 + 28n$$

$$f'_n = f_0 + 42 + 28n$$

where $f_0 = 25501.0$ MHz and $n = \{1, 2, \dots 32\}$. See ERC Recommendation T/R 13-02 (CEPT ECC, 2010) Annex B and Rec. ITU-R F.748 (ITU-R, 2001a) Annex 1.

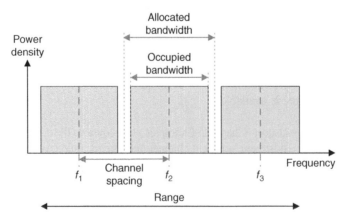

Figure 5.1 Bandwidths and channel spacing

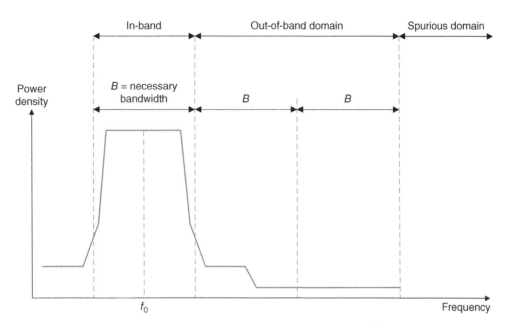

Figure 5.2   Example of in-band, out-of-band and spurious domains

The Radio Regulations (RR) defines the OBW as follows:

1.153 **occupied bandwidth***: The width of a frequency band such that, below the lower and above the upper frequency limits, the mean powers emitted are each equal to a specified percentage β/2 of the total mean power of a given emission.*
Unless otherwise specified in an ITU-R Recommendation for the appropriate class of emission, the value of β/2 should be taken as 0.5%.

The calculation of power in this definition is discussed further in Section 5.3.1.
The RR also define a necessary bandwidth as follows:

1.152 **necessary bandwidth***: For a given class of emission, the width of the frequency band which is just sufficient to ensure the transmission of information at the rate and with the quality required under specified conditions.*

This takes a system performance-based approach rather than the energy-related definition of the OBW. Typically the necessary bandwidth is the channel bandwidth, which in Figure 5.1 is called the **allocated bandwidth** (ABW). For passive services this relates to the bandwidth necessary to make the required observations.

These bandwidths are also used to determine whether interference is:

• **Co-frequency interference**: where there is overlap between the OBW of the wanted and at least one interfering system

- **Non-co-frequency interference**: where there is no overlap between the OBW of the wanted system and any interfering system.

These two can require different approaches to interference analysis. Co-frequency analysis can be simplified to model the carriers using rectangular blocks as described in Section 5.2. However, non-co-frequency analysis needs to consider how the power transmitted and the receiver filter varies in the frequency domain, as described in Section 5.3.

Terms that might be encountered when defining a sharing scenario include:

- **Co-channel interference**: similar to co-frequency, but links the interference to a channel arrangement. This might be appropriate when both the wanted and interfering systems use the same channel plan
- **Adjacent channel interference**: similar to non-co-frequency, though it implies a similar channelisation plan for wanted and interfering
- **In-band interference**: similar to co-frequency interference, though 'band' might imply a frequency band in which the wanted and interfering systems both operate, but not in a co-frequency manner
- **Out-of-band interference**: similar to non-co-frequency, though, as will be seen later, the out-of-band domain is limited to only selected frequency offsets
- **Adjacent band interference**: used instead of non-co-frequency interference, though again 'band' could be taken as a frequency band.

This book will standardise on the term co-frequency where there is overlap between the wanted and interfering system's OBW and non-co-frequency for cases where there is no such overlap.

The co-frequency or non-co-frequency terms define whether the interfering path has bandwidth overlap or not, while the other terms, such as adjacent channel, out of band, etc. can be used to categorise the frequency sharing scenario.

The power densities immediately outside the necessary bandwidth will be determined by the modulation and filtering at the transmitter and typically decreases until it falls below the noise floor. There can be occasional additional signals with large frequency differences from the centre frequency due to harmonics and equipment imperfections. The part immediately outside the necessary bandwidth is called the out-of-band domain and beyond that the spurious domain, and together they comprise **unwanted emissions**.

The divide between the **out-of-band domain** and the **spurious domain** is typically (but not always) 250% of the necessary bandwidth from the central frequency, as shown in Figure 5.2. This also shows a typical spectrum mask, whereby the power density of the carrier in the frequency domain is defined by a set of straight lines (dBW/Hz vs. Hz). The definitions in the RR are:

1.146   unwanted emissions: Consist of spurious emissions and out-of-band emissions.

1.146A   out-of-band domain (of an emission): The frequency range, immediately outside the necessary bandwidth but excluding the spurious domain, in which out-of-band emissions generally predominate. Out-of-band emissions, defined based on their source, occur in the out-of-band domain and, to a lesser extent, in the spurious domain. Spurious emissions likewise may occur in the out-of-band domain as well as in the spurious domain.

1.146B   spurious domain (of an emission): The frequency range beyond the out-of-band domain in which spurious emissions generally predominate.

The ITU-R uses the **designation of emission** to classify a communications system's carrier using a format given in RR Appendix 1 (ITU, 2012a). The first characters give the necessary bandwidth via three numbers with the decimal point replaced by a symbol giving the units as one of:

- Letter H for bandwidths between 0.001 and 999 Hz given in Hz
- Letter K for bandwidths between 1.00 and 999 kHz given in kHz
- Letter M for bandwidths between 1.00 and 999 MHz given in MHz
- Letter G for bandwidths between 1.00 and 999 GHz given in GHz.

The full designation of emission is an alphanumeric string that contains the following information:

- Necessary bandwidth as three numerals and one character
- Carrier definition via three symbols:
    1. Type of modulation of the main carrier
    2. Nature of signal(s) modulating the main carrier
    3. Type of information to be transmitted.
       There are also two optional additional characters:
    4. Details of signal(s).

As most carriers are digital and handle multiple data types, information other than the necessary bandwidth is usually not relevant.

Example 5.2
The necessary bandwidth of a 36 MHz carrier (e.g. for satellite TV transmissions) would be written as 36M0.

Example 5.3
The necessary bandwidth of a 12.5 kHz carrier (e.g. for land mobile voice communications) would be written as 12K5.

5.2   Bandwidth Adjustment Factor

For co-frequency scenarios it is often necessary to adjust for the difference in bandwidths and degree of overlap using the bandwidth adjustment factor A_{BW}. A typical co-frequency scenario is shown in the frequency domain in Figure 5.3.

The check for whether there is overlap and hence co-frequency interference uses the two frequencies and (occupied) bandwidth B:

$$|f_W - f_I| < \frac{(B_W + B_I)}{2} \tag{5.2}$$

The non-co-frequency interference case is

$$|f_W - f_I| \geq \frac{(B_W + B_I)}{2} \tag{5.3}$$

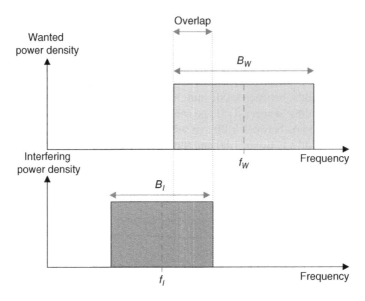

Figure 5.3   Co-frequency overlap scenario

If there is overlap then it is possible that not all the interfering power will enter the receiver. In co-frequency cases the transmit carrier is often modelled as a rectangular block (as in Fig. 5.3) and the receiver as a perfect filter with the same frequency and bandwidth as the wanted signal. In this case the power received would be a fraction of the total interfering power, calculated using

$$Overlap = min \left\{ B_W, B_I, \frac{(B_W + B_I)}{2} - |f_W - f_I| \right\} \tag{5.4}$$

$$A_{BW} = 10\log_{10}\left(\frac{Overlap}{B_I}\right) \tag{5.5}$$

Equation 3.204 then becomes

$$I = P'_{tx} + G'_{tx} - L'_p + G'_{rx} - L_f + A_{BW} \tag{5.6}$$

Example 5.4

The wanted system is a digital TV system with occupied bandwidth = 8 MHz transmitting at a frequency of 706 MHz, while the interferer is a mobile communication system (e.g. LTE) transmitting with bandwidth = 10 MHz at a frequency of 700 MHz. The relevant calculations are

$$Overlap = min \left\{ 8, 10, \frac{(8 + 10)}{2} - |706 - 700| \right\} = 3 \text{ MHz} \tag{5.7}$$

$$A_{BW} = 10\log_{10}\left(\frac{3}{10}\right) = -5.2 \text{ dB} \tag{5.8}$$

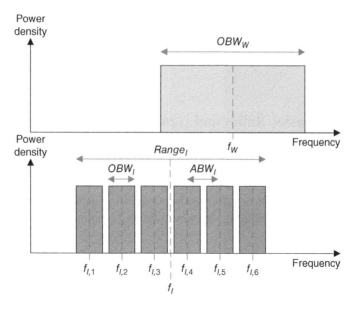

Figure 5.4 Narrowband into wideband scenario

It is possible that the victim system has a much wider bandwidth than the interferer in which case there could be multiple interfering entries as in Figure 5.4. Rather than just considering the occupied bandwidth $OBW = B$, it can be necessary to consider the channel bandwidth or ABW.

In this case there are two possible approaches. Firstly, the calculation could be repeated for each central frequency, so that using index i

$$I_i = P'_{tx} + G'_{tx} - L'_p + G'_{rx} - L_f + A_{BW,i} \tag{5.9}$$

$$I_{agg} = 10\log_{10}\sum_i 10^{I_i/10} \tag{5.10}$$

Alternatively, the multiple channels can be considered a 'super-carrier' with its own centre frequency and frequency range – its 'bandwidth'. The total power is then the power of one carrier multiplied by the number of interfering carriers in the overlap of the super-carrier with the victim OBW. Hence the channel bandwidth or ABW must be used instead of the OBW to calculate the bandwidth adjustment factor as in

$$A_{BW,agg} = 10\log_{10}\left(\frac{Overlap}{ABW_I}\right) \tag{5.11}$$

Then

$$I_{agg} = P'_{tx} + G'_{tx} - L'_p + G'_{rx} - L_f + A_{BW,agg} \tag{5.12}$$

This approach requires that all the other terms (gains, powers, propagation losses, etc.) are identical for each channel, so it should be used with care. One case where such an approach can be appropriate is for GSO satellite coordination, as described in Section 7.5.3.

5.3 Spectrum Masks, Ratios and Guard Bands

5.3.1 Transmit Mask and Calculated Bandwidth

The OBW is defined in the RR to be the frequency range, which contains 99% of the total mean power so that the two extreme ends of the spectrum mask contain 0.5% each, as in Figure 5.5.

The bandwidth of a carrier can be calculated from a mask that defines how the transmit power density varies in the frequency domain. One way to define this **transmit spectrum mask** is as a series of points with index i:

$$[M_{tx,i}, f_i] \quad \text{for} \quad i = (1 \ldots n) \tag{5.13}$$

where

M_{tx} = power spectral density relative to in-band or peak power density in dB
f = frequency

Figure 5.6 shows these mask points as small circles connected via straight lines in (dB, Hz). Each of the straight line segments $M_{tx}(f)$ in decibels can be converted into curved lines in absolute (i.e. linear) $m_{tx}(f)$ where

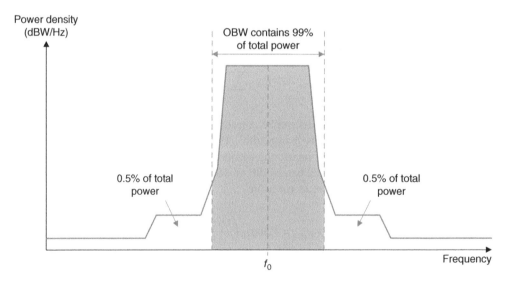

Figure 5.5 Calculation of occupied bandwidth

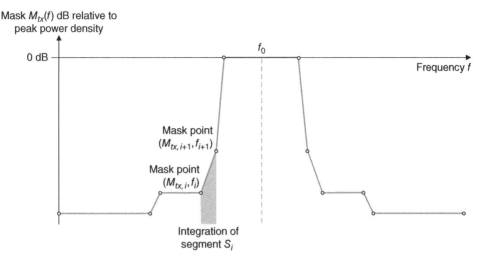

Mask $M_{tx}(f)$ dB relative to
peak power density

0 dB

f_0

Frequency f

Mask point
$(M_{tx,i+1}, f_{i+1})$

Mask point
$(M_{tx,i}, f_i)$

Integration of
segment S_i

Figure 5.6   Transmit spectrum mask with segment

$$M_{tx}(f) = A_i + B_i f \quad f_i \le f \le f_{i+1} \tag{5.14}$$

$$m_{tx}(f) = 10^{M_{tx}(f)/10} \quad f_i \le f \le f_{i+1} \tag{5.15}$$

The intersect and slope parameters (A_i, B_i) can be calculated from the segment's start and end points $(M_{tx,i}, f_i)$ and $(M_{tx,i+1}, f_{i+1})$ using

$$A_i = \frac{M_{tx,i} f_{i+1} - M_{tx,i+1} f_i}{f_{i+1} - f_i} \tag{5.16}$$

$$B_i = \frac{M_{tx,i+1} - M_{tx,i}}{f_{i+1} - f_i} \tag{5.17}$$

The mask $m_{tx}(f)$ gives power relative to peak transmit power spectral density ρ_{tx} (e.g. in watts/Hz); hence the total power in absolute (e.g. watts) is

$$p_{tx} = \int_{f=0}^{f=\infty} \rho_{tx} m_{tx}(f) df \tag{5.18}$$

In practice the transmit mask will only be defined for a range f_{min} to f_{max} and so the integration is

$$p_{tx} = \int_{f_{min}}^{f_{max}} \rho_{tx} m_{tx}(f) df \tag{5.19}$$

This could be integrated numerically, but a more accurate approach is to split the mask into segments where the mask is a straight line in dB. These segments can then be integrated analytically before being summed, so that, given that the peak power density is a constant, the power calculation is

$$p_{tx} = \rho_{tx} \sum_i s_i \tag{5.20}$$

where

$$s_i = \int_{f_i}^{f_{i+1}} m_{tx}(f) df = \int_{f_i}^{f_{i+1}} 10^{M_{tx}/10} df \tag{5.21}$$

Hence

$$s_i = \int_{f_i}^{f_{i+1}} 10^{(A_i + B_i f)/10} df = \int_{f_i}^{f_{i+1}} 10^{A_i/10} 10^{B_i f/10} df \tag{5.22}$$

Now

$$10^x = e^{\ln(10)x} \tag{5.23}$$

So

$$s_i = 10^{A_i/10} \int_{f_i}^{f_{i+1}} e^{\ln(10)B_i/10 f} df \tag{5.24}$$

Using

$$Q_i = \frac{\ln(10)B_i}{10} \tag{5.25}$$

$$s_i = 10^{A_i/10} \int_{f_i}^{f_{i+1}} e^{Q_i f} df \tag{5.26}$$

The integration method depends upon whether $Q_i = B_i = 0$, that is, if the segment is horizontal. If $Q_i \neq 0$ then

$$s_i = \frac{10^{A_i/10}}{Q_i} \left(e^{Q_i f_{i+1}} - e^{Q_i f_i} \right) \tag{5.27}$$

If $Q_i = 0$ then

$$s_i = 10^{A_i/10} (f_{i+1} - f_i) \tag{5.28}$$

Note that both terms are dependent upon the frequency difference rather than absolute frequency and hence the integration will give the same result for any centre frequency.

Equation 5.27 can be rewritten by replacing the variable Q_i as follows:

$$s_i = \frac{10^{A_i/10}}{\ln(10)B_i/10}\left(e^{\ln(10)B_if_{i+1}/10} - e^{\ln(10)B_if_i/10}\right) \tag{5.29}$$

$$s_i = \frac{\left(10^{(A_i+B_if_{i+1})/10} - 10^{(A_i+B_if_i)/10}\right)}{\ln(10)B_i/10} \tag{5.30}$$

$$s_i = \frac{10}{\ln(10)}\frac{m_{i+1}-m_i}{M_{i+1}-M_i}(f_{i+1}-f_i) \tag{5.31}$$

It is therefore possible to analytically calculate the area of each segment of a spectrum emission mask: the total area of the mask can then be calculated by the summation of all the segments.

The process can be reversed to derive the OBW using

$$\int_{f_1}^{f_2} \rho_{tx}m_{tx}(f)df = 0.99p_{tx} = \int_{fmin}^{fmax} \rho_{tx}m_{tx}(f)df \tag{5.32}$$

where

$$OBW = f_2 - f_1 \tag{5.33}$$

The power density term cancels out, implying that values other than peak could be used – for example, the mean in-band power density – and the format of the integration would be unchanged. Then the frequencies f_1 and f_2 can be derived using

$$\int_{f_1}^{f_0} m_{tx}(f)df = \int_{f_0}^{f_2} m_{tx}(f)df = 0.495\,a_{tx} \tag{5.34}$$

where

$$a_{tx} = \int_{fmin}^{fmax} m_{tx}(f)df \tag{5.35}$$

The calculation is likely to involve part of a segment—hence reversing Equations 5.28 and 5.31 to calculate a difference in frequency.

Note how the bandwidth adjustment factor in Section 5.2 used the two parameters (P_{tx}, OBW) where the transmit power was assumed to be wholly in the OBW. However, in this section we have used P_{tx} to represent the total power including that part outside the OBW, indeed using the transmit spectrum mask to calculate OBW as the frequency range with 99% of the total power. Hence there are two transmit powers to consider, the total power P_{tx} and the in-band transmit power P_{ib} connected via their absolute values:

$$p_{ib} = 0.99.p_{tx} \tag{5.36}$$

In general when transmit powers are given, they are P_{tx} but the difference is negligible as

$$P_{tx} - P_{ib} = 0.0436\,\text{dB} \tag{5.37}$$

For co-frequency interference analysis, taking $P_{ib} = P_{tx}$ would slightly overestimate the interference, but as studies are based on assumptions and can contain components with significant uncertainties (in particular propagation models), this difference can usually be ignored. Furthermore, it is generally considered better practice to err slightly on the cautious side for interfering paths.

5.3.2   Standards and Spectrum Emission Masks

When these masks are defined in standardisation documentation from organizations such as ETSI, a range of **spectrum emission mask** formats are used, which, in some cases, differ from the transmit spectrum mask $M_{tx}(f)$ used in the previous section. This section gives some examples and, where necessary, their conversion into $M_{tx}(f)$ format.

Example 5.5
A point-to-point fixed service system with bandwidth 28 MHz and spectrum efficiency class 4H (which implies a 32-state modulation such as 32QAM) is operating in the 28 GHz band. This equipment is covered by the following ETSI standard:

- ETSI EN 302 217-2-2: Fixed Radio Systems; Characteristics and requirements for point-to-point equipment and antennas; Part 2-2: Digital systems operating in frequency bands where frequency co-ordination is applied; Harmonized EN covering the essential requirements of Article 3.2 of the RED (ETSI, 2014c).

The relevant spectrum mask is given in Table 5.1 and shown in Figure 5.7.
The frequency offset is relative to the carrier centre frequency and the mask is defined such that the '0 dB level shown on the spectrum masks relates to the power spectral density at the carrier centre frequency'. It can be seen that in this case it exceeds zero by 2 dB: this relates to the mask being also used for multi-carrier systems, as it might be that the peak power spectral density is not at the carrier centre. It should be noted that having mask values greater than zero is mathematically acceptable as the mask is integrated and then compared to a partial integral of the same mask. It is also worth noting that these masks are envelopes: equipment meeting these

Table 5.1   Transmit mask for 28 MHz
point-to-point fixed link

Frequency offset (MHz)	Mask (dB)
12	2
15	−10
16.8	−33
35	−40
48.3	−50
70	−50

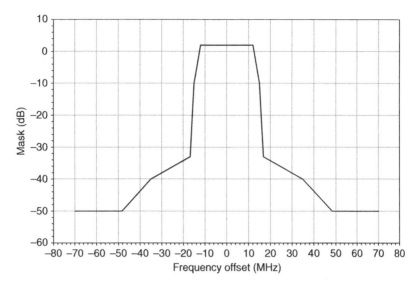

Figure 5.7 Transmit mask for 28 MHz point-to-point fixed link

standards should not exceed the values given and are expected to have equal or lower power densities. However, when they are used to calculate the OBW or mask integration in Section 5.3.3, it is assumed they are exactly met at all points.

The mask can be seen to extend to 2.5 times the 28 MHz channel separation and the floor is 50 dB down from peak power spectral density. The calculated 99% power bandwidth OBW = 27.97 MHz.

Example 5.6

A wide-area base station (BS) is transmitting a 10 MHz carrier with power = 46 dBm in Europe at frequencies above 1 GHz. The relevant mask is given by the 3GPP standard:

- 3GPP TS 36.104: LTE; Evolved Universal Terrestrial Radio Access (E-UTRA); Base Station (BS) radio transmission and reception (ETSI/3GPP, 2014).

This document is also ETSI TS 136 104.

The mask is specified in terms of maximum power density for various offsets Δf from the boundary channel edge frequency (i.e. not central frequency as in Example 5.5) as shown in Table 5.2. This table has been simplified to use a single-frequency offset definition excluding the one that includes the measurement bandwidth.

To convert this to a transmit spectrum mask of the form $M_{tx}(f)$, it is necessary to convert each point using

$$M(f) = R - 10\log_{10}(B_M) - (P_{tx} - 10\log_{10}(B))$$ (5.38)

Here $P_{tx} = 46$ dBm (i.e. using the same power units as R) and the bandwidth $B = 10$ MHz so the converted mask is shown in Table 5.3 and Figure 5.8.

Table 5.2 Emission mask for 10 MHz LTE wide-area base station

Frequency offset Δf	R = requirement	B_M = measurement bandwidth
0 MHz $\leq \Delta f <$ 5 MHz	$-7\,\mathrm{dBm} - \dfrac{7}{5}\Delta f$	100 kHz
5 MHz $\leq \Delta f <$ 10 MHz	$-14\,\mathrm{dBm}$	100 kHz
10 MHz $\leq \Delta f <$ 20 MHz	$-15\,\mathrm{dBm}$	1 MHz

Table 5.3 Transmit mask for 10 MHz LTE base station

Frequency offset (MHz)	Mask $M(f)$ (dB)
5	−33
10	−40
15	−40
15	−51
25	−51

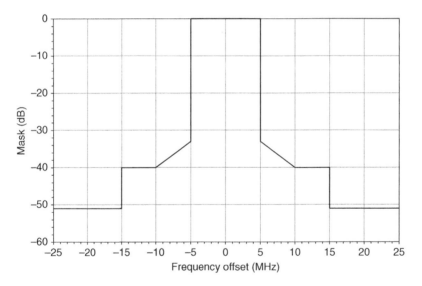

Figure 5.8 Transmit mask for 10 MHz LTE base station

Example 5.7

An 8 MHz bandwidth digital video broadcast (DVB) signal is transmitting at a power of 20 kW on a frequency adjacent to a sensitive (low-power or receive-only) system, and so it must use the critical mask given in the following:

- ETSI EN 300 744 European Standard (Telecommunications series). Digital Video Broadcasting (DVB); Framing structure, channel coding and modulation for digital terrestrial television (ETSI, 2009).

The mask is given in Table 5.4.

The relative level is specified as 'power level measured in a 4 kHz bandwidth, where 0 dB corresponds to the total output power'. A 20 kW = 43 dBW transmitter in 8 MHz has power density = 34 dBW/MHz = 10 dBW/4 kHz. The mask is then the relative level in Table 5.4 plus the total power = 43 dBW minus the in-band power density = 10 dBW/4 kHz as shown in Figure 5.9, extending the mask to 250% of the bandwidth.

Note how this mask is significantly tighter than those shown for the FS point-to-point link in Example 5.5 or the base station (BS) in Example 5.6. This is partly necessary to reduce unwanted emissions from the higher DVB transmit power, but it also relates to equipment cost, as effective filtering is expensive and requires additional on-site space. These factors are particularly an issue for mobile phones that have stringent cost and space constraints, and their transmit masks are typically less tight than the ones given in these examples.

Table 5.4 Emission mask for 8 MHz DVB critical case

Frequency offset (MHz)	Relative level (dB)
−12	−120
−6	−95
−4.2	−83
−3.8	−32.8
3.8	−32.8
4.2	−83
6	−95
12	−120

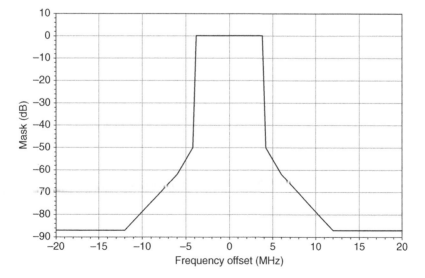

Figure 5.9 Transmit mask for 8 MHz DVB critical case

5.3.3   The Mask Integration Adjustment Factor

The transmit mask $M_{tx}(f)$ describes how the transmit power spectral density varies in the frequency domain. A **receive spectrum mask** can similarly be defined, namely, $M_{rx}(f)$ in dB or $m_{rx}(f)$ in absolute, which gives the frequency response of the receiver. For a received signal before filtering with power density $\rho_{rx}(f)$, the power of the signal actually received in the range δf would be

$$\delta p_{rx}(f) = \rho_{rx}(f)m_{rx}(f)\delta f \tag{5.39}$$

Hence the total received power would be

$$p_{rx} = \int_{f=0}^{f=\infty} \rho_{rx}(f)m_{rx}(f)df \tag{5.40}$$

In this form mask values are negative in dB with 0 dB representing a transparent filter with no attenuation. Note that sometimes the filter response is defined as losses and consequently with positive values.

Consider the configuration in Figure 5.10. Here the radio path in Figure 3.74 has been simplified so that the transmitter is directly connected to the receiver, excluding the antennas, propagation effects and receive line feed.

In this case the receive power density at point (p) in the figure can be specified from the transmit power density and transmit mask as in

$$\rho_{rx}(f) = \rho_{tx}m_{tx}(f) \tag{5.41}$$

Hence it is possible to calculate the power that would be detected at the receiver as

$$p_{rx} = \int_{f=0}^{f=\infty} \rho_{tx}m_{tx}(f)m_{rx}(f)df \tag{5.42}$$

$$p_{rx} = \rho_{tx}\int_{f=0}^{f=\infty} m_{tx}(f)m_{rx}(f)df \tag{5.43}$$

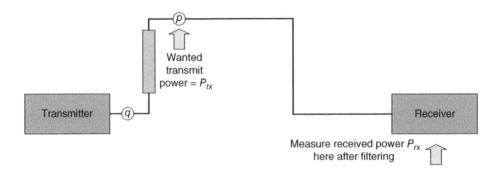

Figure 5.10   Simplified radio path for mask integration calculations

But Equation 5.18 stated that

$$P_{tx} = P_{tx} \int_{f=0}^{f=\infty} m_{tx}(f) df \tag{5.44}$$

The ratio of the power received after filtering to the transmitted power can therefore be calculated as

$$\frac{P_{rx}}{P_{tx}} = \frac{\int_{f=0}^{f=\infty} m_{tx}(f) m_{rx}(f) df}{\int_{f=0}^{f=\infty} m_{tx}(f) df} \tag{5.45}$$

As with the calculation of the bandwidth, the integration of the two masks is over a restricted range, usually the frequencies over which the receive mask is defined. In addition, this frequency range is split into segments where both the transmit mask and receive mask can be defined by straight line segments, as in Figure 5.11.

Writing a_{tx} as the sum of the integration of the transmit segments s_i as in Equations 5.21 and 5.35, this **mask integration adjustment**, A_{MI} is therefore

$$A_{MI} = 10 \log_{10}(a_{MI}) \tag{5.46}$$

where

$$a_{MI} = \frac{\sum_i t_i}{a_{tx}} \tag{5.47}$$

$$t_i = \int_{f_i}^{f_{i+1}} m_{tx}(f) m_{rx}(f) df = \int_{f_i}^{f_{i+1}} 10^{(M_{tx} + M_{rx})/10} df \tag{5.48}$$

Given suitable creation of segments, both masks will be straight lines in (dB,f), and so we can use

$$M_{tx}(f) = A_i + B_i f \tag{5.49}$$

$$M_{rx}(f) = C_i + D_i f \tag{5.50}$$

Hence

$$t_i = \int_{f_i}^{f_{i+1}} 10^{(A'_i + B'_i f)/10} df \tag{5.51}$$

where

$$A'_i = A_i + C_i \tag{5.52}$$

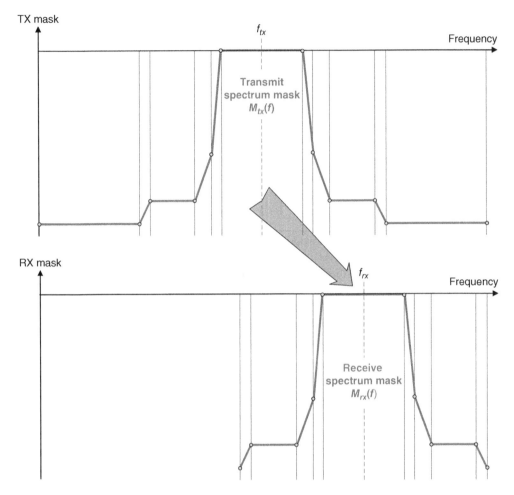

Figure 5.11   Calculating A_{MI} using segments of $M_{tx}(f)$ and $M_{rx}(f)$

$$B'_i = B_i + D_i \tag{5.53}$$

This uses the same notation as for the transmit mask integration in Section 5.3.1 to convert from mask points to lines:

$$A_i = \frac{M_{tx,i}f_{i+1} - M_{tx,i+1}f_i}{f_{i+1} - f_i} \tag{5.54}$$

$$B_i = \frac{M_{tx,i+1} - M_{tx,i}}{f_{i+1} - f_i} \tag{5.55}$$

$$C_i = \frac{M_{rx,i}f_{i+1} - M_{rx,i+1}f_i}{f_{i+1} - f_i} \tag{5.56}$$

$$D_i = \frac{M_{rx,i+1} - M_{rx,i}}{f_{i+1} - f_i} \qquad (5.57)$$

This can be integrated analytically using

$$t_i = 10^{A'_i/10} \int_{f_i}^{f_{i+1}} e^{Q'_i f} \, df \qquad (5.58)$$

where

$$Q'_i = \frac{\ln(10)B'_i}{10} \qquad (5.59)$$

Again, the integration method depends upon whether $Q'_i = B'_i = 0$, that is, if the segment is horizontal.

If $Q'_i \neq 0$ then

$$t_i = \frac{10^{A'_i/10}}{Q'_i} \left(e^{Q'_i f_{i+1}} - e^{Q'_i f_i} \right) \qquad (5.60)$$

If $Q'_i = 0$ then

$$t_i = 10^{A'_i/10} (f_{i+1} - f_i) \qquad (5.61)$$

As for the mask integration, it is possible to simplify these by substituting the A, B, C, D terms with mask values as in Equations 5.52–5.57. An example of the format of the equations after this simplification can be found in the HCM Agreement Annex 3B, 'Determination of the Masks Discrimination and the Net Filter Discrimination in the Fixed Service' (HCM Administrations, 2013).

The A_{MI} is typically integrated over the receive mask bandwidth, while the transmit mask is integrated over the transmit mask bandwidth to calculate a_{tx}. A decision has to be made when there is no overlap between the transmit and receive masks as in Figure 5.12: does this imply the interference will be negligible or should a value be calculated anyhow?

The following options could be considered:

1. Use the last value in the transmit mask extended to the full bandwidth of the receive mask
2. Use another value such as the spurious emissions limit (discussed in Section 5.3.6)
3. Assume there is no interference.

The approach to take will depend upon the interference scenario and the data available.

The A_{MI} can then be used in both the wanted and interfering calculations to derive the receive power taking into account the transmit and receive spectrum masks:

$$C = P_{tx} + G_{tx} - L_p + G_{rx} - L_f + A_{MI}(tx_W, rx_W) \qquad (5.62)$$

$$I = P'_{tx} + G'_{tx} - L'_p + G'_{rx} - L_f + A_{MI}(tx_I, rx_W) \qquad (5.63)$$

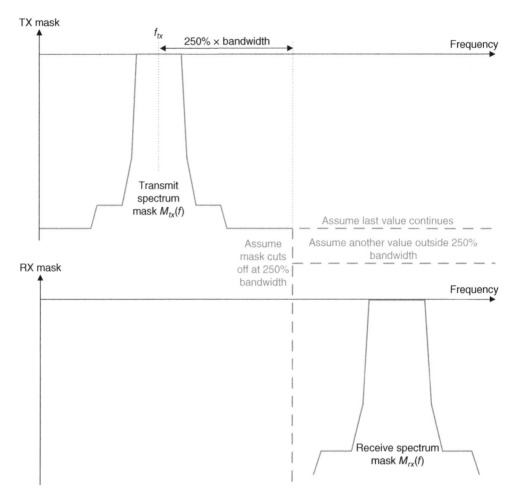

Figure 5.12 Options when no overlap between $M_{tx}(f)$ and $M_{rx}(f)$

In both cases the receive mask comes from the wanted or victim system, while the transmit mask is either taken from the wanted or interferer depending upon whether calculating C or I, respectively. This approach assumes that the TX mask is invariant as the transmit power changes and that the gain and propagation losses are averaged over the relevant bandwidth.

These equations work for both co-frequency and non-co-frequency interference calculations, making them generic and flexible. There is no constraint on the use of similar bandwidths or channels. A single power equation can be used for both wanted and interfering paths and all types of interference.

A constraint in use of the A_{MI} is the need to define the receive filter characteristics. Transmit spectrum masks are readily available via standardisation documents, but these requirements do not generally extend to receivers. This is an area subject to significant debate, and it has been proposed that receiver characteristics, in particular the receive spectrum mask, be included in future standardisation documents.

A receive spectrum mask could be created using one of the following methods:

- Construct a perfect receiver with $M_{rx}(f) = 0$ dB in-band and (say) -999 dB outside the OBW
- Construct a receive mask using $M_{rx}(f) = 0$ dB in-band and outside that use the adjacent channel selectivity (ACS) ratio(s) described in Section 5.3.5 as in Example 5.8
- Use the transmit mask as an approximation to the receive mask as they could have similar characteristics
- Derive the mask using measured data, for example, based upon how the C/I threshold varies by frequency difference
- Use a theoretical receive mask such as the Butterworth or Nyquist filters as described in Example 5.9 and Section 3.4.7
- Use a combination of transmit mask parameters and theoretical filter models as used in ETSI TR 101 854 (ETSI, 2005).

It can be useful to make some assumptions and use them in a study to encourage the relevant system operators to either accept the mask or propose an alternative. There is often reluctance by system operators to provide the receiver filter as it can reveal sensitive information about their network, and this is one reason why other metrics are considered, such as those in the following sections.

Another term that can be useful to consider is how the A_{MI} varies with frequency compared to both the transmitter and receiver having the same centre frequency. This frequency rejection factor can be calculated using

$$F_{FR} = A_{MI}(0) - A_{MI}(\Delta f) \tag{5.64}$$

Note that Example 7.20 describes an analysis comparing the results using A_{BW} with those from A_{MI} when calculating the I/N between terrestrial links and a satellite Earth station (ES).

Example 5.8

Within ITU-R JTG 4-5-6-7, studies analysed sharing between IMT and DTT in the 700 MHz band (ITU-R JTG 4-5-6-7, 2014). The transmit and receive spectrum masks were created from the LTE standard's ACLR and ACS parameters (which are defined in Section 5.3.5) in Table 5.5 and are shown in Figure 5.13.

Note how the masks were extended beyond the 250% bandwidth cutoff for out-of-band emissions into the spurious domain. From these two spectrum masks the mask integration adjustment A_{MI} can be calculated for various frequency offsets as shown in Figure 5.14.

Table 5.5 IMT LTE transmit and DTT receive characteristics

System	IMT LTE transmit	DTT receive
Signal bandwidth	10 MHz	7.6 MHz
Channel bandwidth	10 MHz	8 MHz
ACLR/ACS first channel	$ACLR_1 = 45$ dB	$ACS_1 = 45$ dB
ACLR/ACS second channel	$ACLR_2 = 45$ dB	$ACS_2 = 50$ dB
Spurious domain	54 dB	55 dB

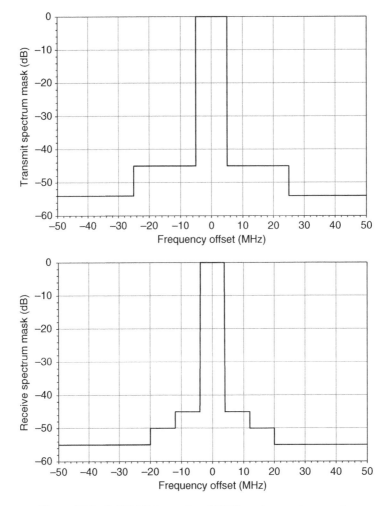

Figure 5.13   IMT LTE transmit and DTT receive spectrum masks

It can be useful to calculate and store tables of how A_{MI} varies with frequency to improve computational performance.

Example 5.9

Example 5.4 calculated the bandwidth adjustment factor $A_{BW} = -5.2$ dB assuming rectangular carriers with wanted system 8 MHz DVB at $f = 706$ MHz and interferer 10 MHz LTE at $f = 700$ MHz. Using the LTE transmit spectrum mask from Example 5.6, what is the A_{MI} assuming the DVB receiver can be modelled via a Butterworth filter with $n = \{2, 3, 4, 5\}$?

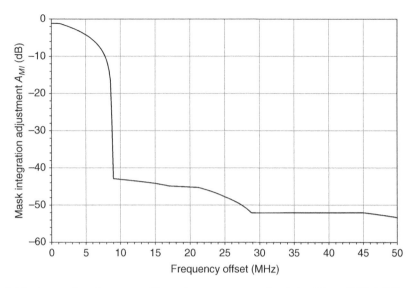

Figure 5.14 A_{MI} against frequency offset using spectrum masks derived from LTE ACLR and DTT ACS values

Table 5.6 Example A_{MI}

DVB filter order	A_{MI} (dB)
$n = 2$	−4.72
$n = 3$	−4.99
$n = 4$	−5.11
$n = 5$	−5.17

The equations for the Butterworth filter are given in Section 3.4.7, and the resulting A_{MI} are shown in Table 5.6 with example frequency plot for the case of $n = 4$ in Figure 5.15. Due to the inclusion of more of the interfering system's power, the A_{MI} is up to 0.5 dB higher than A_{BW} derived using the more simplistic bandwidth overlap calculation.

5.3.4 Frequency-Dependent Rejection and Net Filter Discrimination Terminology

The previous section calculated the A_{MI} factor by integrating the transmit and receive spectrum masks to derive the net effect of both. The difference in A_{MI} between transmitter and receiver operating with the same central frequency was described as the F_{FR} or the frequency rejection factor.

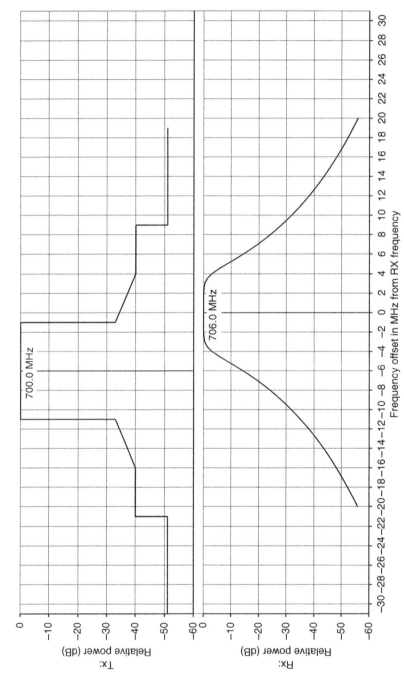

Figure 5.15 Example frequency view showing LTE transmit spectrum mask and receive spectrum mask generated using Butterworth filter with $n = 4$

A number of alternative terminologies are defined in documents such as:

- Rec. ITU-R SM.337: Frequency and distance separations (ITU-R, 2008b)
- ETSI TR 101 854: Technical Report; Fixed Radio Systems; Point-to-point equipment; Derivation of receiver interference parameters useful for planning fixed service point-to-point systems operating different equipment classes and/or capacities (ETSI, 2005)
- HCM Agreement Annex 3B, 'Determination of the Masks Discrimination and the Net Filter Discrimination in the Fixed Service' (HCM Administrations, 2013).

Rec. ITU-R SM.337 (ITU-R, 2008b) uses the **frequency-dependent rejection** (FDR) calculated from the **on-tune rejection** (OTR) and the **off-frequency rejection** (OFR):

$$FDR(\Delta f) = OTR + OFR(\Delta f) \tag{5.65}$$

These are defined as follows (using the notation from the previous section):

$$OTR = 10\log_{10}\left[\frac{\int_{f=0}^{f=\infty} m_{tx}(f)df}{\int_{f=0}^{f=\infty} m_{tx}(f)m_{rx}(f)df}\right] \tag{5.66}$$

$$OFR(\Delta f) = 10\log_{10}\left[\frac{\int_{f=0}^{f=\infty} m_{tx}(f)m_{rx}(f)df}{\int_{f=0}^{f=\infty} m_{tx}(f)m_{rx}(f+\Delta f)df}\right] \tag{5.67}$$

Hence

$$FDR(\Delta f) = OTR + OFR(\Delta f) \tag{5.68}$$

$$FDR(\Delta f) = 10\log_{10}\left[\frac{\int_{f=0}^{f=\infty} m_{tx}(f)m_{rx}(f)df}{\int_{f=0}^{f=\infty} m_{tx}(f)m_{rx}(f+\Delta f)df} \cdot \frac{\int_{f=0}^{f=\infty} m_{tx}(f)df}{\int_{f=0}^{f=\infty} m_{tx}(f)m_{rx}(f)df}\right] \tag{5.69}$$

$$FDR(\Delta f) = 10\log_{10}\left[\frac{\int_{f=0}^{f=\infty} m_{tx}(f)df}{\int_{f=0}^{f=\infty} m_{tx}(f)m_{rx}(f+\Delta f)df}\right] \tag{5.70}$$

So

$$FDR = -A_{MI} \tag{5.71}$$

However ETSI TR 101 854 and the HCM Agreement use the **mask discrimination** (MD) and **net filter discrimination** (NFD) where

$$MD = OTR \tag{5.72}$$

$$NFD = OFR(\Delta f) \tag{5.73}$$

The OTR and OFR introduce additional terms that are not required as the key is to calculate the $FDR = -A_{MI}$. Furthermore, the language of the NFD is not clear, as the words imply the normalised integration of the transmit and receive masks to calculate the net discrimination due to the filters rather than normalised to the transmit mask area. It would be more consistent if the NFD was defined to be the same as the FDR or mask integration adjustment, A_{MI}, and indeed it has been noted that it has been used in this way in studies presented at international meetings.

Therefore, this book has taken an approach based upon first principles that defines the power equation using A_{BW} for co-frequency cases and A_{MI} for non-co-frequency cases, rather than the OTR/MD or OFR/NFD terms that introduce additional complexity or are not, in some cases, well defined and consistently employed.

5.3.5 Adjacent Channel Leakage Ratio, ACS and Adjacent Channel Interference Ratio

One of the most common scenarios for the non-co-frequency case is interference between radio systems operating in adjacent channels. This could be the immediately adjacent channel or one of those after it. For this scenario the calculation of interference is dominated by two paths (as shown in Fig. 5.16):

- Path 1: Interferer's emissions in its assigned channel that are received by the victim in its adjacent channel
- Path 2: Interferer's emissions in its adjacent channel that are received by the victim in its assigned channel.

Note that harmful interference due to path 1 can be described as **blocking**. In this case the A_{MI} can be approximated using two parameters:

- For path 1, the **ACS**
- For path 2, the **adjacent channel leakage ratio** (*ACLR*).

The *ACS* is defined as

$$acs = \frac{Power\ at\ receiver\ from\ given\ source\ in\ assigned\ channel}{Power\ at\ receiver\ from\ given\ source\ in\ adjacent\ channel} \tag{5.74}$$

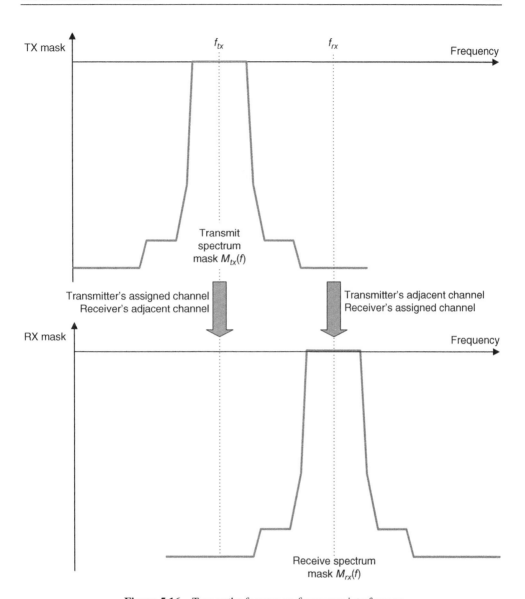

Figure 5.16   Two paths for non-co-frequency interference

The *ACLR* is defined as

$$aclr = \frac{Mean\ power\ transmitted\ in\ assigned\ channel}{Mean\ power\ transmitted\ in\ adjacent\ channel} \tag{5.75}$$

These are usually given in decibel format where both will be positive numbers:

$$ACS = 10\log_{10}acs \tag{5.76}$$

$$ACLR = 10\log_{10} aclr \tag{5.77}$$

Example 5.10
The *ACS* for 10 MHz LTE user equipment (UE) specified in 3GPP TS 36.101 (ETSI/3GPP, 2010)/ETSI TS 136 101 is

$$ACS = 33\,\mathrm{dB}$$

The *ACLR* for 10 MHz LTE BS specified in 3GPP TS 36 141 (ETSI/3GPP, 2011)/ETSI TS 136 141 is

$$ACLR = 44.2\,\mathrm{dB}$$

Together they can be used to derive an approximation for the A_{MI} by adding the power received via each of the two paths. Power must be added in absolute, so, excluding other components of the link calculation, this is

$$p_1 = \frac{p_{tx}}{acs} \tag{5.78}$$

$$p_2 = \frac{p_{tx}}{aclr} \tag{5.79}$$

$$p = p_1 + p_2 = \frac{p_{tx}}{acs} + \frac{p_{tx}}{aclr} = \frac{p_{tx}}{acir} \tag{5.80}$$

Here the ratio of the total interference between adjacent channels is given by the **adjacent channel interference ratio** (*ACIR*). Hence

$$acir = \cfrac{1}{\cfrac{1}{aclr} + \cfrac{1}{acs}} \tag{5.81}$$

Example 5.11
With *ACS* = 33 dB and *ACLR* = 44.2 dB as in Example 5.10 the *ACIR* is

$$ACIR = 32.7\,\mathrm{dB}$$

Typically the *ACIR* is positive and dominated by the lower of the *ACS* and *ACLR*.
In the case where the TX and RX systems are in clearly separated channels, the mask integration adjustment can be approximated using

$$A_{MI} \sim -ACIR \tag{5.82}$$

This approach has the advantage of simplicity in that it is defined via two numbers that can be used as input parameters to interference analysis. In addition, they are more often provided in standardisation documents: in particular the *ACS* is more readily available than the full receive spectrum mask $M_{rx}(f)$. The disadvantage is a loss of accuracy and reliance on frequency offsets defined by a channel plan rather than a more generic approach.

5.3.6 Spurious Emissions and dBc

The previous sections considered the frequencies adjacent to the in-band domain, namely, the out-of-band domain, where the emissions are determined by the modulation and can be modelled using the $M_{tx}(f)$. There will be some power transmitted outside the out-of-band domain due to spurious emissions, hence the spurious domain in Figure 5.2.

The RR define spurious emission as

1.145 spurious emission: Emission on a frequency or frequencies which are outside the necessary bandwidth and the level of which may be reduced without affecting the corresponding transmission of information. Spurious emissions include harmonic emissions, parasitic emissions, intermodulation products and frequency conversion products, but exclude out-of-band emissions.

Levels for spurious emissions are given in RR Appendix 3 'Maximum permitted power levels for unwanted emissions in the spurious domain' often in the form of dBc:

Where specified in relation to mean power, spurious domain emissions are to be at least x dB below the total mean power P, i.e. −x dBc.

For terrestrial services (excluding radar) calculations are made relative to bandwidths that depend upon the central frequency as in Table 5.7. For space services the reference bandwidth is always 4 kHz.

The spurious emission limit in the reference bandwidth is then

$$P_s = P - X_{dBc} \tag{5.83}$$

Values and methods to calculate X_{dBc} are given in a table in RR Appendix 3. For example, for the land mobile service it is

$$X_{dBc} = min\{70, 43 + 10\log_{10}p_{tx}\} \tag{5.84}$$

Rec. ITU-R. SM.329 (ITU-R, 2012a) includes the RR levels as Category A limits together with additional limits for specific countries and services (e.g. Category B is for Europe, while

Table 5.7 Reference bandwidths for terrestrial spurious emission calculations

Frequency range	Reference bandwidth
9 kHz to 150 kHz	1 kHz
150 kHz to 30 MHz	10 kHz
30 MHz to 1 GHz	100 kHz
Above 1 GHz	1 MHz

Category C is for the United States and Canada), which are more stringent. Regional organisations and standardisation bodies also define spurious emission levels, such as in ETSI standards
and ERC Recommendation 74-01 (CEPT, 2011c).

Example 5.12

Example 4.1 described a private mobile radio (PMR) system operating with $P_{tx} = 12$ dBW transmitting at $f = 420$ MHz. The reference bandwidth is 100 kHz and X_{dBc} can be calculated using

$$X_{dBc} = min\{70, 43 + 12\} = 55$$

The spurious emission limit is therefore

$$P_s = 12 - 55 = -43\,\text{dBW}/100\,\text{kHz}$$

The spurious emission limits are not very strenuous, but then for almost all of the spurious
domain, emissions should be significantly lower than these levels. It is only expected that there
are occasional spikes of power that approach the emission limits. In particular, there can be harmonics on multiples of the transmit frequency where there are significantly greater emissions.
These might require modifications to frequency allocations to avoid sensitive services in
other bands.

Example 5.13

At WRC-97 the following footnote was added to the RR:

5.291A   *Additional allocation: in Germany, Austria, Denmark, Estonia, Finland, Liechten-
          stein, Norway, Netherlands, the Czech Rep. and Switzerland, the band 470–494
          MHz is also allocated to the radiolocation service on a secondary basis. This use
          is limited to the operation of wind profiler radars in accordance with Resolution 217.*

One reason why the upper limit of this footnote allocation is 494 MHz was to avoid harmonics
from these high-power wind profiler radars into the sensitive receivers of the mobile-satellite
service (MSS) operating at 1 980–2 010 MHz in the Earth-to-space direction.

5.3.7   Intermodulation

Intermodulation is caused by non-linear behaviour of components, in particular amplifiers
reacting to multiple frequencies. The result is additional or spurious signals that are multiples
of the frequencies involved. Intermodulation is more typically an issue for practical spectrum
management, such as when frequencies are selected for private mobile radio (PMR) systems,
rather than regulatory-style studies. Intermodulation can be an issue for a range of radio systems including satellite services, in particular when the bandwidth is relatively narrow.

If an environment contains a set of frequencies $\{f_1, f_2, \ldots f_n\}$, then intermodulation can lead to
emissions on frequencies:

$$f = k_1 f_1 + k_2 f_2 + \cdots + k_n f_n \tag{5.85}$$

Here k_i are non-zero integers, which results in $f > 0$. Intermodulation is classified via the number of frequencies involved and the order $= O$, where

$$O = |k_1| + |k_2| + \cdots + |k_n| \tag{5.86}$$

Example 5.14
The third-order intermodulation products for a two-frequency scenario where $f_1 = 165.1\,\text{MHz}$ and $f_2 = 165.2\,\text{MHz}$ are

$$f = \{2f_1 + f_2, 2f_1 - f_2, f_1 + 2f_2, -f_1 + 2f_2\}$$
$$= \{495.4\,\text{MHz}, 165.0\,\text{MHz}, 495.5\,\text{MHz}, 165.3\,\text{MHz}\}$$

The principal intermodulation products for the two-frequency case are shown in Figure 5.17. The most critical cases are typically:

- Those products near the transmit frequencies where there is likely to be less rejection from filters
- Multiples of the transmit frequencies, which include harmonics.

The products near the transmit frequencies that contain combinations of the difference in frequency are $\Delta f = f_2 - f_1$ and hence are the odd number harmonics, for example, 3rd, 5th and 7th as in Figure 5.17.

There are three main ways in which intermodulation can cause a degradation to a radio service given in Report ITU-R M.739 (ITU-R, 1986):

1. Unwanted emissions are generated in transmitters
2. Unwanted emissions are generated in elements external to the transmitters
3. Unwanted emissions are generated in the radio-frequency stages of the receiver.

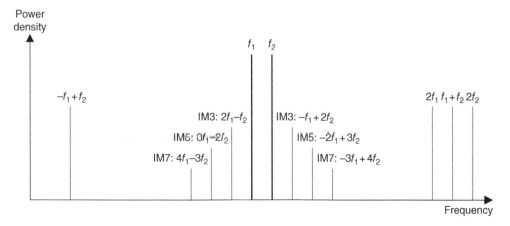

Figure 5.17   Two frequency intermodulation products

Unwanted emissions generated external to the transmitters are sometimes described as passive intermodulation or the 'rusty bolt' scenario. In this case there is a component on a site with multiple transmitters that can reradiate signals leading to intermodulation effects. Passive intermodulation is more of an issue for site management and measurement rather than analytic interference analysis.

The last active stage in most radio systems is the amplifier. If there are multiple frequencies present (either due to a multi-channel wanted system or to other nearby radio systems), then this can lead to transmitter intermodulation.

The impact of transmitter intermodulation can be calculated using two terms:

1. A_c: coupling loss between the two signals, that is, ratio of the power emitted by one system (the unwanted) as measured at the output of another transmitter
2. A_I: the intermodulation conversion loss, the ratio of the unwanted signal to the level of the intermodulated product (e.g. third order).

The power of the intermodulation product transmission can then be calculated using

$$P_{IM} = P_{TX} - A_C - A_I \qquad (5.87)$$

Example 5.15
Two private mobile radio (PMR) systems are located at the same site as in Figure 5.18, one operating at 165.1 MHz and the other at 165.2 MHz. Each is transmitting with power 5 W = 7 dBW, and there are $A_c = 30$ dB of coupling loss between the two systems and $A_I = 10$ dB of intermodulation conversion loss. The two nearest third-order intermodulation product signals will therefore be transmitting with

$$f = \{165.0\,\text{MHz}, 165.3\,\text{MHz}\}$$

$$P_{IM} = 7 - 30 - 10 = -33\,\text{dBW}$$

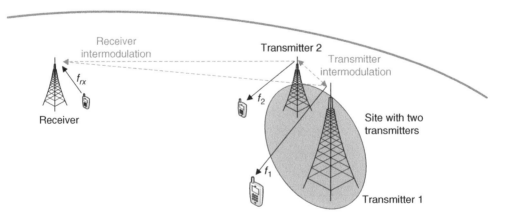

Figure 5.18   Transmitter and receiver modulation

The converse of transmitter intermodulation is receiver intermodulation, where signals from (two or more) interfering stations of frequencies (f_1, f_2) result in additional spurious emissions in the receiver. The two signal third-order intermodulation interference power can be calculated using the following equation from Rec. ITU-R SM.1134 (ITU-R, 2007f):

$$P_{INO} = 2(P_1 - \beta_1) + (P_2 - \beta_2) - A_I \tag{5.88}$$

where

P_{INO} = power of the third-order intermodulation product at the receiver on frequency $f = 2f_1 - f_2$
P_1, P_2 = powers of the interfering signals at frequencies f_1, f_2, respectively
β_1, β_2 = frequency selectivity at offsets Δf_1 and Δf_1
A_I: the intermodulation conversion loss, the ratio of the unwanted signal to the level of the intermodulated product.

The term $(P_i + \beta_i)$ corresponds to Equation 5.63:

$$I = P'_{tx} + G'_{tx} - L'_p + G'_{rx} - L_f + A_{MI}(tx_I, rx_W)$$

Hence

$$P_{INO} = 2I_1 + I_2 - A_I \tag{5.89}$$

A similar equation to calculate the impact of receiver generated intermodulation can be found in the following:

- Australian Communications and Media Authority (ACMA): Radiocommunications Assignment and Licensing Instruction (RALI): Frequency Assignment Requirements for the Land Mobile Service LM 8 (ACMA, 2000).

This gives the following values for two frequency A_I:

Third order $A_I = 9$ dB
Fifth order $A_I = 28$ dB.

Note that both ACMA LM 8 and Rec. ITU-R SM.1134 give FDR terms that could be used in the calculation of intermodulation products. For example, Rec. ITU-R SM.1134 gives the following equation:

$$\beta(\Delta f) = 60\log_{10}\left[1 + \left(\frac{2\Delta f}{B_{RF}}\right)^2\right] \tag{5.90}$$

5.3.8   Block Edge Masks and Guard Bands

The transmit spectrum mask defines the characteristics of a transmitter in the frequency domain and is a useful tool when studying non-co-frequency interference scenarios. However, the impact on a receiver varies in the frequency domain: for a given transmit spectrum mask, greater power can be transmitted when there is a larger frequency separation.

The **block edge mask** (BEM) is a regulatory tool that defines the maximum power that can be transmitted across a boundary between frequency blocks in a way that allows flexibility in selection of the transmit power. It is usually given in the form of a table of values [equivalent isotropically radiated power (*EIRP*) density vs. frequency separation from boundary] with straight lines between data points in dBW versus MHz.

Example 5.16
The BEMs for terrestrial services operating in the 28 GHz area licences auctioned in the United Kingdom are given in Figure 5.19 (Ofcom, 2007). Note how there are different masks for point-to-point and point-to-multipoint systems, with the former defined as cases where the antenna half power beamwidth is less than 5°. In this example negative frequency offsets are within the frequency block and positive ones outside.

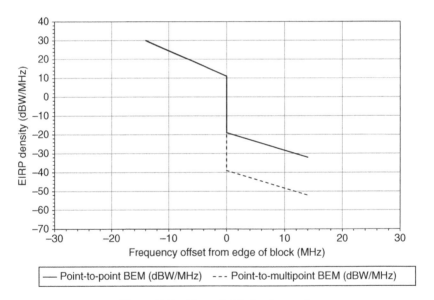

Figure 5.19   Example block edge mask

BEMs are derived from interference studies that used generic assumptions about the systems on either side of the boundary. If there is the same service on both sides, then there is likely to be a reciprocal BEM in the other direction; otherwise it could be different.

The BEM is a very useful regulatory tool as it provides both protection and flexibility. Operators can deploy transmitters in their frequency block as long as they meet this mask, typically without further approval. It is relatively straightforward to calculate what power is permitted and, if necessary, undertake measurements to check whether emissions meet this mask. The transmit power can vary depending upon frequency separation from the boundary, giving significant flexibility to the operator.

They can be included in a licence to define emission rights and obligations without limiting it to a specific standard and hence give a degree of technology neutrality. So a BS could be initially deployed using WCDMA and then upgraded to LTE without requiring regulatory approval if the BS meets the licence's BEM.

Example 5.17
Two fixed links are proposed to be deployed in the 28 GHz band with BEM as in Example 5.17. Both use the 28 MHz mask described in Example 5.5 as shown in Figure 5.7. The frequency offsets from the edge of band and peak EIRP densities are as in Table 5.8. Note that the difference in the two central frequencies is consistent with a 28 MHz channel plan.

It is generally assumed that the FS mask's responses to changes in power are linear, so that increases in the transmit power increase the mask by an equal amount for all frequencies. Hence the EIRP density for various frequency offsets can be calculated using

$$EIRP_{density}(\Delta f) = EIRP_{density}(peak) + M_{tx}(\Delta f) \tag{5.91}$$

The EIRP masks for the two links can be compared against the BEM either analytically or graphically as in Figure 5.20. In this case it can be seen that

- Link A: this exceeds the BEM and so would not be consistent with the operator's obligations
- Link B: this is below the BEM at all points and would be acceptable.

Note that in general regulators do not consider that the BEM gives the operator on one side of the boundary the right to transmit in the adjacent block even if they are operating below the limits defined.

It is often difficult to operate transmitters right up to the boundary due to BEMs such as these. Often there are **guard bands**: extra frequency 'space' included to separate transmitters in one band and receivers in the other as shown in Figure 5.21. In general no transmitters would be permitted to operate within this guard band.

Example 5.18
The channel plan for fixed links in Example 5.17 was modified to include a 2 MHz guard band, increasing the frequency separation of both links from the edge of the block. The resulting EIRP masks of both links meet their obligations as can be seen by Figure 5.22.

Table 5.8 Example point-to-point link parameters

Link	A	B
Frequency offset from block edge (MHz)	−14	−42
Peak EIRP density (dBW/MHz)	−10	15

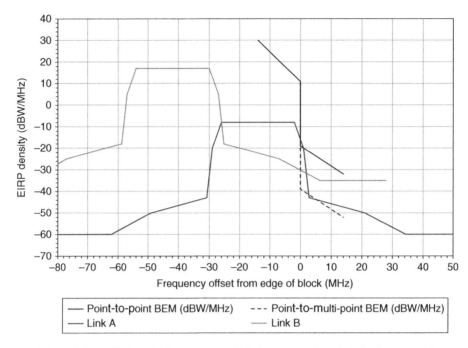

Figure 5.20   EIRP masks for two example links compared against block edge mask

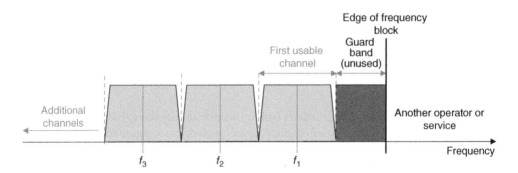

Figure 5.21   Channel plan with guard band

Guard bands are often used at block edges, either on their own or in conjunction with a BEM, as regulatory tools to manage interference between systems at frequency boundaries. Key tasks for interference analysis studies are then to identify the following:

- What would be a suitable BEM?
- What would be a suitable guard band?

An example of the types of analysis that can be used to develop suitable values for a BEM can be found in CEPT ECC Report 131, 'Derivation of a block edge mask (BEM) for terminal

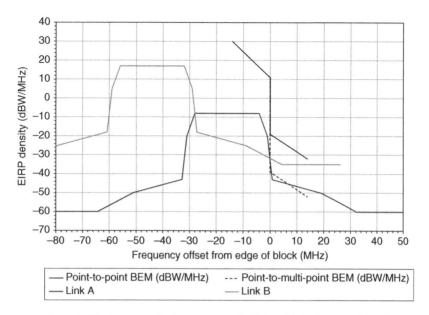

Figure 5.22 EIRP masks for two example links with 2 MHz guard band

stations in the 2.6 GHz frequency band (2 500–2 690 MHz)' (CEPT, 2009). The initial approach was based upon minimum coupling loss (described in Section 6.6) followed by a more detailed Monte Carlo analysis (as in Section 6.9). BEM have been defined in CEPT studies for a wide range of services including UMTS and LTE.

Tighter BEM and larger guard bands give additional protection but at the cost of potentially unused spectrum: some guard bands are significantly larger than the 2 MHz in Example 5.18.

The reduction in efficiency due to unused spectrum near the boundary can be alleviated by permitting these BEMs to be exceeded if there is a coordination agreement between the operators on either side. However this introduces costs and assumes that both operators will be equally interested in reaching an agreement. Another approach is to permit very low power systems on a secondary (no interference) basis.

The BEM doesn't fully protect receivers in adjacent bands: it would have been defined in a way that balances constraints on the transmitters on one side with degree of protection on the receiver side. A BEM that fully protects all receivers on one side of the band edge would likely involve severe restrictions on the ability to deploy transmitters on the other.

Another issue relates to density: the BEM alone puts no constraint on the deployment model for transmitters. The BEM could have been developed assuming a wide-area deployment model with tens of kilometres between transmitters. If the actual deployment involves transmitters spaced less than a kilometre apart then this could result in more locations where there is harmful interference than envisaged when the BEM was developed.

Changes in the density of deployment are particularly an issue if a band has experienced change of use (CoU) from one technology or service to another. An example of this would be Nextel's purchase of fleet despatch frequencies at around 800 MHz in the United States, followed by CoU into a public mobile wireless network. This led to issues relating to harmful interference into commercial and public safety radio systems in adjacent bands, which was only resolved via an FCC-backed frequency replanning exercise (FCC, 2004).

One possible approach to manage density was proposed in a study into technology-neutral spectrum usage rights (Aegis Systems Ltd, Transfinite Systems Ltd and Indepen Ltd, 2006) that suggested an area-based power flux density (PFD) threshold of the form

aggregate field strength not to be exceeded over X% of an area for Y% time

The difficulty with this form of constraint is that it is unclear how to define the area to use for the X% test, and so this metric is not widely used.

5.4 Polarisation

The radio wave has been described via its frequency and power, in particular how the power varies in the frequency domain. Another attribute of the radio wave can also be important, namely, its polarisation.

Radio waves comprise of variations in the electric and magnetic fields as determined by Maxwell's equations. The variation of the electric field defines the polarisation, which is typically one of:

- **Linear**: sin or cos waves with a defined plane. There can be many different planes, but in many geometries these can be identified as **linear horizontal** (LH) or **linear vertical** (LV), as in Figure 5.23
- **Circular**: the plane of the wave rotates around the direction of travel. There are two options: **left-hand circular** (LHC) and **right-hand circular** (RHC), as in Figure 5.24.

The difference between LHC and RHC polarised waves is defined in the RR as follows:

1.154 right-hand (clockwise) polarized wave: An elliptically- or circularly-polarized wave, in which the electric field vector, observed in any fixed plane, normal to the direction of propagation, whilst looking in the direction of propagation, rotates with time in a right-hand or clockwise direction.

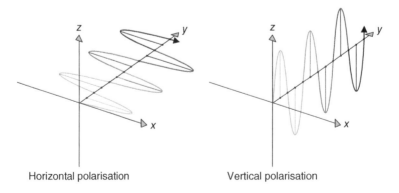

Figure 5.23 Examples of linear polarisation

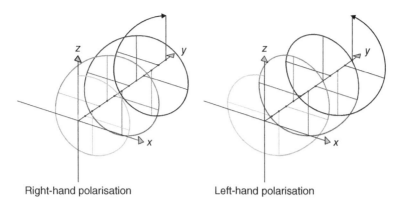

Figure 5.24   Examples of circular polarisation

1.155 left-hand (anticlockwise) polarized wave: An elliptically- or circularly polarized wave, in which the electric field vector, observed in any fixed plane, normal to the direction of propagation, whilst looking in the direction of propagation, rotates with time in a left hand or anticlockwise direction.

An easier way of remembering these two directions is to use either your left or right hand with the thumb extended and fingers curled up. With the thumb pointing in the direction of travel of the radio wave, the curve of the fingers will show the direction it rotates around that line.

There can also be intermediate polarisations, described as elliptical, where one of the circular polarisation directions is greater than the other. In an extreme case this becomes equivalent to linear polarisation.

The polarisation can impact the interference calculation in a number of ways:

- The polarisation can affect the propagation loss as described in Section 4.1
- The polarisation can be changed by the propagation environment, also described in Section 4.1
- The gain pattern of the transmit and receive antennas could vary depending upon the polarisation involved, as described later
- The receiver could be able to reject some of the interfering signal if it is using a different polarisation than that used by the wanted signal
- A system could transmit on two polarisations to either double the link's capacity or for the two wanted signals to be combined together to improve the $C/(N + I)$.

One difficulty in modelling polarisation effects is that the angle of linear polarised waves alters outside the main beam of an antenna, while circular polarised carriers can become increasingly elliptical. It can, therefore, be dangerous to assume that significant benefits can come from including polarisation advantages. For example, Appendix 8 of the RR suggests that the polarisation loss L_{pol} in decibel is as in Table 5.9.

The polarisation loss can then be taken away from the interfering signal as in

$$I = P'_{tx} + G'_{tx} - L'_p - L_{pol} + G'_{rx} - L_f \qquad (5.92)$$

Table 5.9 Polarisation loss in decibels from RR Appendix 8

Polarisation	LH	LV	RHC	LHC
LH	0.0	0.0	1.46	1.46
LV	0.0	0.0	1.46	1.46
RHC	1.46	1.46	0.0	6.02
LHC	1.46	1.46	6.02	0.0

Antennas and receivers are usually designed to generate and receive the required polarisation, and hence there is typically no change to the wanted signal link budget equation.

There can be significant reductions in the interfering signal for linear polarisations if the geometry is suitable. This is particularly the case if the geometry is constant so the radio system can be engineered on that basis, such as for point-to-point fixed links.

Example 5.19

To gain type approval in Europe, gain patterns of class 4 antennas proposing to be operated in the fixed service in bands between 24 and 30 GHz must meet the radiation pattern envelope (RPE) defined in ETSI EN 302 217-4-2 (ETSI, 2010). This includes two patterns, one for the co-polar case and another for cross-polar, as shown in Figure 5.25. It can be seen that the most significant difference is near the main beam, with the difference decreasing as the off-axis angle increases.

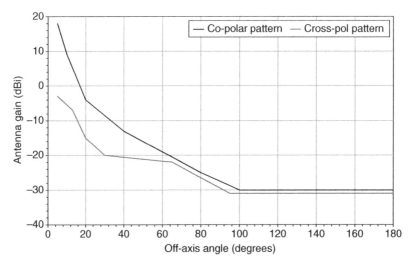

Figure 5.25 Example of class 4 antenna RPE (24–30 GHz)

These two gain patterns can then be used in the interference calculation: the mask to use will depend upon whether the wanted and interfering systems are using the same or different polarisations. Take the scenario in Figure 5.26; there are two cases to consider:

1. Interfering station's transmit antenna LH co-polar gain pattern into victim antenna LV cross-polar gain pattern

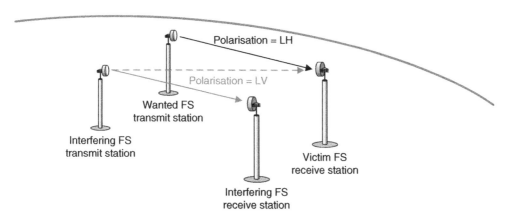

Figure 5.26 Fixed link linear polarisation scenario

2. Interfering station's transmit antenna LV cross-polar gain pattern into victim antenna LH co-polar gain pattern.

This can be considered as two interfering signals, but the other components of the link budget are likely to be the same, and hence it is possible to use a single link budget and add the two gains in absolute (as it is the powers that must be added) as in Rec. ITU-R F.699 (ITU-R, 2006b):

$$G_t(\varphi_t) + G_r(\varphi_r) = 10\log_{10}\left[10^{\frac{G_{tH}(\varphi_t) + G_{rV}(\varphi_r)}{10}} + 10^{\frac{G_{tV}(\varphi_t) + G_{rH}(\varphi_r)}{10}}\right] \tag{5.93}$$

where H implies LH polarisation, V is for LV polarisation and (t, r) are for transmit and receive gains/angles.

The greatest advantage comes when both the wanted and interferer are at boresight. An example of this would be for fixed links when the same geometry (transmit/receive stations) and frequency channel is used twice, once for LH and once for LV polarisations. As all geometry and propagation losses involved should be the same (if the same antennas are used), then the C/I is directly related to the antenna polarisation advantage.

Given the calculation of the wanted and interfering signals,

$$C = P_{tx} + G'_{tx} - L_p + G_{rx} - L_f$$

$$I = P_{tx} + G'_{tx} - L_p + G'_{rx} - L_f$$

Then:

$$\frac{C}{I} = \left(G_{tx} - G'_{tx}\right) + \left(G_{rx} - G'_{rx}\right)$$

The difference between the two gains (co-polar and cross-polar) is the cross-polar discrimination (XPD) defined in Rec. ITU-R F.746 (ITU-R, 2012b) as:

$$XPD_{H(V)} = \frac{Power\ received\ on\ polarisation\ V\ transmitted\ on\ polarisation\ V}{Power\ received\ on\ polarisation\ H\ transmitted\ on\ polarisation\ V} \qquad (5.94)$$

Hence

$$\frac{C}{I} = XPD \qquad (5.95)$$

For the case in Example 5.19, the XPD required by the ETSI standard is 27 dB, which will result in a $C/I = 27$ dB, which should be sufficient in most cases to permit co-channel dual-polar (CCDP) operation.

GSO satellite systems can also use LV versus LH polarisation methods to manage interference, both within a system and to/from another. Care is required as the vertical versus horizontal planes will vary over the large surface of the Earth that can be seen from GSO. More information can be found in Rec. ITU-R S.736 (ITU-R, 1997b).

5.5   Adaptive Systems: Frequency, Power and Modulation

For some radio systems, the frequency, power and modulation used remains constant during its operational lifetime. For example, once a PMR push-to-talk system has been deployed, it is likely to use the initial set of parameters without change for a period of 5–10 years. However, many other radio systems alter their behaviour depending upon circumstances, and the frequency, power and modulation can change automatically, and this section considers these types of adaptive system.

Note that more advanced methods could be used including adaptive gain patterns and geo-databases (discussed in Section 7.10 on white space technology).

5.5.1   Dynamic Frequency Selection

In dynamic frequency selection (DFS), the system senses the radio environment prior to transmitting. It will not use a particular channel if it detects another system already operating on it.

An example of this would be operation of radio local area networks (RLANs) in parts of the 5 GHz band, which, as described in EN 301 893 (ETSI, 2014a), use DFS to:

- Detect interference from radar systems and to avoid co-channel operation with these systems
- Detect other RLANs in order to avoid busy channels and provide a near-uniform loading of the spectrum.

This requires a radio activity sensing mode that can detect other RLANs or radars operating in a channel and definition of a threshold above which the channel is considered busy and hence unavailable.

The benefit of DFS is that it allows new entrants to operate in a band while protecting incumbents and is relatively low cost in that the decisions are made by the equipment automatically, not requiring any infrastructure or human intervention.

However DFS only works if various conditions are met, and this could lead to it being unsuitable for some scenarios. In particular

- **Carrier detection**: it is necessary that the receiver can detect the service to be protected. This is feasible if the service is high power (e.g. radar) or nearby (e.g. another RLAN). It is more difficult if the other service has transmitters far away and uses high-gain antennas to amplify the weak signal. Satellite services in particular often operate with very low power signals that are unlikely to be detected with low-gain receivers. Sometimes measurement campaigns present results showing 'empty' spectrum, which is therefore assumed to be available as it is unoccupied. However it could be heavily used for (say) satellite receivers using parabolic dish antennas.
- **Hidden nodes**: a channel could be busy with another system operating with its receiver nearby but with the transmitter hidden or obstructed from the DFS system. It would therefore conclude that the channel is clear and hence transmit, causing interference into the other system's receiver. An example of this was shown in Figure 3.22, which resulted in a collision in a Wi-Fi network.
- **Delay**: a channel availability check (CAC) is performed by the RLAN at regular intervals (typically 1–10 minutes), but between those times it is feasible for it to operate in a way that could cause harmful interference for a short period of time.

The constraints on DFS mean that its application is generally restricted to specific scenarios, in particular low-power devices such as 5 GHz RLAN sharing with high-power radar systems.

Example 5.20
Examples of the trigger levels for 5 GHz RLAN devices can be found in Rec. ITU-R M.1652 (ITU-R, 2011b); in particular the DFS detection thresholds are:

−62 dBm for devices with a maximum EIRP of under 200 mW
−64 dBm for devices with a maximum EIRP of 200 mW to 1 W averaged over 1 μs.

5.5.2 Automatic Power Control

Radio systems are often designed based upon a specific wanted signal strength or receiver sensitivity level (RSL), described further in Section 5.8. This level will be derived based upon factors such as bandwidth, modulation, margins and noise figures as discussed in Section 5.8. The actual level received will depend upon the various factors in the link budget:

$$C = P_{tx} + G_{tx} - L_p + G_{rx} - L_f$$

Some of these terms could vary in time, such as:

- Transmit gain: if the receiver is moving, then the antenna gain towards it could vary. For example, the gain at a BS towards a mobile within a cell would vary depending upon its location
- Propagation loss: even with no motion of transmit or receive stations, there could be significant changes in the propagation loss due to atmospheric effects. If one of the stations is moving, there could also be changes due to multi-path or variations in terrain and clutter

- Receive gain: as with the transmit gain, this could change if one of the stations is in motion.

These factors result in a received wanted signal level that varies in time. If the transmit power is fixed, then at many times the link will be using additional power over the minimum level needed to achieve the required RSL. However, in systems that involve two-way communication, there is the opportunity to create a feedback loop so that the transmit power is selected, which just achieves the RSL.

Example 5.21
A land mobile system uses a constant transmit power to provide a service. For most of the time this results in a received signal significantly greater than the required RSL. However, during a fade of depth 26 dB, it is insufficient, as shown in Figure 5.27.

If power control were used, the received signal could be kept just above the RSL and additional power transmitted during a fade. Often this will be constrained by the maximum transmit power permitted. The preferred transmit power can be calculated by using a default P_{tx} and calculating the resulting received signal C:

$$C = P_{tx} + G_{tx} - L_p + G_{rx} - L_f$$

The transmit power P'_{tx} is then the one that would ensure the received signal was the target RSL:

$$RSL = P'_{tx} + G_{tx} - L_p + G_{rx} - L_f$$

So

$$P'_{tx} = P_{tx} - (C - RSL) \tag{5.96}$$

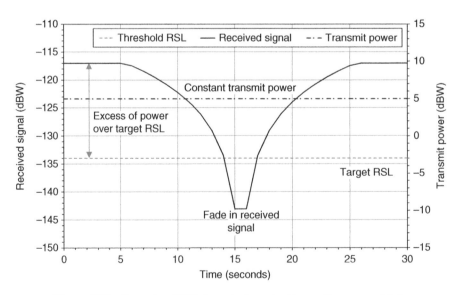

Figure 5.27   Example of RSL impact of constant transmit power and fade

An alternative approach is to calculate the total link loss:

$$L_{total} = G_{tx} - L_p + G_{rx} - L_f \qquad (5.97)$$

The transmit power used would then have to be capped by $P_{tx,max}$, so that

$$P'_{tx} = min[P_{tx,max}, RSL - L_{total}] \qquad (5.98)$$

Automatic power control (APC) has many advantages:

- It is spectrum efficient, as it provides the service required while generating the minimum level of interference
- It is power efficient, reducing environmental impact and extending battery life or maximising capacity for systems with limited power (e.g. satellites)
- It can overcome propagation events by increasing power to avoid loss of service.

In some services it is permitted to increase power above the licensed level for short periods of time to compensate for propagation losses. This might not lead to greater interference if the interfering signal is also subject to propagation loss (i.e. rain loss but not necessarily multi-path loss).

Mobile networks using CDMA typically require use of APC to ensure that the uplink signals from handsets to the BS have similar received powers levels, as part of their intra-system interference management.

APC is also called adaptive transmit power control (ATPC) or just power control.

It is often useful to have the target for APC to be just above the RSL as there will be a delay in response of the power control loop to very rapid changes in the propagation loss (see discussion in Section 5.8). There can also be quantisation in the APC system, for example, the power is increased in 0.1 dB steps. Note also that the APC loop could be based on a target metric other than RSL such as bit error rate (BER), packet error rate (PER) or $C/(N+I)$, though these are harder to model. In particular, targets that depend upon interference levels can result in feedback loops where multiple systems all increase their transmit power to reduce the effect of interference, resulting in greater interference into other systems.

Power control is an important system feature to model when undertaking interference analysis. In some situations it is possible to create an initial simulation and determine either the mean transmit power or the distribution of transmit powers that would be expected for that scenario.

Example 5.22

The land mobile system in Example 5.21 was modified to use APC with target threshold 1 dB above the RSL at −133 dBW and power control range $P_{tx} = [-15 \, dBW, +10 \, dBW]$. The resulting received signal and transmit power used is shown in Figure 5.28. It can be seen that:

- For most of the time the transmit power is lower than that in Example 5.21
- As the fade begins, the transmit power increases while the received signal level is unchanged
- For the deepest fade there is insufficient transmit power and so the received signal drops below the RSL

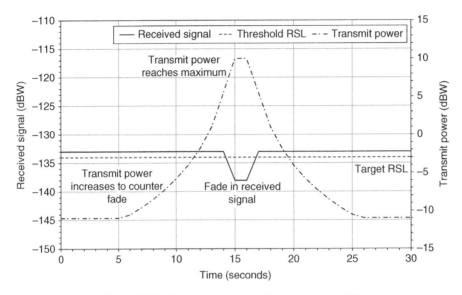

Figure 5.28 Improvement in performance using APC

- The duration of the fade event, when the received signal is less than the target RSL, is reduced compared to Example 5.21.

5.5.3   Adaptive Coding and Modulation

In the system in Example 5.21, use of a constant transmit power led to a significant margin over the target RSL in all but the deepest of fades. This margin could be used by the system to provide a higher data rate by increasing the modulation. Section 3.4.5 described how the target S/N needed to achieve a given BER is greater for higher-order modulations. In addition, Shannon's channel coding theorem, given as Equation 3.30, identified that the maximum capacity of a channel also increases with S/N.

Changes in the modulation can be accompanied by changes in the coding, employing more or less additional bits. This also modifies the S/N required to achieve a specified BER but at the cost of reduced user traffic capacity.

For example, the Wi-Fi standard IEEE 802.11n (IEEE, 2009a) uses the following:

- Modulations: {BPSK, QPSK, 16-QAM, 64-QAM}
- Code rates: {1/2, 2/3, 3/4, 5/6}.

Another example would be for point-to-point links used for mobile network backhaul, which tend to adjust their modulation to the one that provides the highest data carrying capacity for the current propagation conditions.

From an interference analysis point of view, use of **adaptive coding and modulation** introduces additional complexity, for example:

- There could be multiple thresholds to consider, one for each modulation type and coding scheme
- The transmit spectrum mask will be different for each modulation type.

It is often necessary to simplify the analysis, for example:

- When the adaptive modulation system is the interferer, consider the widest of all the spectrum masks
- When the adaptive modulation system is the victim, use the most sensitive threshold.

There is the danger, however, of taking a series of worst-worst-worst-case assumptions. If the system is capable of adjusting its modulation, then the impact of interference can be a decrease in capacity: the system is forced to switch to a lower modulation with a reduction in payload data rate. However that might be necessary to facilitate sharing between services.

5.6 End-to-End Performance

Often interference analysis considers a single radio link, that is, one path from a transmit station to a receive station. However such a link can be part of an end-to-end communication route where it is necessary to consider multiple hops.

A particular example of this is for satellite systems that, when providing a service on the ground, must employ two links, the **uplink** and the **downlink**. The satellite must be able to retransmit the uplink signal as a downlink, usually with a different antenna and centre frequency.

The connection at routing nodes such as the satellite can use one of two techniques:

1. Bent pipe repeaters, that is, the node (e.g. satellite) simply amplifies the signal as received, changes its frequency and then routes the signal to the relevant antenna
2. Regeneration, whereby the node converts the incoming data down to a bit stream, which is then recoded and transmitted.

For terrestrial digital systems it is often convenient to use full regeneration as it allows additional functionality, in particular multiplexing where the traffic from multiple users shares resources such as backhaul.

Satellite systems tend to use bent pipe repeaters as it is transparent and will be less affected (if at all) by changes in the underlying protocols used. This is particularly important given the equipment is inaccessible and the satellite lifespan can be 10–20 years, potentially greater than the lifetime of an air interface standard.

One downside of using a bent pipe amplifier approach is that noise on the uplink is also amplified and transmitted on the downlink, increasing the total link noise. The end-to-end performance of a satellite link including two hops {uplink, downlink} can be calculated from the performance of each hop using **thermal addition**.

Consider the geometry in Figure 5.29. This shows a GSO satellite system with link parameters in absolute as defined in Table 5.10.

The C/N for the uplink can be calculated in absolute using

$$\frac{c}{n} = \frac{p_{tx} g_{tx} g_{rx}}{l_p n} \tag{5.99}$$

Similarly, for the downlink the C'/N' in absolute is

$$\frac{c'}{n'} = \frac{p'_{tx} g'_{tx} g'_{rx}}{l'_p n'} \tag{5.100}$$

But the power for the downlink is equal to the power for the uplink times the amplifier gain at the satellite, γ:

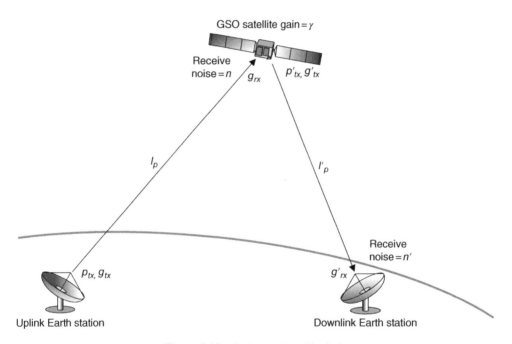

Figure 5.29   End-to-end satellite link

Table 5.10   Notation for Figure 5.29

Direction	Uplink	Downlink
Transmit power	p_{tx}	p'_{tx}
Transmit gain	g_{tx}	g'_{tx}
Path loss	l_p	l'_p
Receive gain	g_{rx}	g'_{rx}
Receive noise	n	n'

$$p'_{tx} = \gamma c = \gamma \frac{p_{tx} g_{tx} g_{rx}}{l_p} \tag{5.101}$$

The total noise of the end-to-end (e2e) link is equal to the noise on the downlink alone plus the noise on the uplink amplified by the satellite and transmitted down along with the wanted signal:

$$n_{e2e} = n\gamma \frac{g'_{tx} g'_{rx}}{l'_p} + n' \tag{5.102}$$

Now

$$\frac{1}{c/n} + \frac{1}{c'/n'} = \frac{l_p n}{p_{tx} g_{tx} g_{rx}} + \frac{l'_p n'}{p'_{tx} g'_{tx} g'_{rx}} \tag{5.103}$$

Using Equation 5.101

$$\frac{1}{c/n} + \frac{1}{c'/n'} = \frac{l_p n}{p_{tx} g_{tx} g_{rx}} + \frac{l_p l'_p n'}{p_{tx} g_{tx} g_{rx} \gamma g'_{tx} g'_{rx}} \tag{5.104}$$

$$\frac{1}{c/n} + \frac{1}{c'/n'} = \frac{l_p n \gamma g'_{tx} g'_{rx} + l_p l'_p n'}{p_{tx} g_{tx} g_{rx} \gamma g'_{tx} g'_{rx}} \tag{5.105}$$

$$\frac{1}{c/n} + \frac{1}{c'/n'} = \frac{n\gamma g'_{tx} g'_{rx}/l'_p + n'}{c'} \tag{5.106}$$

Therefore as the end to end $c_{e2e} = c'$, the overall C/N can be calculated using

$$\frac{1}{c_{e2e}/n_{e2e}} = \frac{1}{c/n} + \frac{1}{c'/n'} \tag{5.107}$$

This equation is described as the thermal addition of the uplink and downlink C/Ns. As any interference on the uplink would also be amplified by the satellite, the following are also true:

$$\frac{1}{c_{e2e}/i_{e2e}} = \frac{1}{c/i} + \frac{1}{c'/i'} \tag{5.108}$$

$$\frac{1}{c_{e2e}/(n_{e2e} + i_{e2e})} = \frac{1}{c/(n+i)} + \frac{1}{c'/(n'+i')} \tag{5.109}$$

Example 5.23
A satellite system has an uplink $C/N = 19$ dB and downlink $C/N = 14$ dB. The overall end-to-end path's $C/N = 12.8$ dB.

For links with regeneration the end-to-end performance is based on the BER of each hop. When error rates are small and independent between hops, the total BER can be calculated by adding each hop's BERs. In many cases the BER of the end-to-end link is dominated by the BER of the worst hop. In this case

$$BER_{total} \sim max\{BER_1, BER_2, BER_3 \dots \} \tag{5.110}$$

Given that the relationship between BER and $C/(N+I)$ is fixed for a particular modulation scheme (see Fig. 3.14), then an approximation for digital systems is

$$CtoNI_{total} \sim min\{CtoNI_1, CtoNI_2, CtoNI_3 \dots \} \tag{5.111}$$

5.7 Modelling Deployment and Traffic

In many radio systems the level of interference created will depend on the behaviour of users, in particular the services requested from the network and their location. In interference analysis this is captured by:

- The **deployment model**: defines where transmit and receive stations are located
- The **traffic model**: defines the type of service requested, how often and for how long plus the required link metrics (e.g. E_b/N_0).

This section describes how to model these factors.

5.7.1 Deployment Range

A key question for aggregate interference studies is how large a deployment is necessary. It should be sufficiently extensive that including additional transmitters would make a negligible difference to the results. However, there can be computational constraints on how many stations can be modelled in a simulation.

The deployment range needs to be considered in both the frequency and geographic domains, driven by whether the scenario is co-frequency or non-co-frequency. A key factor will be the degree of frequency reuse, which will depend upon the access method, as discussed in Chapter 3.

In both the frequency and geographic domains, the range over which to aggregate should take account of:

- The need to continue adding interfering transmitters until including more would make a negligible difference to the results. The definition of negligible will be scenario dependent but could be in the range 0.1–1.0 dB
- Physical, resource and self-interference limits on the number of interfering transmitters. For example, if the model is of a dense urban deployment of BS, it will be limited by the size of the city. In addition, there could be intra-system interference constraints on the maximum feasible number of transmitters.

Consideration of the scenario to be analysed might help identify the aggregation rules. For example, if the interferer is deployed only in urban areas, then there are two scenarios depending upon whether:

1. The victim is in the centre of the urban area: In this case only the urban area around the victim would need to be considered
2. The victim is outside of the urban area: In this case the aggregate interference from the urban area could be considered using techniques as in Section 5.7.5.

5.7.1.1 Co-frequency Analysis

In the frequency domain, the analysis should include all the interfering transmitters that are co-frequency with the victim receiver and take account of the relative bandwidths, which could be:

1. Victim bandwidth larger than the interferer channel bandwidth: In this case there will be an aggregation of interfering channels, noting there could be a cap due to system constraints (e.g. resource management) on the total. In cases where the ratio of bandwidths is not an integer, then the A_{BW} could be used for the partially overlapped channel.

Example 5.24
The victim is a satellite Earth station with receive bandwidth of 30 MHz, while the interferer is 10 MHz LTE. Therefore three channels worth of LTE transmitters could be included, though it might be that in the scenario being analysed there would be a maximum of 20 MHz per base station.

   If the interferer bandwidth is much less than the victim, then this can lead to large numbers of interferers. One way to reduce the computational overhead is to work to a reference bandwidth or do the calculations for one interfering channel and scale to the full victim bandwidth.

2. Victim bandwidth smaller than the interferer channel bandwidth: In this case the aggregation can use a single channel of interferers together with the bandwidth adjustment factor A_{BW}.

5.7.1.2 Non-co-frequency Analysis

This case has the same requirement to consider a sufficient number of interfering transmitters that adding further ones would make a negligible difference to the results, and it is necessary to aggregate in both:

- Frequency: at least one adjacent channel and possibly more
- Distance: it generally is not necessary to consider as large an area as the co-frequency case, but it could involve multiple interfering transmitters for each channel.

Example 5.25
Figure 5.30 shows an analysis of the aggregate interference from a mobile network's grid of base stations (BS) into a DTT fixed receive station. The parameters are given in Tables 5.11 and 5.12. While 46 dBm is typically specified as the peak power for a 10 MHz LTE BS, in this case

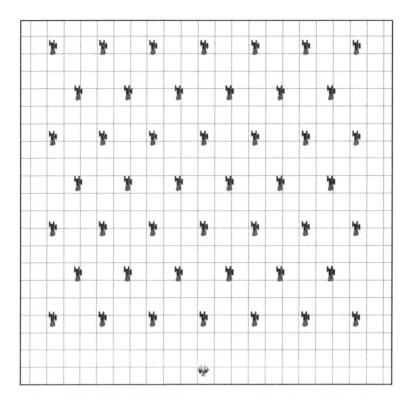

Figure 5.30   Grid of IMT LTE base stations and first victim DTT receiver

Table 5.11   IMT LTE base station parameters

Height	30 m
Sectors	3
Peak gain	15 dBi
Gain pattern	Parabolic with floor −10 dBi
Beamwidths (azimuth × elevation)	120° × 5°
Downtilt	3°
Transmit power	43 dBm
Bandwidth	10 MHz
BS separation	3 km

Table 5.12   DTT fixed receiver parameters

Height	10 m
Gain pattern	ITU-R BT.419
Peak gain	9.15 dBi
Pointing	Away from BS
Receive noise figure	7 dB
Bandwidth	8 MHz
Distance from nearest BS	{3 km, 103 km}[a]

[a] That is, 0 and 100 km from the edge of the IMT LTE coverage.

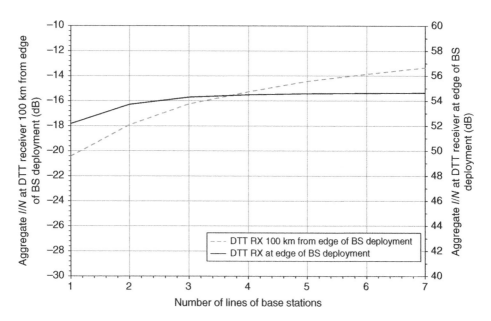

Figure 5.31 Change in aggregate I/N as number of lines of BS increases

the power was reduced by 3 dB. This is because it is unlikely that all BS would be operating at maximum power simultaneously and the average power is assumed to be half the peak power.

The IMT LTE BS were deployed in seven lines with either 6 or 7 BS per row. The I/N was calculated at each digital terrestrial television (DTT) receiver using Rec. ITU-R P.452 as the propagation model with $p = 20\%$ of time fully correlated for {1, 2, 3, 4, 5, 6, 7} rows of LTE BS.

Figure 5.31 shows how the aggregate I/N increases as the number of rows of BS increased from 1 to 7. It can be seen that while the DTT RX near the BS deployment had the highest I/N, the increase due to additional rows of BS was noticeably less (2.5 dB) than for the DTT RX (7.1 dB) that was 100 km away. This is because the relative increase in the distance of the nth row of BS compared to the first is much greater for the nearer DTT RX.

The impact of rows 1–7 compared to rows 1–6 was 0.02 dB for the nearer case and 0.53 dB for the DTT receiver at 100 km away. These deltas are becoming sufficiently small that it might be unnecessary to include additional rows.

5.7.2 Activity Models and Erlangs

A simple traffic model is to define the probability that a user is active, p(active) also called the activity factor (AF) or duty cycle. Then at each simulation sample a random number is generated, and if it is less than p(active), then that user is considered to be active and hence transmitting. The AF could vary depending upon the service, for example, voice or data.

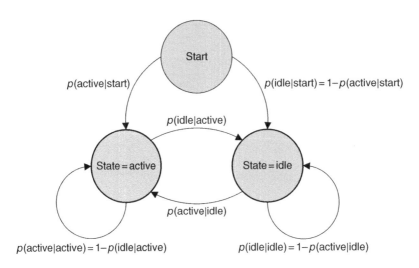

Figure 5.32   Activity state diagram

Example 5.26

A study for Ofcom measured the activity of voice and data users of private mobile radio (PMR) systems in London (Ofcom, 2014) and detected activity factors of around $AF = 0.02$ for voice. Data service measured values were in the range $0.1 < AF < 0.8$ depending upon the service, for example, the frequency of GPS polling.

The AF can be extended to have a separate probability that if the user is active at one sample they are switched off for the next sample. This effectively gives the user two states, active and idle, and a probability of transition between the two. It is also necessary to give a third probability, that the user is initially active. The resulting state machine is shown in Figure 5.32, and in many scenarios this level of detail is sufficient, particularly if combined with other techniques described in this section.

More detailed models can take account of call duration, so that the input parameters are

$$\text{Average duration of a call}: H(\text{in unit time})$$

$$\text{Average call requests}: \lambda(\text{per unit time})$$

These are the raw inputs of the two Erlang traffic models, Erlang B and Erlang C (Rappaport, 1996). The traffic intensity per user in Erlangs is then

$$A_u = \lambda H \tag{5.112}$$

With U_n users to service the total traffic in Erlangs is

$$A_t = U_n A_u = U_n \lambda H \tag{5.113}$$

If these U_n users are able to access C_n channels, then the average traffic intensity per channel is

$$A_{tc} = \frac{U_n A_u}{C_n} \tag{5.114}$$

Note that this represents traffic demand rather than traffic carried, which could be reduced if there is congestion. As the users are assumed to be independent, there is a slight probability that all would want to access a channel at the same time, and if $C_n < U_n$ then some users will be unable to access the channel or be **blocked**.

Note that this usage of the word blocked relates to traffic congestion and not due to a non-co-frequency interference path as in Section 5.3.5.

The probability of blocking can be calculated for the two cases:

1. Erlang B: there is no queueing: if a call is blocked, it is assumed to have failed
2. Erlang C: there is queueing: if a call is blocked, it will wait until the channel is free, which leads to higher probability of blocking as no calls are lost.

For Erlang B the probability of blocking can be calculated using

$$p[blocking] = \frac{\frac{A_t^{C_n}}{C_n!}}{\sum_{k=0}^{C_n} \frac{A_t^k}{k!}} \tag{5.115}$$

For Erlang C the probability that the delay is greater than zero (i.e. some degree of blocking) can be calculated using

$$p[delay > 0] = \frac{A_t^{C_n}}{A_t^{C_n} + C_n!\left(1 - \frac{A_t}{C_n}\right)\sum_{k=0}^{C_n-1} \frac{A_t^k}{k!}} \tag{5.116}$$

In the case of Erlang C the delay length probability can be calculated using

$$p[delay > t] = p[delay > 0]e^{-(C_n - A_t)t/H} \tag{5.117}$$

The average delay D for *all* calls in a queued system is

$$D[all] = p[delay > 0]\frac{H}{C_n - A_t} \tag{5.118}$$

Considering only those that are delayed, the average wait time is

$$D[queued] = \frac{H}{C_n - A_t} \tag{5.119}$$

Note that when a single channel is being shared (i.e. $C_n = 1$), these reduce to

$$\text{Erlang B}: \quad p[blocking] = \frac{A_t}{1 + A_t} \tag{5.120}$$

$$\text{Erlang C}: \quad p[delay > 0] = A_t \tag{5.121}$$

Example 5.27

A single PMR channel is being shared between six independent users. Each makes 4 calls/hour of average duration 18 seconds, which implies their $AF = 0.02$. The probabilities are

$$\text{Erlang B}: \quad p[blocking] = 0.107$$

$$\text{Erlang C}: \quad p[delay > 0] = 0.12$$

These calculations are usually applied to the design of a communication network to ensure that there are sufficient resources but are also of interest during interference analysis to ensure the parameters used are consistent. They are of direct interest to the planning algorithm for PMR systems described in Section 7.2, which treats blocking as interference.

5.7.3 Traffic Type

There are a number of ways that traffic can be categorised including:

- Analogue versus digital
- Voice versus data
- Data type: web browsing, email, video, file transfer, GPS polling, etc.

Which of these is most important will depend upon the scenario. While analogue systems will have a different transmit spectrum mask and thresholds than a digital system, these modulation types are becoming less prevalent.

There can be significant differences between voice and the data services and between data services in terms of the transmit powers, bandwidths, AF, densities, etc. In particular

- Voice call: low data rate but requires low latency
- Video call: high data rate and requires low latency
- Messaging: low data rate and can accept high latency
- File download or upload: possibly high data rate but can accept higher latency
- GPS polling: depends on polling cycle time
- Web browsing: variable depending upon site.

It is not feasible to model all aspects of a network, and so simplifications are necessary. Hence, users could be classified by service type and mapped onto the critical parameters of:

- Transmit power per user, possibly indirectly via a target RSL or E_b/N_0 as in Table 3.7 based upon Rec. ITU-R M.1654 (ITU-R, 2003c)
- Density of users per sector or cell per carrier bandwidth
- Percentage of users indoors rather than outdoors (as this has a significant impact on the propagation model).

This will require an understanding of the multiple access method used (as discussed in Section 3.5) and how that affects the power per user and per BS. The number of users could take account of other constraints, such as LTE resource blocks.

Table 5.13 UE parameters from Report ITU-R M.2292

Indoor user terminal usage	50% for macro rural scenario 70% for macro urban/suburban scenario
Average indoor user terminal penetration loss	15 dB for macro rural scenario 20 dB for macro urban/suburban scenario
User terminal density in active mode to be used in sharing studies	$0.17/5$ MHz/km² for macro rural scenario $2.16/5$ MHz/km² for urban/suburban scenario
Maximum user terminal transmitter output	23 dBm
Average user terminal transmitter output	2 dBm for macro rural scenario -9 dBm for macro urban/suburban scenario

It can be more appropriate to use a separate network-level simulation that takes account of resources, protocols and adaptive features (power, modulation, antenna) to generate simplified parameters used in interference analysis.

Example 5.28
Report ITU-R M.2292 (ITU-R, 2013r) summarises the characteristics of user equipment (UE) with the values in Table 5.13. The approach taken is to specify user densities and average powers rather than model traffic types in detail.

The number of users could also vary during the day, the so-called diurnal variation. For example, peak traffic might only be reached during the early evening around 17:00–18:00. Different traffic types or user types could have different busy hours, reducing the total aggregate interference. For interference from terrestrial systems into satellite systems, the field of view is likely to be so large that it will cover multiple time zones, each with its own busy hour, and so the aggregate interference would vary by time of day.

Example 5.29
The non-GSO mobile-satellite service (MSS) described as LEO-F in Rec. ITU-R M.1184 (ITU-R, 2003a) has a fully loaded beam EIRP of 49 dBW/MHz, as described in Section 6.2.2. It is assumed that the majority of the traffic is voice and the activity factor is 0.4. With over a hundred voice calls within the 1 MHz carried per beam, it could be acceptable to use the voice activity factor to calculate the averaged EIRP as

$$EIRP = 49 + 10\log_{10}(0.4) = 45 \text{ dBW/MHz}$$

5.7.4 Deployment Models

There are two aspects to deployment that must be considered:

1. Fixed deployments, that is, assets such as towers and GSO satellites that can be considered located at a stationary point for a given scenario
2. Mobile deployments, that is, users such as handsets, aircraft and ships that can (or indeed are likely to) change location within a scenario.

These represent two different deployment 'spaces' and care should be taken when considering mixing or convolving them together. Variations in the time domain, such as propagation loss, can be convolved in Monte Carlo or dynamic analysis including mobile deployments but not fixed deployments without justification. An example of when it might be acceptable is if there are sufficient fixed stations that averaging is realistic, as in the aggregate equivalent isotropically radiated power (AEIRP) methodology described in the following section. The mobile deployment could also consider the location variability term in the propagation models.

In general, fixed deployments are a defined input into a simulation rather than something that is varied, though there can be multiple sets of fixed deployments that are considered during sensitivity analysis. An exception is the two-stage Monte Carlo methodology described in Section 6.10 in which an initial run analyzes the fixed deployment space while the second is in the time domain and includes mobile deployment and propagation variation.

Mobile deployments are usually defined as having equal likelihood within an area, specified by a circle, square or hexagon. In reality there are variations within the cell, but these are rarely modelled in order to make the analysis generic rather than specific. In addition, this type of detailed traffic model is likely to be proprietary and valued information to an operator.

A standard deployment for mobile networks involves a grid of BS with sectors and cells as in Figure 5.33. A predefined number of users would be randomly located within each hexagonal sector using a uniform distribution with constant density.

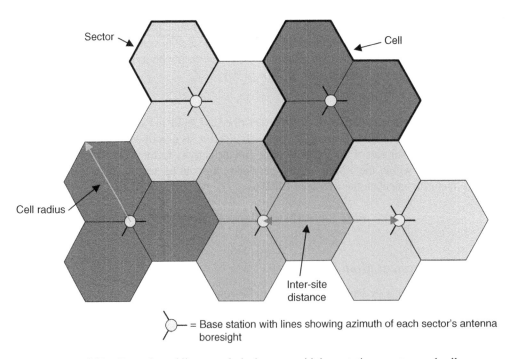

Figure 5.33 Example mobile network deployment with base stations, sectors and cells

5.7.5 Aggregation Techniques

When there are large numbers of stations, it can become computationally challenging to model each one individually. A number of techniques can be used to reduce complexity and create manageable scenarios.

5.7.5.1 Use Adjusted Peak Power

For example, when modelling the BS transmit (downlink) direction of a mobile network, it could be acceptable to model just the BS and not each user in each cell. The worst-case assumption would be to model the BS operating at its maximum power (e.g. 46 dBm in 10 MHz). If there are large numbers of BS, it becomes implausible that each would be operating at peak power simultaneously. Then an average power could be used (say) 3 dB down from peak, that is, 43 dBm in 10 MHz as in Example 5.25.

5.7.5.2 Aggregate N Transmitters

In some cases there can be extremely large numbers of identical transmitters. For example, consider the 5 GHz RLAN into non-GSO MSS feeder links scenario within Rec. ITU-R M.1454 (ITU-R, 2000a). Here the non-GSO satellite is in line-of-sight of both Europe and North America, potentially seeing many millions of RLAN devices. In this case it can be useful to model the interference from a large numbers of actual stations using one simulated station and adjusting its transmit power accordingly.

If every station in the simulation is to represent N actual stations, the transmit power should be adjusted using

$$P'_{tx} = P_{tx} + 10\log_{10}N \qquad (5.122)$$

Where there is an AF or duty cycle, then the number of active stations (rather than total number of stations) can be derived and used in the calculations so that

$$N_{active} = AF \times N_{stations} \qquad (5.123)$$

Then

$$P'_{tx} = P_{tx} + 10\log_{10}N_{active} \qquad (5.124)$$

A similar approach can be used for the EIRP:

$$EIRP'_{tx} = EIRP_{tx} + 10\log_{10}N_{active} \qquad (5.125)$$

5.7.5.3 AEIRP Calculations

Another approach is to determine the AEIRP towards the horizon within, for example, the area served by a single BS. If the size of this area is much less than the distance to the victim station, then the path loss will be similar and the EIRP can be aggregated separately. This can either be a single fixed number or a distribution that can then be used as an input into a Monte Carlo analysis.

An example of this approach is given in two ITU-R Recommendations:

1. Rec. ITU-R F.1766: Methodology to determine the probability of a radio astronomy observatory receiving interference based on calculated exclusion zones to protect against interference from point-to-multipoint high-density applications in the fixed service operating in bands around 43 GHz (ITU-R, 2006d)
2. Rec. ITU-R F.1760: Methodology for the calculation of aggregate equivalent isotropically radiated power (a.e.i.r.p.) distribution from point-to-multipoint high-density applications in the fixed service operating in bands above 30 GHz identified for such use (ITU-R, 2006c).

The approach is to first determine the distribution of AEIRP from all the stations (both BS and user terminals) dispersed over a defined area called a building block. The AEIRP takes into account the following:

- Variations in transmitter and receiver heights
- Variations in station location and hop length
- Variations in antenna azimuth and associated gain
- Variations in transmit power control.

The result is a cumulative distribution function (CDF) $AEIRP(p)$ that models the total interfering potential from that cell and can be used as the input into the aggregate interference calculation.

Note that this approach initially considers deployment likelihood and then convolves in the time domain:

1. The AEIRP distribution defines the likely variation in AEIRP due to variations in the deployment of fixed stations
2. This distribution is used as an input into a Monte Carlo analysis that convolves AEIRP variation with propagation and victim antenna pointing.

This can be an acceptable approach given that large numbers of deployments would have AEIRP distribution as derived in the first stage, though properly a two stage approach should be used as described in Section 6.10.

5.8 Link Design and Margin

A key parameter for interference analysis is the threshold that separates acceptable interference from harmful interference. This threshold will depend on the system involved and in particular the method used to plan the wanted service. This planning can include the calculation of the transmit power and/or the service area to be covered.

To understand or indeed to derive interference thresholds, it is therefore necessary to understand the approach used to plan the wanted service and in particular the margin included for interference. While there can be a range of approaches to link design and interference

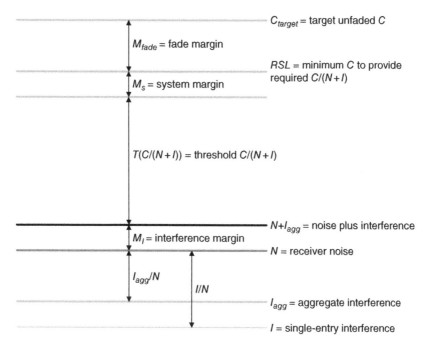

Figure 5.34   Link calculation and interference margin

apportionment (as discussed in Sections 5.9 and 5.10), this section gives one of the most commonly used methods. An alternative approach is used for broadcast coverage prediction as discussed in Section 7.3.

The key concepts are shown in the **power line diagram** in Figure 5.34, which graphically describes power levels at the receiver, with higher levels towards the top.

The starting point for the link design is the following parameters:

N = receiver noise (either specified via a temperature or noise figure)
M_i = interference margin
$T(C/(N + I))$ = ratio of wanted signal to total of noise and interference that is required to provide the target quality of service (QoS)
M_s = system margin (could be zero)
M_{fade} = fade margin (could be zero)

The objective is to calculate the target RSL[1] required to provide the required QoS. This can be done using

---

[1] In some documents the acronym RSL refers to the receive signal level. However in this book the receive signal level symbol is either C if wanted or I if interfering, while RSL refers to the target sensitivity.

$$RSL = N + M_s + M_i + T\left(\frac{C}{(N+I)}\right) \tag{5.126}$$

The system margin is included to cover for any imperfections in implementation and ideally would be zero. An example of what could be included in M_s would be antenna de-pointing or operating away from a satellite beam's boresight.

In many cases there can be propagation variation that has to be taken into account, such as multi-path or rain fade. This could be included in the RSL, but as the fade margin often varies by deployment (e.g. depending on path length or rain zone), it is useful to have separate targets for the faded and unfaded wanted signal, namely, the RSL and C_{target}, respectively.

Fading could be handled by using either APC (as in Section 5.5.2) or constant power with a fixed fade margin. The target unfaded wanted signal is then

$$C_{target} = RSL + M_{fade} \tag{5.127}$$

Having derived a target wanted signal level including fade margin, it is possible to calculate the transmit power required to **close the link** (i.e. meet the RSL objective for the required percentage of time), starting from

$$C_{target} = P_{tx} + G_{tx} - L'_p + G_{rx} - L_f \tag{5.128}$$

Here L'_p is the unfaded propagation loss. We can therefore calculate

$$P_{tx} = C_{target} - \left(G_{tx} - L'_p + G_{rx} - L_f\right) \tag{5.129}$$

Example 5.30
A fixed link system with bandwidth 28 MHz and receive noise figure of 7 dB is to use a 16-QAM carrier with interference and system margins of 1 dB and fade margin of 30 dB. The RSL can be calculated as follows:

$$N = 10\log_{10}(290) + 7 + 10\log_{10}(28) + 60 - 228.6 = -122.5\,dBW$$

The threshold for 16-QAM = 20.5 dB from Table 3.6 so

$$RSL = -122.5 + 1 + 1 + 20.5 = -100\,dBW$$

The target unfaded wanted signal level is then

$$C_{target} = -100 + 30 = -70\,dBW$$

Example 5.31
The fixed link system from the previous example is using antennas with gain 30 dBi and line feed 1 dB over a hop 8 km long at a frequency of 28 GHz. Using the free space path loss in

Equation 3.13 and assuming 1 dB of additional gaseous attenuation, the required transmit power can be derived:

$$L_{fs} = 32.45 + 20\log_{10}(8) + 20\log_{10}(28{,}000) = 139.5\,\text{dB}$$

Hence

$$L'_p = L_{fs} + L_{attenuation} = 140.5\,\text{dB}$$

Therefore from Equation 5.129

$$P_{tx} = -70 - 30 + 140.5 - 30 + 1 = 11.5\,\text{dBW}$$

This is a large transmit power for a fixed link with an EIRP = 41.5 dBW. Fixed links in these bands are often limited to paths with lengths less than 10 km due to large propagation losses, in particular fade loss due to rain. It might be necessary to reduce the modulation or bandwidth, resulting in lower data rates to close the link. More information in the planning and frequency assignment of fixed links is given in Section 7.1.

The fade margin will depend upon the propagation models used and the availability object-ive. For example, for the fixed link the propagation models could be:

- Rec. ITU-R P.525: free space path loss
- Rec. ITU-R P.676: gaseous attenuation
- Rec. ITU-R P.530: multi-path and rain fade.

For a space-to-Earth link it is likely to be a different set, such as:

- Rec. ITU-R P.525: free space path loss
- Rec. ITU-R P.676: gaseous attenuation
- Rec. ITU-R P.618: rain fade.

Where there are availability requirements, these are used to calculate the required margin. For example, the propagation models in Recs. ITU-R P.530 and P.618 have an associated per-centage of time, as discussed in Chapter 4.

Furthermore, safety of life systems could have additional margin to provide enhanced link availability.

In the previous equations the fade margin was used to go from the *RSL* to the target wanted signal, but the equations can be simplified by using a total propagation loss as in

$$RSL = N + M_i + M_s + T\left(\frac{C}{(N+I)}\right) \tag{5.130}$$

$$P_{tx} = RSL - G_{tx} + L_p(\%) - G_{rx} + L_f \tag{5.131}$$

where the total propagation loss for p% of time is $L_p(\%)$.

In some scenarios, the transmit power is fixed (or known) and what is variable (or unknown) is the range. An example of this would be a mobile network with fixed BS power. In this case what is required is the range, which can be derived by calculating where the wanted signal reaches the *RSL* using

$$L_p(\%) = P_{tx} - RSL + G_{tx} + G_{rx} - L_f \tag{5.132}$$

If a fixed fade margin is assumed, it can be useful to split the propagation losses into the margin for fading and the remaining propagation loss, so

$$L'_p = P_{tx} - RSL + G_{tx} + G_{rx} - L_f - M_f \tag{5.133}$$

Note that for many propagation models, such as those in P.530, the fade depth will depend upon the path length, and so in those cases it would be preferable to use Equation 5.132.

Example 5.32
A mobile network is using the parameters in Table 5.14. Assuming the Hata/COST 231 propagation model for an urban environment, what is the coverage range and hence sector radius?
    From Equation 3.60,

$$N = 5 + 10\log_{10}(290) - 228.6 + 10\log_{10}(5) + 60 = -131 \, \text{dBW}$$

From Equation 5.130 and converting from dBW to dBm to be consistent with transmit power,

$$RSL = -131 + (1) + (0) + (-1) = -131 \, \text{dBW} = -101 \, \text{dBm}$$

From Equation 5.133 the total path loss excluding the fade margin is then

Table 5.14 Mobile downlink parameters

Frequency	800 MHz
Bandwidth	5 MHz
Noise figure	6 dB
Interference margin	1 dB
$C/(N+I)$ threshold	−1 dB
Fast fading margin	4 dB
Location variability margin	7 dB
Transmit power	43 dBm
Transmit peak gain	15 dBi
Transmit relative gain	−2.5 dB
Receive gain	0 dBi
Body loss	1 dB
Base station antenna height	10 m
Mobile receive height	1.5 m

$$L'_p = 43 - (-101) + (15 - 2.5) + 0 - 1 - (4 + 7) = 144.5 \, \text{dB}$$

For the Hata/COST 231 propagation model with environment = urban, this equates to a distance or range of about 1.9 km

In this calculation neither the system margin nor receive line feed terms were defined, but there was a body loss that was used in its place. As noted earlier, there is a large variation in how link budgets are calculated depending upon the parameters available or useful in the calculation. Similarly there can be a large variation in how line diagrams are constructed, and there can be additional factors involved in the wanted signal calculation.

5.9   Interference Apportionment and Thresholds

5.9.1   Interference Margin

The previous section worked 'up' from the noise figure and interference margin to calculate the *RSL* and wanted signal and then on to either the transmit power or range. It is also useful to work 'down' from the noise plus interference margin to calculate the aggregate interference and single entry interference allowances and hence thresholds that can be used in interference analysis.

From Figure 5.34

$$(N + I) = N + M_i \tag{5.134}$$

The power addition is in absolute and hence the aggregate interference should be calculated using

$$n + i_{agg} = 10^{(N + M_i)/10} \tag{5.135}$$

So

$$i_{agg} = 10^{(N + M_i)/10} - 10^{N/10} \tag{5.136}$$

Or

$$I_{agg} = 10 \log_{10} \left(10^{(N + M_i)/10} - 10^{N/10} \right) \tag{5.137}$$

It is often useful to consider the ratio of the aggregate interference to the noise, which can be calculated from the interference margin alone as

$$\frac{I_{agg}}{N} = 10 \log_{10} \left(10^{M_i/10} - 1 \right) \tag{5.138}$$

Alternatively, a given *I/N* can be converted into a loss of link margin using

$$M_i = 10 \log_{10} \left(1 + 10^{(I_{agg}/N)/10} \right) \tag{5.139}$$

Some examples of how the ratio of the permitted (or threshold) aggregate interference to noise and DT/T changes according to the interference margin are given in Table 5.15 and shown graphically in Figure 5.35. This loss of link margin is also known as the receiver desensitisation.

An interference margin of 3 dB is sometimes used as the boundary between **noise limited** and **interference limited** scenarios (Flood and Bacon, 2006).

While any interference margin is feasible, an industry standard value is 1 dB, and the resulting I_{agg}/N threshold, identified in this book by the $T()$ notation, is often rounded to

$$T\left(\frac{I_{agg}}{N}\right) = -6\,\text{dB} \tag{5.140}$$

Then the DT/T threshold is rounded to

$$T\left(\frac{DT}{T}\right) = 25\% \tag{5.141}$$

Table 5.15   Example interference margins as aggregate I/N and DT/T

Interference margin (dB)	0.5	1.0	2.0	3.0
Threshold I_{agg}/N (dB)	−9.1	−5.9	−2.3	0.0
Threshold DT/T (%)	12.2	25.9	58.5	99.5

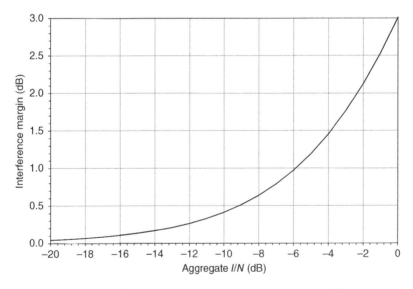

Figure 5.35   Interference margin for a given aggregate I/N

This is an aggregate interference allowance as a ratio of the receiver noise and has to take into account *all* sources including:

- Intra-system interference (i.e. from within a network)
- Inter-system interference from all interferers of the same service
- Inter-system interference from all interferers of different services.

The margin must take into account *all* co-frequency and non-co-frequency sources of interference, including those from primary or secondary allocations and unwanted emissions from services in other bands.

The threshold can be adjusted to take account of other factors. For example, services that are designated as being **safety of life** (see RR Article 4.10) can have 6 dB of additional protection or **safety margin** as in Report ITU-R M.2235 (ITU-R, 2011c).

Another example could be services auxiliary to broadcasting (SAB) and services auxiliary to programme making (SAP) supporting live events with high priority such as the Olympics, for which outages during critical moments could be considered unacceptable.

Other interference margins are feasible and would lead to greater resilience to interference. For example, the planning algorithm for fixed links in the United Kingdom uses 1 dB of margin for most bands but 2 dB in parts of the 6 and 26 GHz bands (Ofcom, 2013).

A disadvantage of additional interference margin is the requirement for higher transmit powers, which itself leads to greater interference into other systems. However, the net result is to facilitate sharing due to the enhanced ability to tolerate interference.

Example 5.33
A scenario is configured in which four interfering entries share a 1 dB interference margin. The interference margin is increased in 1 dB steps up to 8 dB, requiring a corresponding increase in transmit power to close each system's link. However, the additional margin permits more entries without exceeding the aggregate interference threshold, as shown in Figure 5.36. It can be seen that the improved sharing benefits level off as the additional margin is absorbed by the higher transmit power required to close the link.

An example of an interference margin other than 1 dB is given in the IMT LTE link budget in Section 6.9.5 where the margin is 3 dB. This is required because of the significant intra-system interference from users in other cells and sectors: the inter-service threshold between primary services is more typically in the range $T(I/N) = -6$ dB to $T(I/N) = -10$ dB.

The ability of radio systems to increase their transmit power to facilitate a larger interference margin can be restricted due to constraints including:

- Regulatory constraints on *EIRP*, such as the 55 dBW limit for fixed and mobile stations in Article 21.3 of the RR
- Regulatory constraints on off-axis *EIRP* as in Section 7.5.4
- Regulatory constraints on *PFD* (border or area) including field strengths defined in international memorandums of understanding (MoUs) as in Section 5.10.9
- Block edge masks as in Section 5.3.8
- Equipment constraints, include satellite power limits
- Energy costs
- Human exposure limits, described in the boxed text after Section 5.10.9

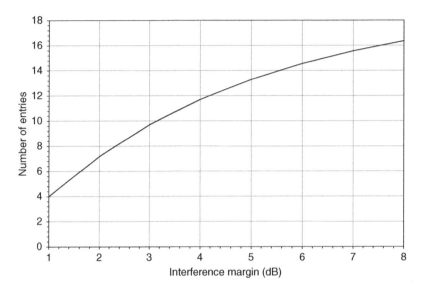

Figure 5.36 Example impact of additional interference margin on sharing

- Increased interference into other systems that are unable to increase their interference margin as they either have already been deployed or are constrained by one of the other bullet points.

Furthermore, a system providing coverage over an area with a fixed transmit power would have its maximum range decreased by an increase in the margin for interference.

A general principle of radio spectrum management is to use the lowest power necessary to provide a service (see RR Article 0.2 in Section 2.4.3.1) and consequentially lowest interference margin. Typically an aggregate 1 dB interference margin is assumed, in particular when setting inter-system and inter-service sharing criteria.

5.9.2   Interference Apportionment

One way to manage interference is to **apportion** the aggregate interference allowance between sources in order to calculate a single entry interference threshold that can be used in a specific sharing scenario. For example, the aggregate interference could be assumed to be the total of n equal single entries, and hence

$$T\left(\frac{I}{N}\right) = T\left(\frac{I_{agg}}{N}\right) - 10\log_{10}(n) \tag{5.142}$$

There can be significant disagreement and debate on how to apportion the interference margin, but commonly used values include those in Table 5.16. Note that the **apportionment** should be done in absolute rather than decibels as it relates to power.

For example, to protect terrestrial broadcasting systems, Rec. ITU-R BT.1895 (ITU-R, 2011a) recommends:

Table 5.16 Example *DT/T* and *I/N* thresholds

Threshold	$T(DT/T)$ (%)	$T(I/N)$ (dB)
Coordination trigger for satellite services	6	−12.2
Common threshold for terrestrial services	10	−10
Co-frequency secondary service	1	−20
Non-co-frequency service	1	−20

3. *That the total interference at the receiver arising from all sources of radio-frequency emissions from radiocommunication services with a corresponding co-primary frequency allocation should not exceed 10% of the total receiving system noise power.*

The single entry interference threshold of $T(I/N) = -10$ dB, which equates to a $T(DT/T) = 10\%$, corresponds to a degradation in the *C/N* margin of 0.414 dB.

This threshold is sometimes further apportioned into multiple services so that (assuming two services):

- The interference threshold for another service should be $T(I/N) = -13$ dB, which equates to a $T(DT/T) = 5\%$.

It is necessary to consider the different levels of aggregation:

- Aggregate interference over all services, systems and transmitters
- Aggregate interference over all systems and transmitters of one service
- Aggregate interference over all transmitters in one system of one service.

The example $T(I/N) = -13$ dB might be the threshold for a single service, but that is likely to involve multiple transmitters. The threshold for a single interfering station would therefore have to be tighter.

Example 5.34

C-band satellite Earth stations must share with a range of services including other satellite networks and fixed links, both with primary allocations. It was argued in JTG 4-5-6-7 that the sharing criteria of $T(I/N) = -10$ dB should be apportioned so that the new entrant of IMT LTE base stations would be permitted a $T(I/N) = -13$ dB. This would be an aggregate over all transmitters in a deployment as discussed in Section 5.7.1.

These apportionment methods have as key input the number of single entry systems that contribute to the aggregate. This value will depend upon the scenario involved and could require preliminary simulation to identify how interference aggregates, such as the analysis in Example 5.25. In general the number will relate to the expected density of transmitters. For example, for low-density point-to-point fixed links, a value of $N = 4$ is typically used by Ofcom in its frequency assignment (Ofcom, 2013).

There is another approach that can be considered for management of the standard 1 dB margin for interference, which is to keep track of the total aggregate interference as part of a first-come, first-served (FCFS) licensing regime. Hence

- The first to apply for a licence will have no interference constraints as there are no victims to consider
- The second to apply will have to protect licensee 1 and can take up to all of its 1 dB interference margin
- The third to apply will have to protect licensees 1 and 2 and will only be able to use that part of licence 1's interference margin not used by licence 2
- Etc.

This would make it increasingly difficult for new entrants, and most licensing regimes would rather give a more even balance between the rights of existing spectrum users with new entrants. Even with apportioning mechanisms, there are still significant benefits to incumbents as they are given continuing rights while new entrants must protect them.

Example 5.35
For satellite systems, an apportionment of interference and definition of interference margin for feeder links is given in Rec. ITU-R S.1432 (ITU-R, 2006h) as

Error performance degradation due to interference at frequencies below 30 GHz should be allotted portions of the aggregate interference budget of 32% or 27% of the clear-sky satellite system noise in the following way:
> *25% for other FSS systems for victim systems not practising frequency re-use*
> *20% for other FSS systems for victim systems practising frequency re-use*
> *6% for other systems having co-primary status*
> *1% for all other sources of interference*

This corresponds to an interference margin of 1 dB for systems practising frequency reuse and 1.2 dB for those not practising frequency reuse.

5.9.3 Short-Term and Long-Term Thresholds

These interference values are such that they could occur all or most of the time and the system would still operate, as long as the fading is not greater than the fade margin. For many systems the fade margin is derived for a very short period of time, for example, only exceeded for 0.01% or even 0.001% of the time. As fading is rarely this deep, there is therefore scope for the system to be able to accept interference above these long-term thresholds, as long as it occurs only for a short period of time.

If the wanted signal was unfaded, then the $N + I$ level can be increased as in Figure 5.37. In this case

$$\frac{I_{agg}(st)}{N} = 10\log_{10}\left(10^{(M_i + M_{fade})/10} - 1\right) \qquad (5.143)$$

In cases where the fade is large, the increase in permissible interference can be estimated using

$$\frac{I_{agg}(st)}{N} \approx \frac{I_{agg}}{N} + M_{fade} \qquad (5.144)$$

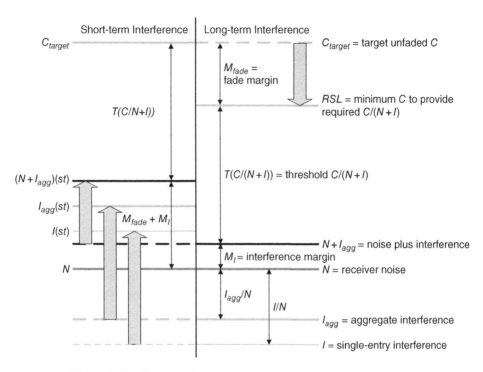

Figure 5.37   Short-term interference versus long-term interference cases

$$\frac{I(st)}{N} = \frac{I}{N} + M_{fade} \tag{5.145}$$

These higher levels of interference are only permitted for short periods of time, hence the (st) notation.

These short-term thresholds are often specified in terms of an associated percentage of time, for example:

$$\frac{I}{N} > Y \text{ dB for no more than } Z\% \text{ of time} \tag{5.146}$$

The implication is that there could be interference greater than the threshold, which would, by definition, lead to a degradation in service, in particular an increase in the unavailability. Note that with N fixed these I/N thresholds can be equivalently written as

$$I > Y + N \text{ dB for no more than } Z\% \text{ of time} \tag{5.147}$$

When protecting services defined across an area, interference thresholds can also be defined for a specified percentage of locations in addition to time, as in

$$I > Y + N \text{ dB for no more than } Z\% \text{ of time and } W\% \text{ of locations} \tag{5.148}$$

An example of this would be the Generic Radio Modelling Tool's (GRMT) spectrum quality benchmark (SQB) as discussed in Section 7.9.

This book will use the following **notation for thresholds and associated percentages of time**:

$T(X)$ = threshold of link metric X
$P_T(X)$ = associated percentage of time for link metric X
$T(X, P(X))$ = threshold of link metric X for specified percentage of time.

In many cases interference thresholds are defined by two points such as (from Rec. ITU-R SM.1448):

- **Short-term**: interference for less than 1% of the time
- **Long-term**: interference for 20–50% of the time.

For short-term thresholds with multiple interferers that are independent of each other, it can be appropriate to apportion in time rather than in interference, as enhancement events are likely to be uncorrelated (see Section 4.8).

An example of interference apportionment based upon I/N can be found in RR Appendix 7 (ITU, 2012a), which defines the permissible short-term interference and associated percentage of time for sharing between satellite ESs and terrestrial services. These can be calculated using

$$P_r(p) = 10\log_{10}(kT_rB) + N_L + 10\log_{10}\left(10^{M_f/10} - 1\right) - W \tag{5.149}$$

$$p(\%) = p_0(\%)/n_2 \tag{5.150}$$

Example 5.36

For a transmitting Earth station in the band 17.7–18.4 GHz with parameters as in Table 5.17, the thresholds to protect digital terrestrial fixed links are

$$P_r(p) = -113\,\text{dBW} \quad \text{in} \quad 1\,\text{MHz}$$

$$p(\%) = 0.0025\,\%$$

Table 5.17 Example Appendix 7 parameters

Aggregate percentage of time	p_0 (%)	0.005%
Number of short-term entries	n_2	2
Link noise contribution	N_L	0 dB
Link performance margin	M_f	25 dB
Adjustment factor for analogue systems	W	0 dB
Receive noise temperature	T_r	1100 K
Bandwidth	B	1 MHz

Rec. ITU-R SF.1006 (ITU-R, 1993) uses the same methodology as Appendix 7 of the RR to derive short-term thresholds but also specifies long-term ones calculated using

$$P_r(p) = 10\log_{10}(kT_eB) + J - W \tag{5.151}$$

$$p(\%) = 20\% \tag{5.152}$$

where

$$J = 10\log_{10}\left(\sqrt{1 + \frac{3}{n_1}} - 1\right) \tag{5.153}$$

SF.1006 also connects the number of short-term interferers with the number of long-term contributions.

Example 5.37
For a transmitting Earth station in the band 17.7–18.4 GHz with parameters as in Table 5.17 and n_1 = number of short-term entries = 5, the calculated long-term thresholds to protect the fixed service are

$$P_r(p) = -144\,\text{dBW} \text{ in } 1\,\text{MHz}$$
$$p(\%) = 20\,\%$$

Example 5.38
A receiver with noise = N dBW has a link budget designed around a 1 dB interference margin and fade margin = M_f dB for a link unavailability of $p\%$ of the time. For a sharing scenario that involves n other systems, a generic set of long-term and short-term interference margins could be as in Table 5.18.

Table 5.18 Example generic interference thresholds

Threshold	Long-term	Short-term
Single entry interference $T(I)$ (dBW)	$N-6-10\log_{10}(n)$	$N-6+M_f$
Associated percentage of time $p_T(I)$ (%)	20	p/n

5.9.4   Thresholds and Bandwidths

If a system is proposed that operates in half the bandwidth of the victim (or is only partially overlapping), then what is the impact on the relevant threshold? Consider the two scenarios

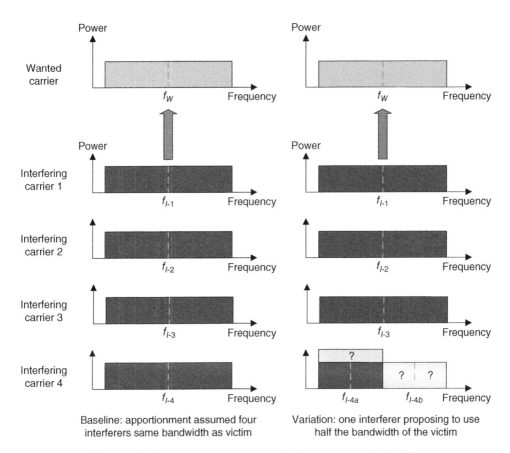

Figure 5.38   Impact on apportionment of using non-equal bandwidths

in Figure 5.38: the baseline assumes four interferers with the same bandwidth as the victim. Assuming four interferers with 1 dB of interference margin using the generic thresholds in Table 5.18, then the long-term threshold is

$$T\left(\frac{I}{N}\right) = -6 - 10\log_{10}(4) = -12\,\text{dB} \tag{5.154}$$

If the same threshold is used, then the 4th interferer would be permitted to operate with double the power density level. But that is only valid if it can be guaranteed that the half bandwidth labelled $f_{I\text{-}4b}$ is not used at a later stage.

If the bandwidths are not identical or there is partial overlap, it can, therefore, be helpful to include a bandwidth adjustment factor (as in Section 5.2) to modify the interference threshold:

$$T\left(\frac{I}{N}\right) = -6 - 10\log_{10}(n) + A_{BW} \tag{5.155}$$

In scenarios when using an *I/N* threshold with differing wanted/interfering bandwidths, it can be simpler to scale all calculations (*C, I, N*) to a reference bandwidth such as 1 MHz, as long as the interferer bandwidth exceeds the wanted or there are interfering carriers across all of the receiver's bandwidth. Alternatively the calculations could be undertaken with respect to one Hertz, that is, to consider I_0/N_0.

Making changes to the transmit power and/or bandwidth (and consequentially to the power spectral density) can assist in interference mitigation, as described in Section 5.11.1.

Example 5.39

In Example 5.34, the aggregate *T(I/N)* for C-band satellite Earth stations (ESs) sharing with IMT LTE base stations was calculated as *T(I/N)* = −13 dB. If the satellite ES is using a 30 MHz bandwidth and the IMT LTE base station transmitting with a 10 MHz carrier, it will be necessary to include aggregation in two dimensions as in Figure 5.39:

- Geographic aggregation, that is, including multiple base stations transmitting on a specific 10 MHz channel

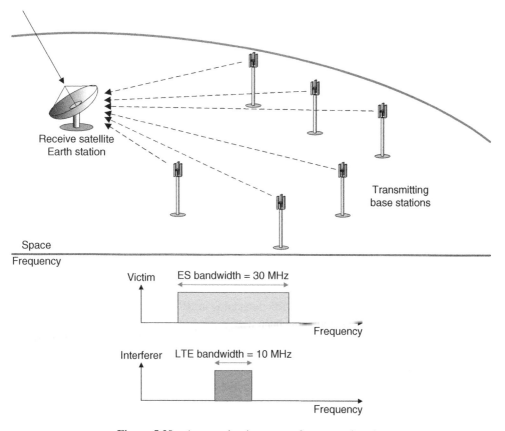

Figure 5.39 Aggregation in space + frequency domains

- Frequency aggregation, that is, considering multiple 10 MHz channels across the ES 30 MHz receiver.

In this scenario it might be acceptable to assume that the IMT LTE operator or operators use the full 30 MHz and hence either increase the aggregate transmit power by $10\log_{10}(3) = 5$ dB or make a similar adjustment to the threshold.

5.10   Types of Interference Thresholds

The previous section described interference thresholds based upon the I/N or simply the interference I at the receiver, with an associated percentage of time that could be for short and/or long-term. However, there are limitations in use of the I/N or I metrics, and a range of other thresholds or metrics can be used.

Consider the sharing scenario in Figure 5.40. Here the primary service is a licensed DTT transmitter serving fixed receivers potentially suffering interference from a nearby white space device (WSD). The wanted signal strength C is greater for receiver DTT Receiver 1 than for DTT Receiver 2, and therefore it is able to accept higher levels of interference from WSD 1 than DTT Receiver 2 could from WSD 2. However, use of either $T(I)$ or $T(I/N)$ would not identify this difference.

Alternative thresholds include:

- C/I or W/U ratios
- Fractional degradation in performance (FDP)
- $C/(N + I)$ and BER or PER
- Increase in unavailability
- Reduction in coverage, range or capacity and consequentially an increase in the number of transmitters required to provide a service or a decrease in the population served
- Field strength, PFD and equivalent power flux density (EPFD)
- Likelihood of observation
- Channel sharing ratios and derived probabilities of blocking.

The implications of a limit being exceeded depends upon the regulatory role of the threshold. For example,

- **Hard limits**, such as the PFD and EPFD thresholds in RR Articles 21 and 22 for satellite systems, must be met unless there has been agreement from the administration involved
- **Coordination triggers**, such as the algorithm for satellite ESs and terrestrial services in RR Appendix 7. In this case if the limit is met, it triggers further analysis that may or may not permit a system to operate
- **Sharing criteria**: In general should be met to prove compatibility. However it can be argued that in cases where the criteria are not met are few (e.g. restricted geometries or very small percentages of time), then sharing might be considered acceptable by spectrum policymakers.

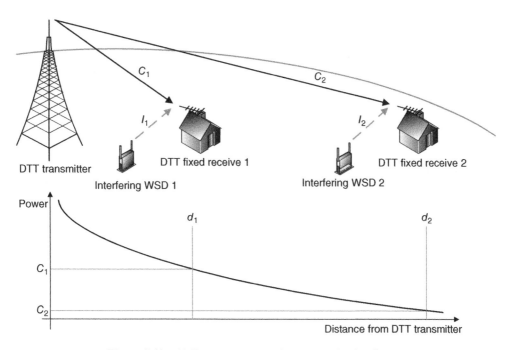

Figure 5.40   Ability to accept interference varying by distance

5.10.1   C/I and W/U Ratios

Sections 5.8 and 5.9 showed how to go from the noise up to the *RSL* and wanted signal and down to the aggregate and single entry interference using margins (interference and fade) and the required $C/(N+I)$. These can be combined to calculate threshold ratios of the carrier (wanted) to interfering (unwanted) signals, that is, the *C/I* or *W/U*.

By definition, in absolute

$$\frac{i}{n} = \frac{i}{c} \cdot \frac{c}{n}$$

(5.156)

So in dB

$$\frac{C}{I} = \frac{C}{N} - \frac{I}{N}$$

(5.157)

An example of this conversion from *I/N* to *C/I* is in Rec. ITU-R S.741 (ITU-R, 1994c), which gives the threshold in *C/I* between fixed-satellite service (FSS) digital carriers as

$$T\left(\frac{C}{I}\right) = T\left(\frac{C}{N}\right) + 12.2\,\text{dB}$$

(5.158)

The figure of −12.2 dB for the *I/N* comes from a $DT/T = 6\%$, which is the coordination trigger value. The $T(C/N)$ can include link margin, as discussed in Section 7.5.3.

While the *C/I* is derived from the *I/N*, there are benefits in including the strength of the wanted signal in the threshold. In particular a satellite system's received signal will vary across the surface of the Earth. At the boundary of its coverage, the minimum *C/N* will be reached, but at other locations, particularly towards the beam boresight, the *C* will be greater. Hence the ability to accept interference using *C/I* would be higher, so this metric is used for detailed coordination between satellite networks, as discussed in Section 7.5. In addition, the *C/I* metric is used in calculations for the 'planned bands' described in Appendices 30, 30A and 30B of the RR (ITU, 2012a).

Example 5.40
A GSO satellite network has a receiver noise = −133.8 dBW. With a *DT/T* = 6% the *I/N* = −12.2 dB so the interference threshold is −146 dBW. If the target *C/N* = 12.9 dB, then the edge of coverage (−4 dB contour) wanted signal *C* = −120.9 dBW. Therefore at the centre of the coverage area, the wanted signal *C* = −116.9 dBW. With a *T(C/I)* = 12.9 + 12.2 = 25.1 dB, this means the maximum interference at the centre of the coverage area is −142 dBW, 4 dB higher than would be permitted using the *I/N* or *DT/T* approach.

The *C/I* ratio is sometimes called the wanted to unwanted or *W/U* ratio, for example, as thresholds for point-to-point fixed links in ETSI TR 101 854 (ETSI, 2005). For these systems, however, there is little difference conceptually compared to using the *I/N* or simply *I* thresholds. As was seen in Figure 5.34 the wanted signal and interfering signals are directly connected via the various margins, and there is no variation in wanted signal across a coverage area. *W/U* ratios can be directly converted into *I/N* ratios using the interference margin, fade margin and *C/(N + I)* target.

Planning methodologies such as OfW 446 (Ofcom, 2013) give *W/U* ratios that vary by carrier type and frequency offset, but these are taking account of differing $T(C/(N + I))$ and A_{MI} and not representing additional information or ability to accept interference. As such, it is generally more transparent to use *I* or *I/N* ratios as they directly link into interference apportionment assumptions and the interference calculation.

There can be benefits in using *C/I* ratios when analysing point-to-area systems such as PMR or broadcasting that use constant transmit power rather than power control. A key question is how the point-to-area system was planned; in particular whether:

- The coverage was planned including an interference margin, similar to those in Sections 5.8 and 5.9, and then the *T(C/I)* can be calculated as in Equation 5.161
- The coverage was planned without an interference margin so that any interference would result in a reduction in coverage. This approach is discussed further in Sections 5.10.5 and 7.3.

Example 5.41
A PMR system is transmitting with power 10 dBW in 12.5 kHz at 420 MHz. Assuming the receiver noise figure is 7 dB and the propagation loss can be modelled with a constant 110.5 dB and slope 37.2 dB, then the signal strength can be calculated against distance. With a *C/(N + I)* threshold = 20 and 1 dB margin for interference, the coverage range is about 8.5 km. The *T(C/I)* is calculated using

$$T\left(\frac{C}{N}\right) = T\left(\frac{C}{N+I}\right) + 1 = 21 \, \text{dB} \tag{5.159}$$

$$T\left(\frac{I}{N}\right) = 10\log_{10}\left(10^{M_i/10} - 1\right) = -6 \, \text{dB} \tag{5.160}$$

$$T\left(\frac{C}{I}\right) = T\left(\frac{C}{N}\right) - T\left(\frac{I}{N}\right) = 27 \, \text{dB} \tag{5.161}$$

The resulting permissible interference varies by distance as shown in Figure 5.41: it can be seen that it increases towards the centre of the coverage area.

Example 5.42
The PMR system in Example 5.41 is planned without an interference margin, and the maximum range increases to 9 km. The amount of interference the system can accept depends upon the strength of the wanted signal and can be calculated using

$$M_i = \frac{C}{N} - T\left(\frac{C}{N+I}\right) \tag{5.162}$$

$$I = N \cdot \frac{I}{N} = N \cdot 10\log_{10}\left(10^{M_i/10} - 1\right) \tag{5.163}$$

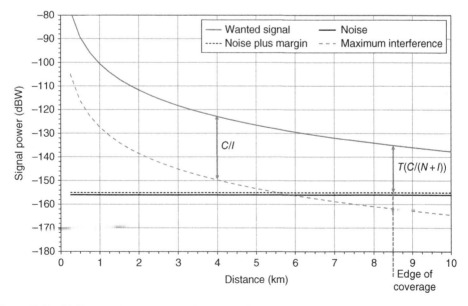

Figure 5.41 PMR wanted signal and maximum interference against distance planning with interference margin

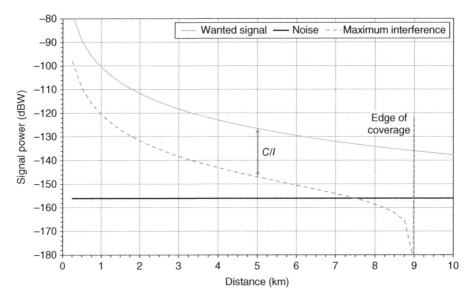

Figure 5.42 PMR wanted signal and maximum interference against distance planning without interference margin

The maximum interference that this system could permit while just meeting the $T(C/(N+I))$ is shown together with the wanted signal and constant noise in Figure 5.42. In this case the $T(C/I)$ is not constant but varies by distance. It is approximately the same as the $T(C/(N+I))$ for the first couple of kilometres, but as the wanted signal approaches the noise floor, interference is rapidly squeezed out until the $C/N = T(C/(N+I))$ when no interference can be tolerated and the $T(C/I)$ becomes infinite.

These two areas, near the transmitter and at the distance where the C/N meets the $T(C/(N+I))$, are often associated with two **protection ratios** (PRs) that are determined via measurement:

1. The $PR(C/I)$ or $PR(W/U)$ is measured using a reference wanted signal C significantly above the minimum threshold value C_{min} (e.g. see (DTG Testing, 2014)). Hence the noise-like interfering signal I and $(N+I)$ are approximately the same, and what is being measured is $T(C/(N+I))$. This ratio is applicable for scenarios where interference dominates the total $(N+I)$ or **interference limited** scenarios
2. The $PR(C/N)$ is measured by identifying the ratio between the receiver noise N and minimum signal strength C_{min} that gives the required performance. Effectively what is being measured is again the $T(C/(N+I))$, in this case in the absence of interference, and hence is relevant for **noise limited** scenarios.

Therefore, in many cases the measured $PR(C/I) = PR(C/N) = T(C/(N+I))$ (e.g. see Tech 3348 (EBU, 2014)), as shown in Figure 5.43. Each ratio is valid within its own zone but not over the full range of distances from the transmitter.

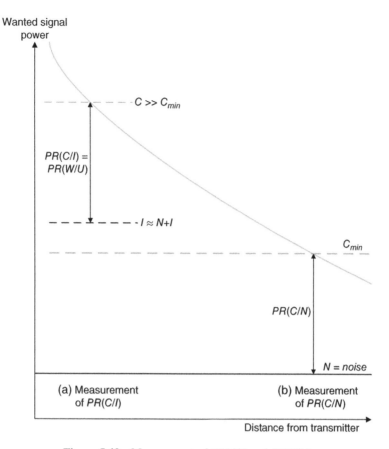

Figure 5.43 Measurement of $PR(C/I)$ and $PR(C/N)$

There are limitations in using $T(C/I)$, namely:

- Using $T(C/I)$ rather than $T(I/N)$ introduces additional complexity into the analysis as it is necessary to model the wanted system as well as the interferer
- It can be more complex to apportion interference
- Many systems use power control so that the received signal is constant across the service area
- A system can meet its C/I requirement but still be unable to provide a service if the $C/(N+I)$ is below the minimum required. For example, if the $C/I = +30$ dB, that is usually a good sign, but it could be because the $C/N = -10$ dB and the $I/N = -40$ dB and hence the link would actually fail. The C/I is not enough on its own to identify if the system is able to provide a service or not
- Often co-frequency analysis is undertaken at the boundary where C is at its minimum and hence there is no benefit in using C/I or W/U compared to use of I/N thresholds.

5.10.2 FDP

The analysis in Section 5.9 took account of the interference at two specific percentages of time, namely, short-term and long-term. However no account was made of the interference levels for intermediate percentages of time.

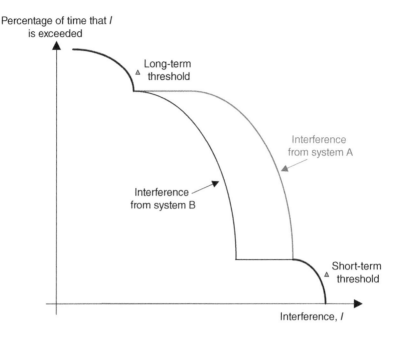

Figure 5.44   CDF plots of interference versus percentage of time for systems *A* and *B*

Consider the case in Figure 5.44 that shows the CDF of interference, *I*, against percentage of time that *I* is exceeded, as generated by two systems, A and B. Both systems create exactly the same interference level for the short-term and long-term thresholds but are very different for intermediate values. In particular it can be seen that system A would cause significantly greater interference than system B, but this information is not being captured when considering the short-term and long-term thresholds alone.

It can, therefore, be useful to use additional mechanisms to analyse the impact of interference, and one of these is the time-averaged I/N or FDP. This measure was developed for scenarios where the wanted system fading is linear in log(percentage of time). An example of this is the multi-path fade model in Rec. ITU-R P.530 (see Section 4.3.3) in which the fade depth A is linear to $10\log_{10}(p)$. In other words

$$a = \frac{k}{p} \tag{5.164}$$

where k is a constant and

$$a = 10^{A/10} \tag{5.165}$$

For systems like fixed links the received signal c in absolute can be calculated using

$$c = \frac{p_{tx} g_{tx} g_{rx}}{l_{fs} a} \tag{5.166}$$

The link will operate for a percentage of time p when the c/n is equal to the threshold $t(c/n)$, that is,

$$\frac{p_{tx}g_{tx}g_{rx}}{l_{fs}n}\frac{p}{k} = t\left(\frac{c}{n}\right) \qquad (5.167)$$

or when

$$p = qn \qquad (5.168)$$

where q is a constant taking in all the other terms.

The effect of interference is therefore to increase the unavailability (i.e. decrease the availability) by

$$\Delta p = q(n+i) - qn = qi \qquad (5.169)$$

As a fraction of the unavailability without interference, this is then

$$\frac{\Delta p}{p} = \frac{i}{n} \qquad (5.170)$$

If the interference varies in time, then this can be integrated so that the *FDP* is

$$FDP = \frac{i_{average}}{n} \qquad (5.171)$$

Note that the time averaging must be done in absolute not in dB and the measure is restricted to scenarios where the performance is limited by fading linear in $\log_{10}(p)$.

The *FDP* and its use are described further in the following documents:

- RR Appendix 5 Annex 1: Coordination thresholds for sharing between MSS (space-to-Earth) and terrestrial services in the same frequency bands and between non-GSO MSS feeder links (space-to-Earth) and terrestrial services in the same frequency bands and between RDSS (space-to-Earth) and terrestrial services in the same frequency bands (ITU, 2012a)
- Rec. ITU-R F.1108: Determination of the criteria to protect fixed service receivers from the emissions of space stations operating in non-geostationary orbits in shared frequency bands (ITU-R, 2005a)
- Rec. ITU-R M.1143: System specific methodology for coordination of non-geostationary space stations (space-to-Earth) operating in the mobile-satellite service with the fixed service (ITU-R, 2005c).

The *FDP* was used as a trigger for coordination in Rec. ITU-R M.1143, which relates to interference from the non-GSO MSS in bands shared with the FS as in Figure 5.45. For the bands identified, the trigger for coordination using the methodology in Rec. ITU-R M.1143 for digital FS links is given in Rec. M.1141 (ITU-R, 2005b) as

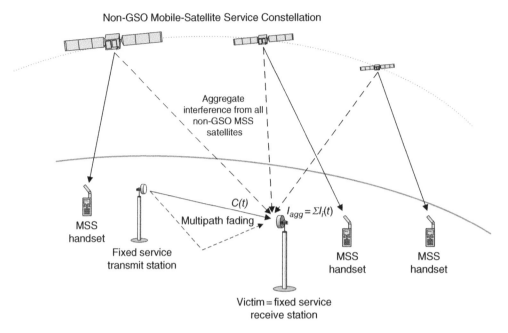

Figure 5.45 Sharing scenario with interferer non-GSO MSS into point-to-point FS link

$$T(FDP) = 25\% \tag{5.172}$$

This is based on an interference margin of 1 dB.

Example 5.43
A non-GSO MSS system is proposing to operate in bands shared with the FS, where the parameters of each system are given in Section 6.2.2. Long-term and short-term I/N thresholds were generated by assuming that the aggregate interference margin was shared between four systems equally and hence were as in Table 5.19 using the generic methodology in Table 5.18.

Analysis of the I/N from the system suggests that the short-term and long-term thresholds were exceeded as can be seen by the CDF in Figure 5.46.

However, the FDP was calculated as being $FDP = 14.9\%$ – less than the threshold for coordination.

The reason behind the apparent discrepancy – one set of thresholds saying there is an issue and the other saying there isn't – is that the FDP is an aggregate threshold, while the I/N thresholds shown here were single entry. If the I/N CDF was compared against the aggregate I/N thresholds, then they would be met. It could therefore be argued that the FDP should be apportioned, with only part of the total unavailability increase being permitted by one system.

Note how the shape of the I/N CDF curve for very short percentages of time is parabolic, being driven primarily by the main beam gain pattern of the victim FS receiver. In addition the curve is smooth for all percentages of time down to 0.001%, suggesting there have been

Table 5.19   FS short-term and long-term single entry I/N thresholds

Threshold	Long-term	Short-term
Percentage of time	20%	0.01%
I/N	−12 dB	+10 dB

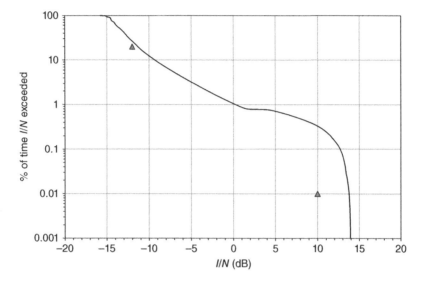

Figure 5.46   CDF of I/N for non-GSO MSS system into FS link

sufficient samples and the time step is small enough for accurate results, as discussed in Section 6.8.

5.10.3   C/(N + I) and BER

The previous section considered one approach to time-varying interference, which was to calculate the average I/N or FDP. Another approach is to fully model how the interference and wanted signals vary in time using the $C/(N+I)$ metric. This has the advantage that it is the $C/(N+I)$ that actually drives the QoS. In particular for digital systems there is a direct relationship between $C/(N+I)$ and the BER. Therefore, a QoS requirement for a threshold BER can be converted into a $C/(N+I)$ threshold via the E_b/N_0: for example, Equation 3.27 can be used for BPSK modulation.

Example 5.44

For the scenario in Example 5.43 of a non-GSO MSS system as interferer and FS as victim the C/N, C/I and $C/(N+I)$ CDFs were generated and are shown in Figure 5.47. It can be seen that the impact of the C/I is to shift the C/N curve left to create the $C/(N+I)$. The C/N, which is

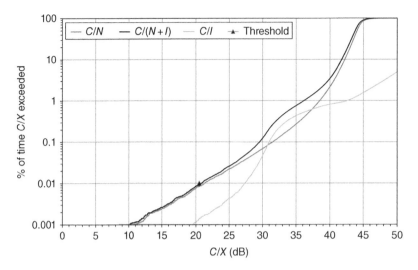

Figure 5.47   CDFs of *C/X* for non-GSO MSS into FS example

driven by fading, dominates the $C/(N+I)$ for long and short periods of time with the C/I dominating the intermediate zone. In this case the $C/(N+I)$ threshold is just met, with a probability of interference of $p = 0.00929\%$.

The methodology used is specified in Rec. ITU-R M.1319 (ITU-R, 2010a) and involves the calculation of the aggregate I_{agg} from all beams of all non-GSO MSS satellites using

$$I_{agg} = 10\log_{10}\left[\sum_{i=satellites}\sum_{j=beams} 10^{I_{ij}/10}\right] \tag{5.173}$$

$$I_{ij} = P'_{tx,ij} + G'_{tx,ij} - L'_{p,ij} + G'_{rx,ij} - L_f + A_{BW,ij} \tag{5.174}$$

In addition

$$C = P_{tx} + G_{tx} - L_{fs} - L_{530}(p\%) + G_{rx} - L_f \tag{5.175}$$

$$N = F_N + 10\log_{10}T_0 - 228.6 + 10\log_{10}B \tag{5.176}$$

Hence it is possible to calculate the $\{C/I_{agg}, C/N, C/(N+I_{agg})\}$ at each time step of a dynamic simulation (as described in Section 6.8) in order to calculate statistics including the CDFs.

Note, again, that the victim had an interference margin of 1 dB that was sufficient, but only just, to ensure it meets its 0.01% unavailability target. In practice there would be additional sources of interference, and this is one drawback of the $C/(N+I)$ approach: to be realistic it must include all sources of interference, which can lead to excessively large and complex simulations. $C/(N+I)$-based analysis can, however, be useful for system design studies in which external interference is well defined and the objective is to optimise a network's performance.

Another approach that can be used when combining interference and wanted distributions $I(t)$ and $C(t)$ is to look at apportionment of unavailability as in the following section.

The *BER* statistics generated from $C/(N+I)$ distributions can be used to generate other network-related statistics. For example, for a given *BER* and packet size it is possible to generate approximate *PER*. In a study for Ofcom on 'Evaluating spectrum percentage occupancy in licence-exempt allocations' (Aegis Systems Ltd and Transfinite Systems Ltd, 2004), the following methodology was used to estimate the *PER* from the *BER*:

$$PER \cong N_{packet}.8.BER \qquad (5.177)$$

where N_{packet} is the number of bytes in the packet and *BER* is assumed to be small. From a *PER* the required *BER* was derived and hence $T(C/(N+I))$. The threshold was then

$$\frac{C}{N+I} > T\left(\frac{C}{N+I}\right) \text{ for } Z\% \text{ of time and } W\% \text{ of locations} \qquad (5.178)$$

5.10.4   Unavailability

While interference analysis is based upon the powers of radio signals, in terms of providing a service, the important factors are the *BER* and the availability objectives. One approach to management of interference is therefore to apportion the unavailability allowance between sources. An example of this can be found in Rec. ITU-R F.1094 (ITU-R, 2007a), which apportions unavailability for fixed links as in Table 5.20.

The decrease in availability (i.e. increase in unavailability) is a generally useful measure to consider in interference analysis, and an example would be

Interference can only increase the unavailability by 10% over the unavailability for the system without interference

For a given distribution of time-varying wanted signal $C(t)$ and given N, there will be an infinite number of distributions of $I(t)$ that results in a given increase in unavailability of the full convolution of $\{C(t), N, I(t)\}$, so this approach is only used to check proposed $I(t)$ distributions such as *EPFD* limits described in Section 7.6.

With two distributions $C(t)$ and $I(t)$ that are independent and have unavailabilities that are small, the total unavailability U can be estimated as the total of the unavailability due to fading of the wanted signal plus that due to interference above the short-term threshold, that is,

$$U(total) \cong U\left(Fade > M_{fade}\right) + U(I > T_{st}(I)) \qquad (5.179)$$

Table 5.20   Unavailability apportionment in Rec. ITU-R F.1094

Percentage	Element	Includes
89	X	The fixed service portion (intra-service sharing), which includes degradations due to equipment imperfections
10	Y	Frequency sharing on a primary basis (inter-service sharing)
1	Z	All other sources of interference

The unavailability due to interference as a ratio of the total unavailability is then

$$\Delta U = \frac{U(I > T_{st}(I))}{U(total)} \tag{5.180}$$

This is supported by Rec. ITU-R S.1323 (ITU-R, 2002c), which constrains the increase in unavailability for satellite networks, recommending that interference should:

5.1 *be responsible for at most 10% of the time allowance for the BER (or C/N value) specified in the short-term performance objectives of the desired network and corresponding to the shortest percentage of time (lowest C/N value)*

The benefit of unavailability analysis is the strong connection to QoS, but at the cost of complexity in simulations and metrics. Unavailability is sometimes used as a method to derive I/N thresholds, which are more readily used in interference analysis.

Example 5.45
For the scenario in Example 5.44 the unavailability statistics are as in Table 5.21. It can be seen that the target that 10% of the unavailability is due to interference is met, though as noted earlier this is an aggregate value not single entry.

Table 5.21   Example increases in unavailability

Target total unavailability	0.01%
Unavailability allowance for interference	10%
Threshold unavailability increase due to interference	0.001%
Unavailability include fading	0.00837%
Unavailability from interference and fading	0.00929%
Increase in unavailability due to interference	0.00092%

5.10.5   Coverage, Range and Capacity

For radio systems providing a service over an area, interference can lead to a reduction in coverage, and this can be the basis of a threshold. Consider the scenario of interference into an IMT WCDMA deployment as in Figure 5.48. The impact of interference is to raise the noise at the IMT WCDMA BS.

While urban cells are noise limited, rural cells are coverage limited and hence will be vulnerable to increases in the $N + I$ that lead to a consequential reduction in range. The result of a reduction in range is to increase the number of BS required to meet coverage targets. Another way to assess interference could therefore be to specify what would be an acceptable increase in the number of BS.

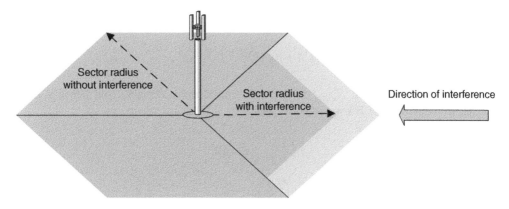

Figure 5.48   Reduction in coverage due to interference

Example 5.46

In Rec. ITU-R M.1654 (ITU-R, 2003c) a requirement that interference from broadcasting-satellite service (BSS) (sound) systems would not lead to an increase in the number of base stations of more than 10% was specified and then converted into a $T(I/N)$. The starting point for this conversion was the observation that interference decreases link margin according to

$$\Delta L = 10\log_{10}\left[1 + 10^{(I/N_{tot})/10}\right] \tag{5.181}$$

The noise for the uplink is a combination of the system noise and thermal noise N. The noise increase on a WCDMA uplink can be calculated using the approach in Section 3.5.5 as N_R; hence

$$\frac{I}{N_{total}} = \frac{I}{N}\frac{N}{N_{total}} = \frac{I}{N}\frac{1}{N_R} \tag{5.182}$$

So

$$\Delta L = 10\log_{10}\left[1 + 10^{\left(\frac{I}{N}-N_R\right)/10}\right] \tag{5.183}$$

This reduction in link margin will result in a corresponding decrease in range. The propagation model used to calculate range is assumed to be Hata/COST 231, which has as its range element the following term from Equation 4.19:

$$L_{50} \sim [44.9 - 6.55\log_{10}(h_{tx})]\log_{10}(d_{km}) \tag{5.184}$$

For a transmit BS height of 30 m (appropriate for rural areas), this simplifies to

$$L_{50} \sim 35.22\log_{10}(d_{km}) \sim 10\log_{10}\left(d_{km}^{3.522}\right) \tag{5.185}$$

The change in area is driven by the distance squared, so

$$\Delta A = \left[1 + 10^{\left(\frac{I}{N}-N_R\right)/10}\right]^{-2/3.522} \tag{5.186}$$

For rural cells a value of $N_R = 1$ dB could be considered acceptable, and hence a 10% increase in the number of BS would correspond to

$$T\left(\frac{I}{N}\right) = -6.4\,\text{dB} \tag{5.187}$$

$$T\left(\frac{DT}{T}\right) = 23\% \tag{5.188}$$

This approach is interesting as it links interference to its impact in terms of cost, in this case additional equipment and deployment expenditure. Indeed cost could be considered an interference metric in its own right, a converse to the value of spectrum in a market-based approach to spectrum management.

Urban cell uplinks are limited by the noise rise. The impact of interference is therefore to decrease the capacity of the cell. For Example 3.10 the impact of a noise rise of 1 dB would lead to a reduction in capacity of about five voice calls.

It should be noted that in many other cases the threshold used for inter-service interference into IMT is an $I/N = -10$ dB.

A similar methodology was used to derive an interference threshold for RLANs in the 5 GHz band as described in Rec. ITU-R M.1739 (ITU-R, 2006f), which converts a reduction in range of 5% to a $T(I/N) = -6$ dB. Furthermore, Section 7.7 describes how a 6% reduction in radar range can be converted to a $T(I/N) = -6$ dB using the approach in Rec. ITU-R M.1644 (ITU-R, 2003b).

A service will always operate better without interference, and the impact of interference is a reduction in margin, capacity or range and is an expected result of sharing between services. The key question is what degree of degradation is acceptable, which is influenced by other factors including economics and policy objectives. It can be appropriate for regulators to assume the 1 dB margin for aggregate interference as an even-handed common baseline between services.

Another example of the use of coverage as an interference metric for broadcasting can be found in Report ITU-R BT.2265 (ITU-R, 2012i). In Annex 2 there is a methodology that describes:

how the degradation to the DTTB reception location probability, (ΔRLP) can be determined when the interfering stations (single or multiple) of other services/applications are implemented ('after') compared to the DTTB reception location probability when the interfering stations of other services/applications are not implemented ('before').

This methodology uses Monte Carlo methodology and takes account of the location variability term in the propagation model. The net result is an assessment of the decrease in coverage of a broadcast network due to interference. The decrease in coverage could also be represented as a degradation to the reception location probability.

While this can be a useful tool for assessing interference, the difficulty is in deciding what would be an appropriate threshold. Any planning algorithm should acknowledge that the system will not operate in an interference-free environment. The questions are then, what would be an acceptable decrease in coverage, and how should it be included in the planning process?

Coverage analysis can be extended to model the percentage of population covered. This requires access to a database of population per location, for example, within a pixel. This could be defined in a number of ways including resident population or daytime population, depending upon requirement and service. The impact of interference could then be assessed in terms of reduction in population covered by a specific service. As these databases are location specific, this approach is less applicable for generic scenarios, where use of an average density gives a direct relationship between area covered and population.

For systems using adaptive modulation, a useful metric is the total traffic carried, calculated by integrating the achievable data rate over a reference time period. The achievable data rate can be derived from the C/(N+I), so the impact of interference would be to reduce this capacity.

5.10.6   Observation Duration and Locations

Thresholds for the radio astronomy service (RAS) are given in Rec. ITU-R RA.769 (ITU-R, 2003d). These are in the form of interference at the receiver or *PFD* for various bands and each of the following observation types:

- Continuum
- Spectral line
- Very long baseline interferometry (VLBI).

What is different from most other types of threshold is that they are defined for an integration time of 2000 seconds. This leads to the requirement to average the (say) gain or power over that period.

Thresholds for remote sensing applications can have constraints on measurement locations. For example, in Rec. ITU-R RS.2017 (ITU-R, 2012h) there are constraints on availability such as 99.99% data availability of measurements on the Earth surface within a square with area 2,000,000 km^2.

Example 5.47
In November 2009, the European Space Agency (ESA) launched the Soil Moisture and Ocean Salinity (SMOS) satellite. This carried a single payload, the Microwave Imaging Radiometer with Aperture Synthesis (MIRAS), operating within the Earth exploration-satellite service (EESS) passive band at 1 400–1 427 MHz. This mission has experienced harmful interference from services operating in adjacent bands, such as wireless camera monitoring systems, TV radiolinks and radars.

5.10.7   Radar and Aeronautical Thresholds

Radar systems can be used for critical safety of life applications, such as air traffic control. Operationally radar systems use metrics such as probability of detection and probability of false alarm (Skolnik, 2001). However, the calculations involved in these measures are, in general, too detailed for most interference analysis. In a similar way that models of mobile networks avoid the details and complexities involved in protocols, radar thresholds are typically based

upon *I/N*. For example, Rec. ITU-R M.1460 (ITU-R, 2006e) gives the threshold for radiodetermination radars in the 2 900–3 100 MHz band as

$$T\left(\frac{I}{N}\right) = -6\,\text{dB} \tag{5.189}$$

The Recommendation goes on to comment:

*This represents the tolerable aggregate effect of multiple interferers; the tolerable I/N
ratio for an individual interferer depends on the number of interferers and their geometry,
and needs to be assessed in the analysis of a given scenario.*

This is therefore the same approach as in Section 5.9. However, it should be noted that this
threshold should be met in each azimuth direction and not averaged over the total operating
zone. Some aeronautical applications (e.g. communication services for air traffic control) could
require protection of a volume, such as a cylinder defined by central point, radius and altitude.
    More information about modelling radar systems including a derivation of the radar equation
is given in Section 7.7.

5.10.8   Channel Sharing Ratio

The methodology for frequency allocation of PMR systems described in Section 7.2 uses metrics based on two factors, namely, the interference level and degree of channel sharing. The
second term is important for the case where a single radio channel is used by multiple systems
at the same location as in Figure 5.49.

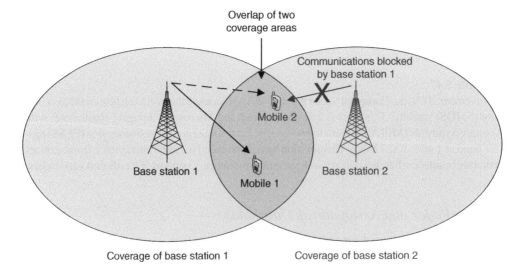

Figure 5.49   Overlap between PMR systems

In this scenario both PMR systems are using the same frequency or channel and have coverage areas that overlap. In the overlap area there can be blocking as communication from either the BS or the mobile of system 1 can prevent access to the radio channel by system 2. The larger the degree of overlap, the greater the likelihood of the channel being blocked.

There could be multiple overlapping areas, and the resulting figure of merit identifies the degree of sharing of the channel averaged over the coverage area of each system. The degree of sharing that is permitted will depend upon the level of traffic or *AF* and the target grade of service identified in the system's licence. The probability of channel blocking can be extended to identify the likelihood of a call being completed successfully.

More information about the calculation of the **channel sharing ratio** is given in Section 7.2.

5.10.9   *Field Strength,* PFD *and* EPFD

Interference analysis is about the power of radio signals, which depends upon the strength of the electric field. The strength of the wanted and interfering signals are often specified by a *PFD*, with conversions between them and the signal received at an isotropic receiver given in Section 3.11.

One of the key strengths of these metrics is the ability to measure them independently of receiver characteristics. This is particularly useful when defining power levels that can trigger regulatory actions if they are met or are exceeded. If an organisation, operator or regulator, has concerns that agreed levels are being breached, they can undertake measurements to check the actual powers.

For this reason the field strength and *PFD* levels are often specified in regulatory texts and also for some sharing scenarios. An example would be Article 21 of the RR, which contains *PFD* levels that satellite systems must meet to avoid harmful interference into terrestrial services, stating:

> 21.16   § 6 1) *The power flux-density at the Earth's surface produced by emissions from a space station, including emissions from a reflecting satellite, for all conditions and for all methods of modulation, shall not exceed the limit given in Table 21-4.*

A satellite system that does not meet these *PFD* limits will not receive a favourable finding by the ITU BR for their filing.

PFD limits are also defined within the table of allocations via numerous footnotes. One such is in RR 5.430A (ITU, 2012a), which gives a *PFD* limit of -154.5 dBW/m^2/4 kHz for 20% of time measured 3 m above terrain. This is used as a regulatory tool to allow entry of IMT mobile services in selected countries in bands currently used by receive satellite ESs. The satellite services in one country will be protected from potential interference due to the deployment of IMT in an adjacent country as long as the *PFD* limit is met. The derivation of this *PFD* limit is given in Example 5.48.

Another example of the use of a *PFD* or field strength on boundary is in a bilateral agreement and MoU between two countries. Typically this is a trigger for coordination: no action is required if the threshold is met, and it could be exceeded with the agreement of the other party. For example, the MoU between the United Kingdom and France for the band 46–68 MHz

(Ofcom and ANFR, 2004) includes constraints on the field strength along the UK coast from French transmitters so that:

The maximum interfering field strength measured in a 12.5 kHz bandwidth for 50% locations and 10% of the time at a height of 10 m above ground shall be:
* *30 dBµV/m for horizontally polarised broadcasting emissions*
* *12 dBµV/m for vertically polarised broadcasting emissions*

Constraints on field strength or *PFD* can be either single entry or an aggregate from multiple transmitters. They can be the same level between the two parties, or they can agree to have frequencies at which priority is given to one or other of the administrations. This **preferential frequencies** approach can take the form of:

1. A higher PFD or field strength along the border line
2. Permission to exceed the PFD along the border with the trigger at a distance = d from the boundary inside the adjacent country.

Figure 5.50 shows an example of using the secondary line for preferential frequencies in the case of checking coordination triggers for PMR systems. This approach is a widely used method included in bilaterals and the HCM Agreement (HCM Administrations, 2013).

Field strength can also be used during service planning including interference analysis. For example, field strengths are used for:

* Broadcasting, as in Recs. ITU-R BT.1368 (ITU-R, 2014d) and BT.2036 (ITU-R, 2013b)
* Private land mobile, as in Ofcom's TFAC OFW 164 (Ofcom, 2008b).

Rec. ITU-R BT.1368 includes equations to calculate the field strength and convert to a receive signal level (see Appendix 1 to Annex 2).

There are, however, three constraints on the use of field strengths compared to signal strength (C or I) at the receiver:

1. Frequency-dependent values. For example, Ofcom's OFW 164 contains over 20 different field strength values for blocking, while only one is required for the equivalent signal level (−116 dBm)
2. Lack of directivity: field strengths generally do not include the receive antenna gain. In many cases this can be a significant factor, with antennas giving 50 dB+ discrimination. Even for broadcast fixed receivers there can be 16 dB of discrimination
3. Complexity of addition of interfering signals received from different directions where each path could have a different receive gain. In practice what is required is to convert field strengths to signal at the receiver, add and then convert back to a field strength.

For these reasons it is generally better to undertake interference analysis using signal strength rather than field strength or *PFD*. As will be seen in Example 5.48 it is possible to convert between the two by making assumptions about antenna gains.

One approach to define a regulatory tool that includes antenna gain and the ability to add multiple interfering signals is the **EPFD**, which is defined in Article 22 of the RR (ITU, 2012a) as

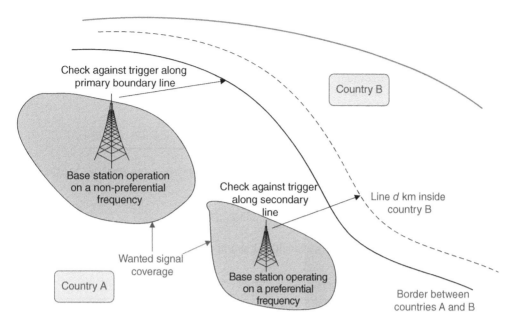

Figure 5.50 Secondary boundary for preferential frequency checks

$$epfd = 10\log_{10}\left[\sum_{i=1}^{N_a}10^{P_i/10}\frac{g_{tx}(\theta_i)\,g_{rx}(\varphi_i)}{4\pi d_i^2}\frac{1}{g_{rx,max}}\right] \tag{5.190}$$

This is effectively the *PFD* multiplied by the receive relative gain and is used as a regulatory hard limit to protect GSO satellite systems from harmful interference from non-GSO FSS systems and is described further in Section 7.6. This metric is very similar to interference at a receiver, with a fixed offset depending on antenna peak gain and frequency.

A similar metric is used in Rec. ITU-R M.1583-1 (ITU-R, 2007b), which addresses non-co-frequency scenarios involving non-GSO constellations and radio astronomy sites. In this case it can be useful to include an additional term to take account of the transmit and spectrum masks, such as A_{MI}.

Example 5.48
During WRC-07 footnote 5.430A was added to the Radio Regulations (see Table 2.1) giving a *PFD* on boundary threshold = −154.5 dBW/m²/4 kHz at a height of 3 m above ground. IMT networks in one country would have to meet this limit on the border of adjacent countries to protect receive satellite Earth stations. This figure was derived using the parameters in Table 5.22 and the following equation:

$$T(PFD,20\%) = T\left(\frac{I}{N},20\%\right) + 10\log_{10}(kT_{rx}B) - G(10°) - A_{e,i} = -154.5\ \text{dBW/m}^2/4\,\text{kHz}$$

$$\tag{5.191}$$

Table 5.22 Parameters used to derive PFD limit in 5.430A

Frequency	f_{MHz}	3 400 MHz
RX temperature	T_{rx}	100 K
Reference bandwidth	B	4 kHz
Percentage of time	p	20%
I/N threshold for $p\%$ of time	$T(I/N, p)$	−10 dB
Elevation angle (deg)	θ	10°
Gain pattern	$G(\theta)$	Rec. ITU-R S.580

The last term comes from Equation 3.227 and the Rec. ITU-R S.580 (ITU-R, 2003e) gain pattern can be modelled as

$$G(\theta) = 29 - 25\log_{10}(\theta) \qquad (5.192)$$

At WRC-15 this PFD threshold was included in a number of additional footnotes in bands from 3 400 to 3 700 MHz.

Human Exposure Limits

A special case of the *PFD* limits is to protect humans from possible harmful effects of electromagnetic radiation. **Human exposure limits** were developed by the International Commission on Non-Ionising Radiation Protection (ICNIRP) for two categories:

- Members of the general public
- Specialists who are exposed due to their occupation and hence will be trained to take precautions and be aware any potential risk.
  The *PFD* limits in W/m^2 for bands between 10 MHz and 300 GHz are shown in Table 5.23.

Table 5.23 Human exposure limits between 10 MHz and 300 GHz in W/m^2

Frequency range	General public exposure	Occupational exposure
$10\,\text{MHz} \le f_{MHz} \le 400\,\text{MHz}$	2	10
$400\,\text{MHz} \le f_{MHz} \le 2\,\text{GHz}$	$f_{MHz}/200$	$f_{MHz}/40$
$2\,\text{GHz} \le f_{MHz} \le 300\,\text{GHz}$	10	50

These are assumed to be averaged over a 6 minute period and levels 1000 times higher (i.e. +30 dB) are permitted for short periods of time. More information is available in *ICNIRP Guidelines for Limiting Exposure to Time-Varying Electric, Magnetic and Electromagnetic Fields (Up to 300 GHz)* (International Commission on Non-Ionizing Radiation Protection (ICNIRP), 1998).

5.10.10 Margin over Threshold

The difference between a link attribute (whichever metric is used) and its threshold is called the **margin over threshold**. Margin over threshold is positive when the system is performing better than the threshold level and negative when it is worse. While for some metrics performance is better when numerically larger (such as C/N), there are other metrics that are better when smaller (such as I/N), and so the margin over threshold equation takes one of two forms.

Examples of the calculation of the margin over threshold for cases where the metric is better when higher include

$$M(C) = C - T(C) \tag{5.193}$$

$$M\left(\frac{C}{N}\right) = \frac{C}{N} - T\left(\frac{C}{N}\right) \tag{5.194}$$

$$M\left(\frac{C}{N+I}\right) = \frac{C}{N+I} - T\left(\frac{C}{N+I}\right) \tag{5.195}$$

$$M\left(\frac{C}{I}\right) = \frac{C}{I} - T\left(\frac{C}{I}\right) \tag{5.196}$$

However the I and I/N metrics are better when lower so

$$M(I) = T(I) - I \tag{5.197}$$

$$M\left(\frac{I}{N}\right) = T\left(\frac{I}{N}\right) - \frac{I}{N} \tag{5.198}$$

For the *PFD* case it depends upon whether the value being measured is wanted or interfering; hence

$$M(PFD_W) = PFD_W - T(PFD_W) \tag{5.199}$$

$$M(PFD_I) = T(PFD_I) - PFD_I \tag{5.200}$$

The general criterion that operation is feasible, and in particular for scenarios including interference that two systems or services are compatible, is that for the required metric X there is **positive margin** over the threshold, namely:

$$M(X) \geq 0 \tag{5.201}$$

If the margin over threshold is negative, then it will be necessary to consider interference mitigation techniques.

But should the test be '>' or '≥'? Generally, interference thresholds represent the level that should not be exceeded. Hence, the value itself is generally acceptable – though numerically incredibly unlikely to occur.

For example, in Article 22 of the RR there are various *PFD* and *EPFD* limits that are defined in the form 'shall not exceed the limits given'. Therefore for non-GSO FSS systems in the band 10.7–11.7 GHz with an $EPFD = -160\,\text{dBW/m}^2$ (exactly) in 40 kHz for less than 0.003% of the time would be acceptable. Similarly the coordination threshold for GSO satellites in Appendix 5 is if the DT/T 'exceeds' 6%. Hence if that value exactly is calculated, then coordination would not be required. Rec. ITU-R SF.1006 (ITU-R, 1993) gives interference thresholds that are the level of 'maximum permissible interference', which also implies that the exact value would be acceptable.

For many scenarios there is a time dimension to the interference threshold, and this leads to another margin to be considered, namely, in time. Consider Figure 5.51, which shows an example CDF where the metric is I/N, with threshold and interference level of $T(I/N)$ with associated percentage of time $p_T(I/N)$.

The two margins over thresholds are

$$M\left(\frac{I}{N}\right) = T\left(\frac{I}{N}\right) - \frac{I}{N}\Big|_{p = p_T(I/N)} \tag{5.202}$$

$$M_p\left(\frac{I}{N}\right) = p_T\left(\frac{I}{N}\right) - p\left[\frac{I}{N} \leq T\left(\frac{I}{N}\right)\right] \tag{5.203}$$

In either case the test is that the margin over threshold not be negative, namely:

$$M(X) \geq 0 \tag{5.204}$$

$$M_P(X) \geq 0 \tag{5.205}$$

Note that if one test is true, then the other must be true as well.

It is often simpler to understand the interference margin over threshold as that clearly shows the change in signal(s) (wanted or interfering) required. However some analysis, in particular Monte Carlo methods described in Section 6.9, has as their output percentages of time or probabilities. In these cases it is necessary to use the margin over threshold in terms of time instead, while the interference threshold is used within the methodology to generate the output statistics.

5.11   Interference Mitigation

When analysis suggests that the interference levels a system generates would be above the agreed threshold, then there are two alternatives: either stop transmitting or investigate ways to mitigate the interference. The mitigation techniques available will depend on the scenario involved, in particular whether the analysis is of a generic scenario or of specific locations and stations.

There is no single approach that can be used in all situations: rather mitigation represents a toolkit of options to be investigated to identify the benefit it brings (in terms of interference reduction) and cost (reduction in wanted service, coverage, financial expenditure, etc.). A good starting point is to consider the fields in the wanted and interfering link budgets and how each could be changed to mitigate interference:

$$C = P_{tx} + G_{tx} - L_p + G_{rx} - L_f + A_{MI} \tag{5.206}$$

$$I = P'_{tx} + G'_{tx} - L'_p + G'_{rx} - L_f + A_{MI} - L_{pol} \tag{5.207}$$

$$N = 10\log_{10}(kTB) \tag{5.208}$$

Sometimes the interference threshold is a trigger for further analysis – in particular relating to the coordination process. Some of the mitigations described here can be used to reduce

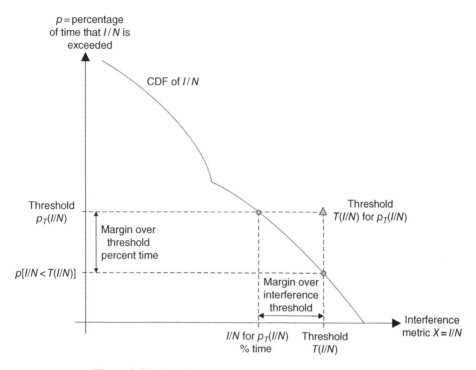

Figure 5.51 Margin over thresholds in interference and time

coordination requirements or within the coordination process to achieve a mutually satisfactory agreement.

5.11.1 Transmit Power and Bandwidth

The simplest way to reduce interference is to decrease the transmit power of the interferer, but this clearly has an impact on the service that can be provided. The result is one or more of:

- A reduction in coverage or range, as the $C/(N+I)$ will be decreased for a given distance
- A reduction in capacity, as the modulation must be reduced or coding increased as the target $C/(N+I)$ is reduced
- A reduction in the *BER* achievable
- A reduction in availability, as the fade margin will protect the system for a lower percentage of time.

One method to reduce the target $C/(N+I)$ while keeping the range and modulation unchanged is to spread the signal over a wider bandwidth as described in Section 3.5.5. However this is effectively reducing the overall capacity and leads to aggregation effects, particularly as there will be more cases of frequency overlap with other systems.

If there is only frequency overlap for part of the interfering system's total bandwidth, then it could be feasible to compensate for lower power in the overlap region by increasing the power outside it.

If APC is not being employed, then its use could help by reducing the average power level. There will still be peaks of power, but if they occur at times when the interference is reduced, it could be acceptable. For example, a satellite ES could increase its power to compensate for rain at its location fading the uplink, but this fading is likely to also reduce the interference into adjacent satellites, as described in Section 4.4.2.

It can be helpful if there are multiple bands available and multiple types of transmitters. For example, a LTE network operator is likely to have access to multiple bands such as 800, 1 800, 2 600 MHz, etc. and also require multiple types of transmitters, ranging from:

- High-power macro BS covering a large area in rural environments
- Low-power small cell operating below the clutter level in dense urban environments.

The operator has the ability to match transmitters to an appropriate band, taking into account propagation conditions and also the interference environment. Bands shared with other services would be more appropriately used for the low-power urban small cell than a high-power rural BS.

Some systems provide a range of services with differing powers and bandwidths. For example, a FSS GSO satellite could support both direct-to-home TV and very small aperture terminals (VSATs). One approach to facilitate coordination would be to design a frequency plan that uses different ends of the available spectrum for each service. It is generally easier to share like with like, so if possible it would be better to:

- Coordinate TV services with an adjacent satellite also providing TV services
- Coordinate VSAT services with an adjacent satellite also providing VSAT services.

5.11.2   Antenna Gain Patterns

One of the best ways to reduce interference is to modify the antenna gain pattern. In particular, using a larger antenna with higher gain benefits as:

- Lower transmit power is required to close the link
- Interference into other systems is limited to a smaller range of geometries
- Interference from other systems is limited to a smaller range of geometries.

However, there are clear costs. In particular, larger antennas are more expensive and take up more space, and in some scenarios it is not possible to increase the antenna size. For example, mobile handsets cannot use highly directional parabolic dishes due to physical constraints.

A number of alternative to simply increasing the antenna size can be considered:

- For the devices that are too small to use parabolic dishes to provide directivity, there can be benefits in use of multiple small antennas or phased arrays to create electronically steerable antennas
- Additional antennas can also be used to adapt to multiple channels between the transmit and receive stations using multiple-input/multiple-output (MIMO) techniques
- Adaptive antennas could create nulls in the direction of the interference
- Multiple antennas can also be used to mitigate fading. For example, the FS can use diversity to reduce multi-path fading as in Rec. ITU-R F.1108 (ITU-R, 2005a)
- Multiple antennas can also mitigate interference. If the wanted signals are combined, then there will on average be a reduction in interference due to the difference in phase between the two paths. Rec. ITU-R F.1108 suggests that for phase difference ϕ (which could be modelled as random) the wanted and interfering signals are transformed in absolute as follows:

$$c \to 2c \tag{5.209}$$

$$i \to 2i\cos^2\left(\frac{\phi}{2}\right) \tag{5.210}$$

- Another approach is to reduce the side-lobes for the critical region. As noted in Section 3.7.5, antenna designers have some flexibility in its design, choosing to optimise its beam or aperture efficiency. There can be the ability (say) to ensure the far sidelobes are as low as possible or that the first sidelobe level is reduced. An example for GSO satellite systems is to use ES antennas that meet Rec. ITU-R S.580 (ITU-R, 2003e) rather than S.465 (ITU-R, 2010b) as this will give 3 dB better off-axis performance
- For specific geometries such as sharing between GSO satellites, it can be beneficial to squeeze the gain pattern in the direction along the GSO arc at the expense of the direction perpendicular to it, as discussed in Section 3.7.6.

The power and antenna gain patterns are often combined via constraints on the off-axis EIRP density. For example, satellite ESs must meet the off-axis EIRP constraints in Rec. ITU-R S.524 (ITU-R, 2006g) such as those in Table 5.24 and in addition often agree satellite and ES off-axis EIRP constraints as part of coordination, as discussed in Section 7.5.4.

Table 5.24 Off-axis EIRP constraints in Rec. ITU-R S.524 for the 27.5–30 GHz band

Angle off-axis within $3°$ of the GSO arc	Maximum EIRP
$2° \leq \varphi \leq 7°$	$19 - 25\log_{10}(\varphi)$ dBW/40 kHz
$7° < \varphi \leq 9.2°$	-2 dBW/40 kHz
$9.2° < \varphi \leq 48°$	$22 - 25\log_{10}(\varphi)$ dBW/40 kHz
$48° < \varphi \leq 180°$	-10 dBW/40 kHz

5.11.3   Antenna Pointing

One effective mechanism to facilitate sharing is to restrict the pointing angles of antennas. The greater the isolation between interfering transmit antenna and victim receive antenna, the lower the interference.

An example of this is shown in Figure 5.52 where a satellite ES is sharing spectrum with BS of an IMT LTE network. One method to facilitate sharing is to only use sectors that point away from the ES. So in this figure the two BS nearest the ES have the antenna pointing at the ES removed. The BS further away has a degree of downtilt (typically 3°–5°): this lowers the EIRP towards the horizon, reducing the potential for interference.

Antenna pointing and gain patterns can be used together as part of interference mitigation. For example, a BS could replace a fixed sectorial antenna with a phased array that creates spot beams that point directly at UEs.

Another example was used by the SkyBridge non-GSO FSS system proposed in the 1990s. The design ensured that it would only use those beams that did not align with GSO ESs pointing at the GSO arc, as in Figure 5.53. This involves use of the α angle and EPFD metric as described in Section 7.6.

In many scenarios involving terrestrial services, one key parameter is the gain towards the horizon. If this can be reduced by including downtilt (e.g. at the base station), it can reduce interference into other systems.

It should be noted that these mitigation methods have costs. For example,

- The mobile network in Figure 5.52 will need to deploy additional base stations plus there is likely to be an exclusion zone around the ES
- The non-GSO network in Figure 5.53 will require additional satellites to provide continuous coverage.

However, these are the types of factors that can be analysed during a network's design phase, so that costs can be reduced and service levels optimised. In some scenarios, such as GSO satellite coordination, the key driver of service is often what can be coordinated with other GSO networks rather than what satellite technology and financing permits. Interference analysis can become a critical factor – if not *the* critical factor – in business planning, determining whether multibillion-dollar enterprises are successful or not.

5.11.4   Locations, Zones and Separation Distance

An effective mechanism to reduce interference is to introduce a degree of geographic separation between interferer and victim. This is a generic sharing methodology, applicable to both co-frequency and non-co-frequency sharing. The disadvantage of this approach is it reduces the area over which a service can be provided.

An example is shown in Figure 5.52, where there is a minimum distance between the satellite ES and the nearest active co-frequency base station. This also shows an example of how other mitigation methods are often used to reduce the separation distance required between victim and interferer, in this case antenna de-pointing.

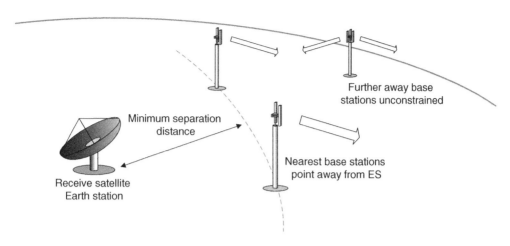

Figure 5.52   Restricting BS antenna pointing to protect receive ES

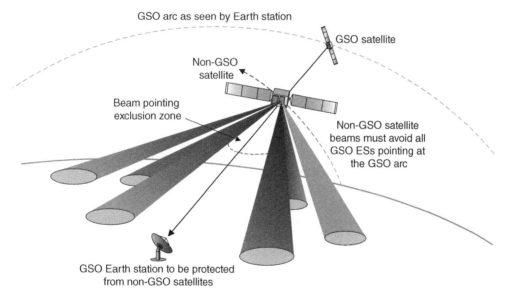

Figure 5.53   Non-GSO satellite avoiding pointing its beams when in line between any GSO ES and GSO arc

Similar separation distances can be required for satellite coordination: if GSO satellites are very close on the geostationary arc, it can be difficult for both to operate at the same geographic location without there being harmful interference.

Note there are at least three different types of zones that can be used to create separation distances:

1. **Exclusion zone**: no transmitters of the interfering system can be deployed in this area
2. **Coordination zone**: transmitters of the interfering system might be able to be deployed in this area subject to passing a coordination process

3. **Interference zone**: locations where a transmitter of the interfering system would cause harmful interference.

A mitigation methodology, therefore, would be to make the interference zone either an exclusion zone or coordination zone as appropriate. This would depend upon factors such as:

1. Availability of mitigation factors: interference from fixed links can vary significantly depending upon pointing angles, and this can be taken into account during coordination
2. Control of deployment: licence-exempt devices are often hard to control, making options beyond simply excluding their operation infeasible
3. Cost of coordination: if there are many potential systems to consider, then the cost of coordinating each one can become excessive.

5.11.5   Deployment Likelihood

Another dimension of probability is deployment likelihood. A worst-case analysis could be undertaken assuming there is in-line, or close to in-line, geometry. For example, an FS antenna could be pointing directly at the interfering/victim satellite ES. If the deployment of the fixed link and satellite ES is independent, then the likelihood of direct alignment is low.

This can be used in a number of ways:

- Use real location and pointing information to identify which specific systems would be affected
- Analyse actual link data to identify the distribution of azimuth and elevation angles and use this to calculate the likelihood of interference.

It could be that the likelihood of interference due to deployment is very low and hence could be acceptable. This can become a policy decision as the victim system could claim (with some justification) that such a change to the regulatory regime would lead to a degradation of the existing interference environment. The policy decision is then whether the benefits of the entrant outweigh the costs of increased interference to incumbents.

For example, introducing Earth stations on mobile platforms (ESOMPs) in parts of the 28 GHz band could cause harmful interference into the terrestrial FS, in particular point-to-point links aligned with a major air traffic route. While it was accepted that there could be interference, it was argued (successfully) in European discussions that this would only occur for a small number of links.

5.11.6   Noise, Feed Loss and Interference Margin

There can be interference mitigation benefits in either increasing *or* decreasing the receiver noise. For example:

- A higher receiver temperature means that for a given interference level I, the I/N is lower. The cost, however, is that the transmit power must be increased, leading to greater interference into other systems and might not be feasible for power limited systems

- A lower receiver temperature means that a lower transmit power is needed to close the link, leading to less interference into other systems. The downside, however, is greater susceptibility to interference, as for a given I the I/N will be higher.

A related change would be to increase the margin for interference, for example, from 1 to 2 dB. However, as noted earlier, this would require a consequential increase in the transmit power, which could lead to interference issues. Similarly increasing the feed loss will decrease the interference at the receiver but at a cost of requiring additional power to close the link.

It is both good engineering practice and an obligation from the RR that radio systems should use the least power necessary to provide the service and in consequence the lowest noise, feed loss and interference margin that would be appropriate.

5.11.7   Receiver Processing

In some cases processing at the receiver can assist in interference mitigation. For example, if the interferer is a radar system, it could be feasible for a radio system to use the gap between pulses (plus between sweeps) to communicate or listen to another signal such as radiolocation. This requires there be a good understanding of the sharing environment at timescales significantly less than 1 second and for the victim to be able to manage with short gaps in availability (possibly leading to a decrease in capacity, performance or accuracy). As the radar system will itself be vulnerable to interference, it would be difficult to use this technique to share with a high power service. Modelling of the impact of pulsed radio signals is non-trivial and often requires measurements to verify actual performance rather than rely solely on simulation.

The receiver can experience a range of additional harmful effects including saturation if the interfering signal received is significantly greater than expected levels.

5.11.8   Time and Traffic

Many radio systems only operate for specified parts of the day and are less busy at (say) 3 in the morning. This means that there is more likely to be spectrum available at that time ,which could be used without causing harmful interference. The downside is that most end customers for radiocommunication systems want services at their busy hour, not at 3 am.

While popular with spectrum managers looking to increase spectrum efficiency, time sharing is limited as a mitigation technique in its own right. An example of where it could be useful is in heavily congested PMR bands where a regulator could allow a channel to be shared between:

- A taxi despatch company, typically quiet in the early morning
- An overnight stock update system, which can be programmed to use this quiet time.

A more typical approach to use of time in spectrum management is to identify how the interference varies in time and compare it against short- and long-term thresholds. It might be that interference levels can at the worst case be very high, but if those levels are only reached for very small percentages of time, then that could be acceptable. However, extremely high levels of interference can lead to serious degradation at the receiver, from loss of synchronisation to more permanent failings, and so there would have to be a cap even for very short periods of time.

Another aspect to the time domain is for shared channels as used by Wi-Fi and some PMR services. In this case spectrum efficiency can be increased by reducing the time taken for a transmission. Hence if Wi-Fi uses the highest modulation feasible, then the time required to transmit a given amount of data decreases, leading to more time being available to other users. For PMR systems, increased sharing of a channel can be achieved if the users have a low AF, that is, occasional voice traffic rather than continuous polling of a fleet of vehicle's GPS position (see Section 7.2).

Traffic levels can be used to reduce the worst effects of aggregation. Rather than assuming all transmitters are at their maximum power, it can be helpful to consider the distribution of user traffic and identify actual power levels, which for some services (e.g. mobile) can be significantly less than maximum.

5.11.9   Polarisation

One way to reduce interference is to operate with the opposite polarisation to the victim. This could be using the same type of polarisation, that is,

- Victim operates LH, so the interferer can use LV (or vice versa)
- Victim operates LHC, so the interferer can use RHC (or vice versa).

Alternatively, it could be to use a different type of polarisation, so that:

- Victim operates linear, so interferer can use circular (or vice versa).

As noted in Section 5.4, there can be significant de-polarisation effects that can make it problematic to use polarisation as a mitigation technique. It is also the case that many systems want to use both polarisations to maximise their own capacity: forcing the wanted or interfering system to only use one polarisation is a constraint that is effectively sharing capacity between them.

However, it can be a significant factor in some scenarios, particularly where interference is near main beam, such as:

- FS, as described in Example 5.19
- FSS, where the interfering system is close on the GSO arc to the wanted system and hence will be close to main beam, leading to greater antenna discrimination.

Sometimes systems are unable to use both polarisations at the same time at the same location. For example, a GSO satellite system might manage intra-system interference by having:

- North beam: LH polarisation
- South beam: LV polarisation.

Therefore, when coordinating with other satellites, the operator could suggest that they arrange their beams so that:

- North beam: LV polarisation
- South beam: LH polarisation.

When there is no main beam alignment, it can still be useful to consider having an alternate type of polarisation, that is, linear versus circular, particularly where there are multiple inter-ferers. Each particular interference path could be de-polarised to a degree, but there will be averaging effects close to 1.5 dB.

5.11.10 Antenna Height

For many terrestrial systems, interference can be decreased by lowering the height of the trans-mitter. Lower heights result in lower ranges and hence lower levels of interference. This is par-ticularly beneficial if the height can be reduced to below the adjacent clutter height as it will ensure additional diffraction losses that could be significant.

The downside is that the range over which a service could be provided is reduced. If the wanted service range is small anyhow, this can be beneficial – for example, IMT LTE urban small cell, Wi-Fi hot spots, etc., as frequency reuse increases.

Height can have an indirect but significant impact on the allocations, in particular as to whether aeronautical services are included or not. Many mobile allocations have a note exclud-ing aeronautical mobile due to the increased range over which interference can occur when transmitting from aircraft height.

5.11.11 Operate Indoors

There can be clear benefits in only operating a service indoors. For example, parts of the 5 GHz band are proposed to be used by RLANS, but these bands are shared with other services includ-ing non-GSO MSS feeder links. One of the mitigation methods was to permit these devices subject to their being used indoors. While some will be operated illegally outside – or legally close to windows with minimal indoor/outdoor loss – the net effect is a decrease in aggregate interference.

Operating indoors can be facilitated by frequency-selective surfaces that can create add-itional building attenuation at specified frequencies.

5.11.12 Improved Filtering and Guard Bands

Frequency scenarios involving non-co-frequency sharing can employ the other techniques described here but additionally:

- Transmit spectrum masks: if this was tightened, then it would reduce interference into sys-tems operating nearby in frequency
- Receive filter mask: by employing additional filtering, it would reduce interference from sys-tems operating nearby in frequency.

There can be disadvantages in both cases due to additional equipment costs and a slight reduction in the wanted signal. The latter particularly can be an issue with systems that have a very low received signal: an example would be radars that are listening for reflections off remote (and possibly small) vehicles. There is a balance to be reached between:

- Maximising received signal by using minimal (if any) filtering
- Minimising interference by using highly effective filters.

An additional issue is that the mask integration factor is driven by the least effective of the transmit and receive masks, as shown in Example 5.11. Hence there is a limit to how much benefit can come from one system introducing extra filtering beyond that used by the other system.

If there is still harmful interference in non-co-frequency cases after taking into account filtering, then one solution is to increase the frequency separation. This requires the use of guard bands as described in Section 5.3.8.

Note the importance of defining what is meant by **frequency separation**, in particular whether it is:

- Separation between the centre frequencies of the wanted and interfering carriers
- Separation between edges of the wanted and interfering carriers.

These two separations are shown in Figure 5.54. Either approach could be appropriate depending upon scenario, as long as the definition is made clear.

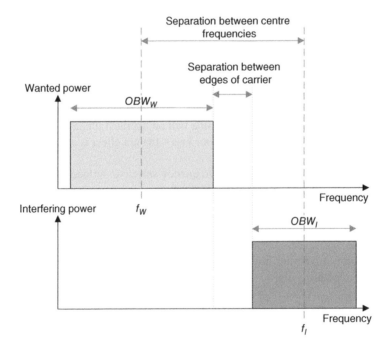

Figure 5.54   Definitions of frequency separation

5.11.13 Site Shielding

Specific locations can be protected by building structures that shield transmit or receive antennas from nearby systems, such as in Example 4.9 where there was calculated to be 18 dB of shielding. Site shielding requires a fixed location with sufficient space around it for construction. The downside of site shielding is cost, but it can bring significant gains.

A typical use might be to protect specific, well-defined and sensitive sites from interference, such as radio astronomy observatories or deep-space communication receivers.

Site shielding can be a mechanism used and paid for by an incoming service to protect the incumbent. The cost of building site shielding is typically born by the party that can utilise the consequential spectrum efficiency gains.

5.11.14 Spectrum Sensing and Geodatabases

One of the difficulties of interference analysis is that the answer to most questions depends upon the assumptions used. The more information is available, the more accurately it is possible to identify when, where and how a system can operate without causing harmful interference. A number of techniques can be implemented in equipment to allow it to make decisions about whether it can transmit and, if so, on which frequencies and at what powers. These include:

- Spectrum sensing or listen before transmit systems as described in Section 5.5.1
- Geodatabases, such as used by WSD described in Section 7.10
- Beacons, whereby each receiver transmits a signal that identifies its location and channel(s) of interest.

The idea of beacons is to avoid the 'hidden node' problem of listen before transmit systems. One issue is that additional spectrum would be required for the beacon signal, and recently interest has focused on use of geodatabases for WSD-type systems.

5.11.15 Wanted System Modifications

In the majority of cases the interference mitigation considers first how to reduce interference generated by the new entrant to protect the incumbent. There is a generally accepted right that an existing system should be permitted to continue to operate uninterfered after the entry of a new system or service.

However this is not to be taken as an unlimited right to stop new services and systems: the existing system or service should have included interference in its planning process, and this implies there will be some margin. The difficulty is in identifying how much (if any) interference margin is left after it has been apportioned between other systems and services. This can become a policy decision by spectrum managers that takes account of wider objectives (e.g. social inclusivity, environment, etc.).

One approach to facilitate change within the wider spectrum economy is to create procedures where new entrants could financially compensate existing users to make those changes that would facilitate sharing. These could include:

- New entrant building site shielding around existing vulnerable receivers
- New entrants paying to upgrade existing service's receive filters or replace an existing antenna with one that is larger and hence more directive
- New entrant paying to relocate existing services to other bands
- New entrant buying spectrum rights off the existing user, either exclusive or as an underlay to existing usage.

Some modifications can involve dramatic changes to the wanted system architecture. For example, it has been proposed that radar systems use stations at separate locations to transmit or receive, which can lead to significantly reduced bandwidth requirements.

5.11.16 Modelling Methodology

In some cases the interference issue is the result of assumptions in the modelling process. Complex scenarios can be simplified to facilitate their analysis, and it is generally a good principle to err on the side of caution. However, there is a danger in using worst-worst-worst-case assumptions that are very unlikely to actually occur.

Some examples of simplified and more detailed analysis are shown in Table 5.25.

The next chapter describes in more depth the range of modelling methodologies that can be used from basic static configurations to more detailed Monte Carlo simulations.

Table 5.25 Examples of simplified and more detailed modelling

Simplified methodology	More detailed methodology
GSO coordination trigger using DT/T	Detailed coordination using C/I metric
Coordination contour assuming worst-case geometry (victim pointing at interferer)	Detailed assessment using terrain database and actual pointing angles
Flat Earth model	Spherical Earth model
Smooth Earth model	Faster roll-off of propagation loss using Hata (generic) or P.452 (specific locations) with terrain data
Low-resolution terrain data	Terrain data + clutter or high-resolution surface database
All interferers on maximum power	Transmit power dependent upon traffic levels
Static geometry at worst case or minimum coupling loss (MCL)	Area, dynamic or Monte Carlo modelling with varying geometry
Static term for non-co-frequency bandwidth adjustment	Integration of the transmit and receive masks
Assume interferer and victim use the same polarisation	Use actual polarisations
Use of mask gain patterns from sources such as ITU-R Recommendations	Use of measured gain patterns

5.12 Further Reading and Next Steps

This chapter has described the interference calculation, building upon the basic link equation from Chapter 3 and using the propagation models from Chapter 4. In particular this section has considered factors that will affect the interfering signal including:

- Bandwidths for co-frequency scenarios
- Transmit and receive masks or adjacent channel ratios for non-co-frequency scenarios
- Spurious emissions including intermodulation products
- Polarisation effects
- Adaptive systems including frequency, power and modulation
- End-to-end performance
- Deployments and traffic.

Having calculated the interfering signal, the key question is what thresholds to use. Different types of metric were considered, including I, I/N, PFD, $EPFD$, C/I and $C/(N+I)$. These were discussed in the context of link design, in particular the interference margin, and it was noted that apportionment of interference is critical.

Finally, this chapter discussed possible interference mitigation techniques. The next chapter moves on to use the interference calculation and thresholds discussed here to review the different types of methodology that can be used for interference analysis.

6

Interference Analysis Methodologies

The previous chapter developed the background and theory of the interference calculation. It described how to calculate a number of interference related metrics including I, C/I, $C/(N + I)$, DT/T, I/N, PFD, $EPFD$ and $E_{\mu V}$. To determine whether the interference levels would be acceptable, it was necessary to consider how to derive thresholds taking into account, in particular, the interference margin during the design of the wanted link budget. Ways to include factors such as differing bandwidths or polarisations were described plus ways to calculate interference by integrating the transmit and receive spectrum masks.

Earlier chapters described underlying concepts required by this interference calculation, in particular:

- The geometric framework to determine locations, distances and angles
- Antenna models to calculate the gain at transmit and receive stations
- The carrier model to derive thresholds for the wanted service
- Propagation models to derive the path loss between stations.

This chapter describes how different types of methodology can use the interference calculation as part of interference analysis. These methodologies include static, minimum coupling loss (MCL), input variation, analytic, area, dynamic and Monte Carlo approaches as described in overview in Section 6.1.

In order to give examples of each of these methodologies, they will be applied to two scenarios involving compatibility between a terrestrial service and a satellite service, namely:

1. IMT base stations into receive satellite Earth stations at C-band
2. A non-GSO MSS network into terrestrial point-to-point fixed link at S-band.

Interference Analysis: Modelling Radio Systems for Spectrum Management, First Edition. John Pahl.
© 2016 John Wiley & Sons, Ltd. Published 2016 by John Wiley & Sons, Ltd.
Companion website: www.wiley.com/go/pahl1015

The regulatory background to these scenarios is described in Section 6.2.

Principles that can be used in selecting the preferred methodology are described in Section 6.12. Guidance on possible working practices when undertaking study work is given in Section 6.13 along with a description of how multiple methodologies can be applied to gain an improved understanding of a sharing scenario.

The following resource is available:

Resource 6.1   The spreadsheet 'Chapter 6 Examples.xlsx' contains example static analysis and MCL calculations plus the IMT link budgets and derivation of the time steps size for dynamic analysis.

In addition a number simulation files are available as resources as described in the succeeding text.

6.1   Methodologies and Studies

Why are there multiple methodologies for interference analysis? The reason for this is that there are different questions that must be answered, which relates to there being a range of motivations for undertaking interference analysis, as discussed in Chapter 2. In particular, Section 2.11 identified the following types of interference analysis:

- System interference analysis: to ensure a proposed system is designed in a way to minimise interference into itself or other services, which could be to assist a licence application or propose changes to the regulations
- Regulatory interference analysis: to undertake studies to change the RR or develop new Recommendations and Reports
- Frequency assignment interference analysis: during the frequency assignment process, often as part of licensing procedures
- Coordination interference analysis: to work with other organisations to agree methods to ensure radio systems from each organisation can operate without causing harmful interference into the other.

In some of these cases, the approach to use will be determined by circumstances. For example, to check whether coordination is required for a new satellite Earth station, the methodology to use is defined in Appendix 7 of the RR, as described in Section 7.4, while between GSO satellites, the methodology is in Appendix 8, as described in Section 7.5.

In other cases, the methodology will not be defined; rather it will be necessary to analyse a scenario involving two or more systems or services. The requirements will vary and could involve wider commercial, regulatory and policy issues where the objective of the analysis is to provide input to a wider audience regarding interference-related issues.

This requires the ability to not just calculate an interfering link budget but be able to answer questions such as:

- What is the interference level between transmitter A and victim receiver B?
- How much separation (frequency and/or distance) should there be between system A and system B to avoid harmful interference?
- How would the level of interference vary by distance?

- How sensitive would be the level of the interference to changes in assumptions?
- Is it possible to calculate interference metrics directly without simulation?
- How would the interference vary geographically?
- How would the interference vary as stations move?
- What is the likelihood of interference?
- What is the capacity loss of a cellular network due to interference?

To answer these questions requires different types of methodology, as identified in Table 6.1 together with the section that describes them further. These methodologies can be also split into those that consider the variability of interference in time and those that just consider a single instance.

There are also further methodologies that could be considered, such as two-stage and area Monte Carlo as described in Section 6.10.

These represent different generic types of methodology, but there can also be specific versions of each. For example, fixed link planning uses multiple, highly specific combinations of

Table 6.1 Interference analysis questions and methodologies

Type of question	Interference analysis methodology	Includes time variation?	Described further in section
What is the interference level between transmitter A and victim receiver B?	Static analysis	No	6.3
How would the level of interference vary by distance? How sensitive would be the level of the interference to changes in assumptions?	Input variation analysis	No	6.4
How would the interference vary geographically over an area?	Area analysis	No	6.5
How much separation (frequency and/or distance) need there be between system A and system B to avoid harmful interference?	Input variation analysis	Not generally or indirectly	6.4
	Minimum coupling loss		6.6
Is it possible to directly calculate interference metrics without simulation?	Analytic analysis	Not generally	6.7
How would the interference vary in time as stations move?	Dynamic analysis	Yes	6.8
What is the likelihood of interference?	Dynamic analysis	Yes	6.8
	Monte Carlo analysis		6.9
	Probabilistic analysis		6.10
What is the capacity loss of a cellular network due to interference?	Monte Carlo	Yes	6.9

static analysis. Understanding the various methodologies described in this chapter will provide greater understanding into the specific algorithms described in Chapter 7.

6.2 Example Scenarios

The methodologies will be demonstrated by their application to one of these scenarios:

1. Analysis of the potential for interference from IMT systems into receive satellite Earth stations (ESs)
2. Analysis of the potential for interference into fixed links from non-GSO mobile-satellite service (MSS) downlinks.

To widen their applicability, these have been selected so that both involve terrestrial and satellite services, with the first involving terrestrial propagation and the second space to Earth paths.

6.2.1 IMT Sharing with Satellite ES

6.2.1.1 Regulatory Background

The motivation to study this sharing scenario was the rapid growth of mobile broadband traffic. Regulatory frameworks should be not just reactive but also predictive, ensuring that suitable allocations and regulations are ready in place to support industry developments. Analysis undertaken in the ITU-R documented in Report ITU-R M.2290 (ITU-R, 2013q) suggested that the spectrum requirements for International Mobile Telecommunications (IMT) for the year 2020 to be in the range of 1 340–1 960 MHz. It was noted that this would vary depending upon country and environment, but there would be the greatest demand in dense urban areas.

This forecast demand significantly exceeded the supply of suitable spectrum at the time of the study in the Report, and hence WRC 2012 included the following item in the agenda for the following Conference:

1. *to consider additional spectrum allocations to the mobile service on a primary basis and identification of additional frequency bands for International Mobile Telecommunications (IMT) and related regulatory provisions, to facilitate the development of terrestrial mobile broadband applications, in accordance with Resolution 233 (WRC-12);*

The primary focus was on bands below 6 GHz, and this led to the requirement to consider sharing scenarios with a very wide range of different services. As these services were the remit of multiple ITU-R Study Groups (as described in Section 2.4.5), it was decided to set up a Joint Task Group (JTG), namely, JTG 4-5-6-7. This group also considered Agenda Item 1.2, which covered studies relating to the use of the band 694–790 MHz within Region 1 by mobile services excluding aeronautical mobile.

This group met six times and the main output was the Chairman's Report, document JTG 4-5-6-7/715 (ITU-R, 2014a). This included 36 annexes, which described the work done, parameters used and output including text for the CPM Report and Draft Reports.

The parameters for the satellite ES and IMT networks using in this chapter are taken from Annex 17 of the Chairman's Report (ITU-R, 2014b):

• Draft New Report ITU-R [FSS-IMT C-BAND DOWNLINK]: Sharing studies between International Mobile Telecommunication-Advanced systems and geostationary satellite networks in the fixed-satellite service in the 3 400–4 200 MHz and 4 500–4 800 MHz frequency bands in the WRC study cycle leading to WRC-15.

The part of the Table of Allocations relevant for the 3 400–4 200 MHz band prior to WRC-15 is shown in Table 2.1. This issue was keenly discussed at WRC-15 where it was agreed to add allocations for mobile and identification for IMT, mostly via footnotes, especially in parts of L-band and C-band.

Note that while the ITU uses the generic IMT-Advanced term, many of the studies were based upon the Long-Term Evolution (LTE) standard. Hence the parameters in this section are described as IMT LTE, a term which is used through-out this book.

6.2.1.2 Satellite ES Parameters

Satellite ES parameters for this scenario are given in Table 6.2.

It was additionally noted that:

• Smaller antennas (1.8–3.8 m) are commonly deployed on the roofs of buildings or on the ground in urban, semi-urban or rural locations, whereas larger antennas are typically mounted on the ground and deployed in semi-urban or rural locations
• 5° is considered as the minimum operational elevation angle.

The response of the ES in the frequency domain was defined via the values in Table 6.3.

Table 6.2 Satellite ES parameters

Parameter	Typical value
Range of operating frequencies	3 400–4 200 MHz, 4 500–4 800 MHz
Antenna diameters (m)	1.2, 1.8, 2.4, 3.0, 4.5, 8, 16, 32
Antenna reference pattern	Rec. ITU-R S.465
Range of emission bandwidths	40 kHz to 72 MHz
Receiving system noise temperature	100 K for small antennas (1.2–3 m)
	70 K for large antennas (4.5 m and above)
Earth station deployment	All regions, in all locations (rural, semi-urban, urban)

Table 6.3 Satellite ES frequency response parameters

Bandwidth	30 MHz
ACLR/ACS first channel	45 dB
ACLR/ACS second channel	50 dB
Spurious	55 dB

The interference thresholds were based on I/N and derived from Rec. ITU-R S.1432 (ITU-R, 2006h) and in particular:

Long-term interference criteria:

- In-band sharing studies: $T(I/N) = -12.2$ dB for 100% of the worst month or $T(I/N) = -10$ dB corresponding to the aggregate interference for $p = 20\%$ of any month. It was noted that this was an aggregate for all other services, and the aggregate from one particular other service should be 3 dB lower, that is, $T(I/N) = -13$ dB
- Adjacent band sharing studies: $T(I/N) = -20$ dB for aggregate interference from all other sources of interference for 100% of the time.

Short-term interference criteria:

- In-band sharing studies: $T(I/N) = -1.3$ dB that may be exceed by up to $p = 0.001667\%$ time (single entry).

6.2.1.3 IMT LTE Parameters

IMT LTE parameters for this scenario are given in Table 6.4.

Table 6.4 IMT LTE base station and cell deployment parameters

Base station characteristics/ cell structure	Macro suburban	Macro urban	Small cell outdoor	Small cell indoor
Cell radius/ deployment density	0.3–2 km (typical figure to be used in sharing studies 0.6 km)	0.15–0.6 km (typical figure to be used in sharing studies 0.3 km)	1–3 per urban macro cell <1 per suburban macro site	Depending on indoor coverage/ capacity demand
Antenna height	25 m	20 m	6 m	3 m
Sectorisation	Three sectors	Three sectors	Single sector	Single sector
Downtilt	6°	10°	n/a	n/a
Frequency reuse	1	1	1	1
Antenna pattern	Rec. ITU-R F.1336	Rec. ITU-R F.1336	Rec. ITU-R F.1336	Rec. ITU-R F.1336
Horizontal 3 dB beamwidth	65°	65°	Omni	Omni
Antenna polarisation	Linear	Linear	Linear	Linear
Indoor base station penetration loss	n/a	n/a	n/a	20 dB (3–5 GHz)

It was also noted that:

- Outdoor small cells would typically be deployed in very limited areas in order to provide local capacity enhancements. Within these areas, the outdoor small cells would not need to provide continuous coverage since there would be an overlaying macro network present
- If the IMT-Advanced network consists of three cell layers – macro cells, small outdoor cells and small indoor cells – they will not all use the same carrier. Two layers may use the same carrier, although separate carriers in the same or different bands are also possible.

The link parameters are given in Table 6.5.

The response of the IMT LTE BS in the frequency domain was defined via the values in Table 6.6.

Table 6.5 IMT LTE base station link parameters

Base station characteristics/cell structure	Macro suburban	Macro urban	Small cell outdoor	Small cell indoor
Below rooftop base station antenna deployment	0%	50%	100%	N/a
Feeder loss	3 dB	3 dB	N/a	N/a
Maximum base station output power (5/10/20 MHz)	43/46/46 dBm	43/46/46 dBm	24 dBm	24 dBm
Maximum base station antenna gain	18 dBi	18 dBi	5 dBi	0 dBi
Maximum base station output EIRP/sector	58/61/61 dBm	58/61/61 dBm	29 dBm	24 dBm
Average base station activity	50%	50%	50%	50%
Average base station EIRP/sector taking into account activity factor	55/58/58 dBm	55/58/58 dBm	26 dBm	21 dBm

Table 6.6 IMT LTE BS frequency response parameters

Bandwidth	10 MHz
ACLR/ACS first channel	45 dB
ACLR/ACS second channel	45 dB
Spurious	55 dB

For simplification, only the base station transmit case was considered, that is, excluding the user equipment (UE) transmit case.

It should also be noted that different radio communities may prefer to use different units. In particular for this example, IMT LTE mobile systems tend to use dBm for power while satellite operators use dBW. For simplicity, only a single power unit will be used throughout this chapter, and to be consistent with the PFD threshold, it will be dBW. This also avoids confusion with the dB unit of area compared to one metre squared, namely, dBm^2.

6.2.2 Sharing Between Non-GSO MSS and FS

6.2.2.1 Regulatory Background

In the early 1990s, a number of non-GSO MSS systems were under development. They proposed to use constellations of satellites in low Earth orbit (LEO) rather than GSO to reduce the delay and path loss to mobile earth stations (MES). These were sometimes split into the 'little LEOs', which aimed to provide messaging services and the 'big LEOs', which focussed on voice services. The latter group required additional new allocations in bands within the 1–3 GHz range, often sharing with existing primary allocations for the fixed service (FS), in particular:

- 2 170–2 200 MHz, proposed to be used for the service (i.e. satellite to handset) downlink by ICO Global Communications, described in the ITU-R as LEO-F
- 2 483.5–2 500 MHz, proposed to be used for the service downlink by Globalstar, described in the ITU-R as LEO-D.

For GSO satellite networks, where the geometry is fixed, terrestrial services could be protected via a power flux density (PFD) mask, such as those found in Article 21 of the RR. However as these systems proposed to use non-GSO constellations, it was suggested that a different approach would be required, in particular to protect FS using digital links.

Given that the band was used for terrestrial services in a very large number of countries, it was considered not feasible to coordinate with each one individually. Therefore an approach was developed whereby:

- The coordination trigger used the fractional degradation in performance (FDP) metric described in Section 5.10.2 calculated using the standard computation program (SCP) in Rec. ITU-R M.1143 (ITU-R, 2005c)
- Detailed coordination, if required, would use the more advanced $C/(N + I)$ methodology in Rec. ITU-R M.1319 (ITU-R, 2010a) as in Example 5.44.

6.2.2.2   Satellite Parameters

The non-GSO MSS system parameters were taken from reference network LEO-F in Rec. ITU-R M.1184 and are given in Table 6.7. The key direction for sharing with the FS is the service downlink operating in the 2 170–2 200 MHz band using 121 spot beams arranged in a hexagonal pattern as in Figure 6.1. While it would be more appropriate to use Rec. ITU-R S.1528 (ITU-R, 2001d) as the gain pattern for a non-GSO system, this Recommendation was not available during the 1990s as it was approved in 2001, and hence the similar Rec. ITU-R S.672 was used.

Note also the discussions in:

- Section 6.8 relating to the implications of the orbit height and suggested alteration
- Example 5.29 relating to traffic levels and aggregate EIRP.

6.2.2.3   Reference FS Parameters

While a wide variety of FS systems operate in this band, the parameters that were used in the analysis are those given in Table 6.8.

In many cases, the gain pattern used in interference analysis involving the FS is that in Rec. ITU-R F.699 (ITU-R, 2006b), as used in Section 7.1. This pattern was design for static analysis

Table 6.7 Reference non-GSO MSS parameters

Orbit type	Circular
Orbit height	10,355 km
Orbit inclination	45°
Number of planes of satellites	2
Number of satellites/plane	5
Number of service link beams	121
Peak gain of beam	32.1 dBi
Beam pattern	Hexagons
Beamwidth	4.4°
Beam separation	4.0°
Gain pattern	Rec. ITU-R S.672-4 side-lobe −25 dBi
Beam EIRP/MHz	49 dBW/MHz
Polarisation	Circular
Voice activation factor	0.4
Outer two ring beams' traffic load	50% of central beams' traffic load
Frequency re-use	1 in 7
Traffic/central beam	1 MHz

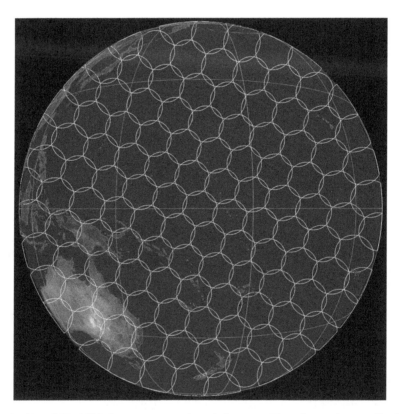

Figure 6.1 Non-GSO MSS beam pattern viewed from satellite. *Screenshot credit: Visualyse Professional. Overlay credit: NASA Visible Earth*

Table 6.8 Reference FS parameters

Latitude	51°N
Azimuth	90°
Heights	20 m
Length	21 km
Antenna peak gain	30 dBi
Antenna beamwidth	5°
Antenna feed loss	2 dB
Gain pattern	Rec. ITU-R F.1245-1
Frequency	2 GHz
Bandwidth	7 MHz
Polarisation	Linear
Noise figure	7 dB
Interference margin	1 dB
Fade margin	22 dB
$C/(N+I)$ threshold	20.5 dB
Unavailability target	0.01%

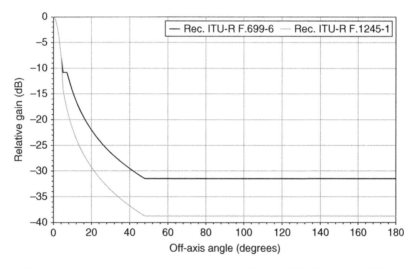

Figure 6.2 Comparison of the gain patterns in Recs. ITU-R F.699 and F.1245

involving a small number of interferers, potentially just the one. Hence it has a conservative sidelobe pattern assuming a fixed geometry where interference could occur continuously at any off-axis angle. However, for this scenario, there are multiple sources of interference, which move, so that it is necessary to have a sidelobe model that describes the average performance. If the interferer is moving, then while there will be angles at which the off-axis gain is higher or lower, the understanding is that statistically it will be the average that is important. In these cases, it is more appropriate to use the antenna gain pattern in Rec. ITU-R F.1245 (ITU-R, 2012c) as shown in Figure 6.2, which compares it against the pattern in Rec. ITU-R F.699 for a peak gain of 30 dBi and beamwidth of 5°.

6.3 Static Analysis

The starting point for most interference studies is to create a **static analysis**. This involves the identification and selection of all of the parameters necessary to derive the required metrics for a single instance and then configuring the simulation tool accordingly. The scenario could even be simplified to the point that it could be calculated in a spreadsheet.

It is often useful to create a diagram that describes the scenario in the space plus frequency domains. The figure can then help create a checklist of all the parameters that will be needed, such as:

- Locations: what are the positions of the stations involved and which geometric system should be used (as in Section 3.8)?
- Antennas: what are the various antennas at each station, in particular the gain patterns and pointing directions (as in Section 3.7)?
- Links: what are the various link attributes, such as frequency, bandwidth, transmit power and receiver noise temperature needed to complete the link budget (as in Section 3.11)?
- Propagation models: what model should be used to calculate how the radio wave propagates for both the wanted and interfering paths (as in Chapter 4)?
- Thresholds: what metric is to be calculated and what would be a suitable threshold (as in Section 5.10)?

These can be used to create the wanted and/or interfering link budgets together with the various metrics such as I/N, C/I, PFD, etc.

For some studies, this can be sufficient. Typically, however, static analysis is used as a reference or baseline. Even if ultimately much more complicated methodologies are used, starting with a basic static analysis is extremely helpful for a number of reasons:

1. To ensure the simulation is complete, that all parameters have been identified
2. To 'sanity check' the calculation, as the simplicity of the simulation means errors are more obvious. This creates a known and trusted baseline to build upon
3. To get an order of magnitude idea of the issues involved. If margins are exceeded by 30 dB, then there is a serious problem; if by only 1 or 2 dB, then there is a high likelihood that sharing will be feasible
4. To compare against other studies. As complexity increases, it becomes ever more difficult to compare results, particularly if different software tools are used. Having a simple scenario that is agreed can be a good test case to identify any potential differences in assumptions or algorithms, noting that even small (and valid) differences can lead to discrepancies (see boxed text 'Testing Against Standards').

> **Testing Against Standards**
>
> I came across an example of the difficulties of comparing implementations of the gain pattern in Rec. ITU-R S.1428 (ITU-R, 2001c), which includes the side-lobe:
>
> $$G(\vartheta) = 29 - 25\log(\vartheta) \text{ for } 95\frac{\lambda}{D} < \vartheta \le 33.1° \tag{6.1}$$
>
> $$G(\vartheta) = -9 \text{ for } 33.1° < \vartheta \le 80° \tag{6.2}$$
>
> We noticed small differences and the divergence was tracked down to the break point where Equations 6.1 and 6.2 meet, which is when
>
> $$\vartheta = 10^{(29+9)/25} = 33.113° \tag{6.3}$$
>
> Hence the break point specified in the Recommendation is only approximately at 33.1°. One implementation was using this rounded value while the other a more precisely calculated break point. Which was right given that both approaches could be considered 'correct'?
>
> In this case, the difference is a negligible 0.004 dB, but it shows how differences in calculations can come from decisions made in implementing complex algorithms, and how there can be arguments supporting both sides. Having a clearly defined scenario with all other parameters and calculations well understood makes it easier to identify areas where further discussion is required.

Example 6.1

The starting point for static analysis is to define one complete parameter set from the range of available alternatives. The scenario to analyse in this example is shown in the space + frequency diagram in Figure 6.3.

Using this figure as a guide, the parameters in Tables 6.9, 6.10 and 6.11 were selected from the descriptions in Section 6.2.

Note how even parameters that are set to zero, such as the feed loss, are listed for clarity and completeness.

These parameters were used to calculate the interfering link budget as shown in Table 6.12 together with information on the calculation steps. The following resource is available:

Resource 6.2 Visualyse Professional simulation file 'Static analysis example.sim' was used to generate the link budgets in this section.

The resulting I/N is significantly above the threshold leading to a negative margin of -38.9 dB, suggesting this is a difficult sharing problem to analyse (indeed, it has been the subject of multiple WRC cycles).

It is worth raising the aggregation issue, as this represents the I/N from:

- A single IMT LTE 10 MHz carrier, but the satellite ES bandwidth is 30 MHz, so there could be a carrier aggregation factor of $\times 3 = +4.8$ dB
- A single IMT LTE base station sector, whereas it is likely to have others (although pointing away from the ES)
- A single IMT LTE base station, though there could be many in the vicinity of the ES.

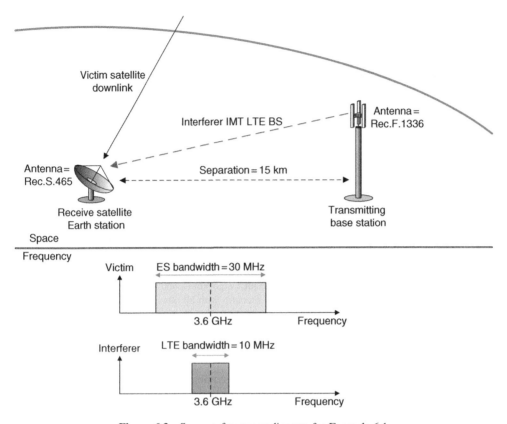

Figure 6.3   Space + frequency diagram for Example 6.1

Table 6.9   Satellite ES parameters

Parameter	Static analysis value
Receive frequency	3 600 MHz
Bandwidth	30 MHz
Antenna gain pattern	Rec. ITU-R S.465
Antenna diameter	2.4 m
Antenna efficiency	0.6
Antenna height	5 m
Antenna elevation angle	10°
Receiving system noise temperature	100 K
Receive feed loss	0 dB

Static analysis can consider aggregation issues by including multiple sources of interference, and this can be one way to build on the starting simulation. Other enhancements that could be considered include comparing co-frequency scenarios with non-co-frequency cases (by changing the frequency offset) or to include terrain data.

Table 6.10   IMT LTE BS parameters

Parameter	Static analysis value
Transmit frequency	3 600 MHz
Bandwidth	10 MHz
Antenna gain pattern	Rec. ITU-R F.1336-4 Recommends 3.1.2
Antenna peak gain	18 dBi
Antenna horizontal beamwidth	65°
Antenna height	25 m
Antenna downtilt	6°
Transmit power at input to antenna	13 dBW

Table 6.11   Scenario parameters

Parameter	Static analysis value
Geometric framework	Spherical earth
Earth model	Smooth
Propagation model	Rec. ITU-R P.452
Percentage of time	20%
Delta N	45 N units/km
Sea level surface refractivity N0	325
Separation distance	15 km
Station deployment geometry	ES pointing towards the IMT LTE BS
	IMT LTE BS pointing towards the ES

Table 6.12   Static analysis resulting interfering link budget

Field	Symbol(s)	Value	Calculation
Frequency (GHz)	f_{GHz}	3.6	—
Victim bandwidth (MHz)	B_V	30	—
Interferer bandwidth (MHz)	B_I	10	—
Interferer power (dBW)	P_{tx}	13	—
Interferer peak gain (dBi)	$G_{max}(tx)$	18	—
Interferer relative gain (dB)	$G_{rel}(tx)$	−18.4	Using F.1336-4
Propagation model	—	P.452	—
Separation distance (km)	d	15.0	—
Percentage of time (%)	p	20	From S.1432
Propagation loss for p % of time (dB)	$L_p(p)$	127.5	Using P.452
Receive peak gain (dBi)	$G_{max}(rx)$	36.9	Using Equation 3.78
Receive relative gain (dB)	$G_{rel}(rx)$	−29.9	Using S.465
Receive feed loss (dB)	L_f	0	—
Bandwidth adjustment factor	A_{BW}	0	Using Equation 5.5
Interference (dBW)	I	−107.9	Using Equation 5.6
Receiver temperature (K)	T	100	—
Noise in victim bandwidth (dBW)	N	−133.8	Using Equation 3.43
I/N (dB)	I/N	25.9	$I - N$
Threshold I/N (dB)	$T(I/N)$	−13	—
Margin (dB)	$M(I/N)$	−38.9	$T(I/N) - I/N$

Example 6.2

An IMT LTE BS with the parameters from Example 6.1 is located 15 km from a border, which must be protected to the *PFD* limit = −154.5 dBW/m^2/4 kHz from footnote 5.430A (see Example 5.48). The path profile between it and the border is given in Figure 6.4: would this base station meet or exceed the limit?

This required the following changes to the baseline static analysis:

1. Add terrain data (in this case from SRTM)
2. Change the height of the ES to that in 5.430A (i.e. 3 m)
3. Calculate the interference into the required reference bandwidth of 4 kHz
4. Convert the interfering signal at an isotropic antenna into a *PFD*.

The resulting calculation is shown in Table 6.13. It can be seen that the margin, while negative, is very close to zero, which suggests there is the potential to meet the relevant criteria. While there are factors that could increase the *PFD* (in particular aggregation as described in Example 6.1), there are also potential mitigation strategies, such as:

- Power: the power could be reduced by 0.1 dB to meet the *PFD* limit
- Pointing: either point away from the ES or increase the downtilt to reduce the gain towards the horizon
- Height: often reducing the height decreases the interference
- Clutter: the analysis was undertaken using terrain but not including clutter loss
- Antenna gain pattern: as identified in Example 5.48, the *PFD* threshold was derived using the ES gain pattern in Rec. ITU-R S.580 rather than S.465. The use of Rec. ITU-R S.580 would reduce interference by 3 dB.

The main limitation of static analysis is it only considers one set of parameters, and the problem domain can be large. With n variables each with N_i possible values, the total number of static analysis combinations is

$$N_{total} = N_1 N_2 N_3 ... N_n \qquad (6.4)$$

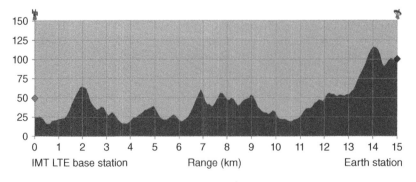

Figure 6.4 Path profile between IMT LTE BS and ES on border

Table 6.13 PFD calculation

Field	Symbol(s)	Value	Calculation
Frequency (GHz)	f_{GHz}	3.6	—
Reference bandwidth (kHz)	B_V	4	—
Interferer bandwidth (MHz)	B_I	10	—
Interferer power (dBW)	P_{tx}	13	—
Interferer peak gain (dBi)	$G_{max}(tx)$	18	—
Interferer relative gain (dB)	$G_{rel}(tx)$	−18.6	Using F.1336-4
Propagation model	—	P.452	—
Separation distance (km)	d	15.0	—
Percentage of time (%)	p	20	From 5.430A
Propagation loss for p % of time (dB)	$L_p(p)$	165.4	Using P.452
Signal at isotropic receiver	S	−153.0	Using Equation 3.204
Area of isotropic antenna (dBm^2)	$A_{e,i}$	32.6	Using Equation 3.227
PFD in 10 MHz (dBW/m^2)	$PFD_{10\,MHz}$	−120.4	$S - A_{e,i}$
Bandwidth adjustment factor (dB)	A_{BW}	−34.0	Using Equation 5.5
PFD in reference bandwidth (dBW/m^2/4 kHz)	PFD	−154.4	$PFD_{10\,MHz} + A_{BW}$
Threshold PFD value (dBW/m^2/4 kHz)	$T(PFD)$	−154.5	From 5.430A
Margin (dB)	$M(PFD)$	−0.08	$T(PFD) - PFD$

For example, with 20 input parameters each with 5 possible values, the total number of per-mutations is $5^{20} = 9.53674 \times 10^{13}$. This is clearly computationally infeasible, but a number of approaches are available that can assist in managing complexity:

- **Visualisation**, such as showing charts, which display graphically how outputs change due to variations in one or more inputs, as discussed in Section 6.4 'Input Variation Analysis' and Section 6.5 'Area and Boundary Analysis'
- **Derived metrics**, such as identifying the distance or frequency separation where a threshold is just met, reducing the number of dimensions of the problem by one. This is discussed further in Section 6.6 'Minimum Coupling Loss and Required Separation Distance'
- **Convolve multiple input variables** by assigning each a probability and then using Monte Carlo methodologies to generate likelihood of various outputs, as discussed in Section 6.9
- **Scenario definition,** in which one or more specific deployments of systems are defined via sets of parameters.
- **Using domain knowledge** to identify those parameters or sets of parameters that can be excluded (e.g. due to market, regulatory or technical constraints).

The first step to manage this complexity is typically to consider input variation analysis as described in the following section.

Static analysis also fails to inform about the variability of interference, such as:

- Geographic variation: for example, over what area could there be interference?
- Station dynamics, for example, motion of aircraft, ships and satellites can lead to interference exceeding a threshold, but how long is an event?
- General variability: for example, if there is mobility of users, variability in traffic and propagation, what is the likelihood of interference?

To be able to answer these types of questions, it is necessary to consider more advanced methodologies, such as area analysis (AA), dynamic simulation and Monte Carlo modelling, as discussed in the following sections.

Note that Example 3.33 contains another scenario that used static analysis.

6.4   Input Variation Analysis

One of the most useful (and hence often encountered) type of methodology is **input variation analysis**, sometimes called **sensitivity analysis**. In this case, an input variable is altered to see what the impact would be on the chosen output metric (e.g. I, C/I, I/N, PFD, etc.). It can build on the starting static analysis to ask questions such as:

- How does the interference vary with separation distance?
- How does the interference vary with frequency separation?
- How does the interference vary with antenna dish size?
- How does the interference vary with antenna pointing angles including downtilt?
- How does the interference vary with antenna height?

This approach can be very useful, as:

- It is clear as to what scenario is being modelled: the configuration is relatively straightforward and hence can be defined, explained and understood readily
- It is clear as to what outputs are being produced: the result is the required metric against the variable that has been changed.

Some questions can be answered without detailed analysis: for example, changes in the input power have a linear impact on the resulting received signal level.

Input variation analysis can also be employed to derive the answers to useful questions, such as:

- What is the distance separation at which the threshold is just met?
- What is the frequency separation at which the threshold is just met?

This type of analysis is often called MCL as described in Section 6.6.

Input variation analysis can be repeated to give a view as to the problem domain in higher numbers of dimensions. For example, a run of interference vs. distance could be repeated for a set of frequency offsets to understand the variation of interference in the (distance, frequency offset) domains.

The following resources are available:

Resource 6.3   Visualyse Professional simulation file 'Static analysis – extend to vary distance.sim' was used in Example 6.3.

Resource 6.4   Visualyse Professional simulation file 'Static analysis – extend to vary frequency.sim' was used in Example 6.4.

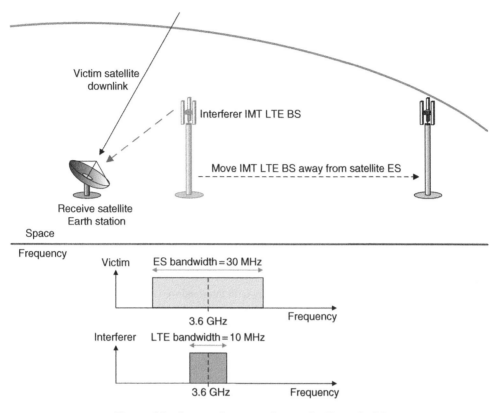

Figure 6.5   Space + frequency diagram for Example 6.3

Example 6.3
The static analysis in Example 6.1 was extended to consider the variation of interference in distance, as shown in the space + frequency diagram in Figure 6.5. The separation distance between the receive satellite ES and the IMT LTE base station was varied from 10 to 100 km. The result was a plot of I/N against distance, as in Figure 6.6, and it can be seen the threshold was met at around 45 km.

Example 6.4
The static analysis in Example 6.1 was extended to consider the variation of interference in the frequency domain, as shown in the space + frequency diagram in Figure 6.7. The transmit and receive spectrum masks were derived from the ACS using the methodology from Example 5.8 and then used to calculate the A_{MI}. The frequency offset between the receive satellite ES and the IMT LTE base station was varied between 0 and 100 MHz. The resulting plot of I/N was as shown in Figure 6.8, and it can be seen the threshold was met when the frequency separation was around 20 MHz.

Note that Example 7.20 describes an analysis comparing the results using A_{BW} with those from A_{MI} when calculating the I/N between terrestrial links and a satellite Earth station.

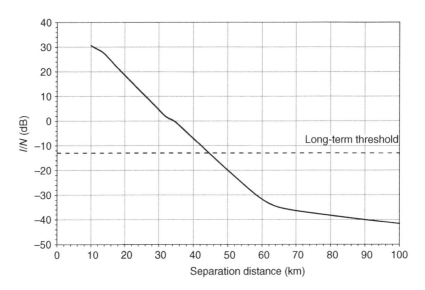

Figure 6.6   I/N vs. distance for long-term threshold for co-frequency case

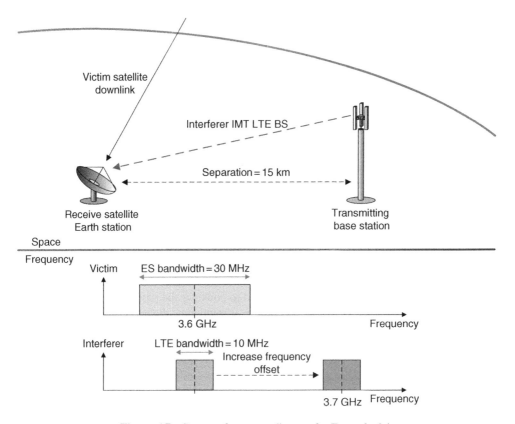

Figure 6.7   Space + frequency diagram for Example 6.4

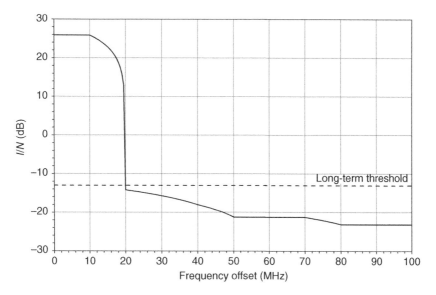

Figure 6.8   I/N vs. frequency offset when separation distance = 15 km

These two examples show the strength of input variation analysis: it gives clear information that assists in understanding a scenario and identifying possible solutions. However, only the distance and frequency were varied, and there are many other factors that would have to be considered, including the short-term threshold.

Input variation analysis can be undertaken in multiple dimensions to produce plots, for example, of *I/N* against both frequency and distance. For geographic variation, an alternative methodology that can examine more combinations of parameters is to analyse interference along a line or in an area or within a volume, as described in the following section.

6.5   Area and Boundary Analysis

One of the most important questions about interference is often which locations could be affected? In most cases, this involves checking locations either across an area or along the boundary of a polygon, as discussed in the succeeding text.

6.5.1   Area Analysis

AA involves one or more stations being positioned at each of a set of points within an area and at each location undertaking a static analysis. Any of the values derived in the calculations at that point can then be shown graphically either through the use of colour-coded pixels (blocks) or by drawing contour lines. The parameter displayed can be any of the link metrics $X = \{C, I, C/I, C/N, I/N, PFD, EPFD,$ etc.$\}$ or one of the intermediate values such as antenna gains or propagation loss. Usually the area to analyse is a rectangle, with pixels calculated in a grid, though it is also possible to create an AA using lines radiating out from a central point.

Example 6.5
The static analysis in Example 6.1 was extended to an AA as shown in Figure 6.9. The IMT
LTE BS location was varied around the receive satellite ES, and at each pixel, the I/N was
plotted as a greyscale and as a contour, with threshold $I/N = -13$ dB, as in Figure 6.10.

Grid lines are spaced 10 km apart and the following resource is available:

Resource 6.5   Visualyse Professional simulation file 'Static analysis – extend to area ana-
lysis.sim' was used in this example.

This AA shows the classic 'keyhole' shape, in this case directed towards the south, due to the
ES antenna pointing in this direction. The pointing methods and assumptions made when creating the AA will significantly impact the resulting graphic. In particular, for this scenario:

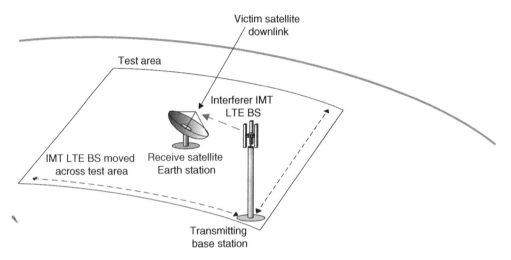

Figure 6.9   Example area analysis

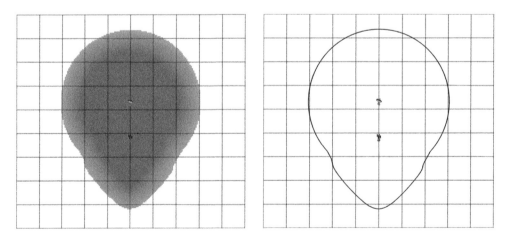

Figure 6.10   Example area analysis showing block plot (left) and contour plot (right)

- Is the ES pointing fixed or does its azimuth vary depending upon the test location of the IMT LTE BS?
- Is the IMT LTE BS antenna pointing fixed or does its active sector azimuth vary depending upon the test location?

Having antennas that always point at the other station is likely to result in larger interference areas than pointing being fixed because there will be less opportunity antenna discrimination to reduce interference levels.

In this case, having the ES point at the IMT LTE BS and also the IMT LTE BS point at the ES makes the only variation between pixels the separation distance between stations, and hence the AA becomes symmetric around the ES (i.e. a circle), with interference only varying by distance as for Figure 6.6.

What is more interesting is an answer to the question:

What is the area around an ES within which an IMT LTE BS should not be located to avoid interference exceeding the threshold?

This is best answered by keeping the ES pointing fixed. While the IMT LTE BS is assumed in this case to have the worst-case assumption of a sector pointing directly at the ES, it could be helpful to consider the alternative mitigated case where the sector pointing directly at the ES is switched off.

If both the ES and IMT LTE BS pointing is kept fixed, as per the static analysis (ES towards the south and IMT LTE BS towards the north), then the shape of the resulting inter-ference zone will be more complex, taking into account the off-axis gain at the base station as in Figure 6.11.

One of the benefits of AA is the ability to visualise the problem and also to identify any unexpected behaviour. As noted previously, in most cases, interfering antennas always pointing at the victim will result in higher interference levels and hence larger interference zones. However, in this case, the zone is wider east–west for the case with fixed north pointing

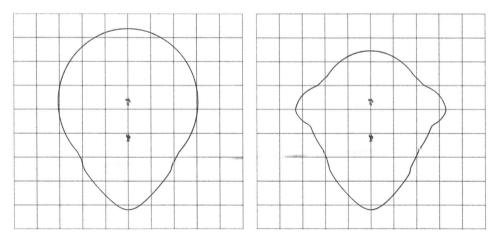

Figure 6.11 Example AA contours generated assuming IMT LTE BS pointing at ES (left) or fixed pointing due north (right)

than always pointing at the victim ES. This relates to the use of downtilt by the base station, which means that when it is pointing 'directly' at the ES, there is still 6° of off-axis discrimination at the antenna, resulting in significant reduction in the gain towards the horizon. This downtilt effect reduces towards azimuth ±90° away from the base station's antenna boresight: while the gain also decreases, it is a lesser effect as the azimuth beamwidth is wider than that in the elevation direction.

A good technique to analyse these types of scenarios is to combine AA with static analysis and move the reference station to the point of interest and examine the link budgets. Another effective use of an AA is to calculate the size of the interference zone, for example, in km².

In this simple case, the AA can be generated by moving either station and the results should be similar, though mirror images of each other. Compare, for example, the exclusion zones on the left in Figure 6.11 with that on the left in Figure 6.45.

For more complex scenarios, in particular aggregate ones involving many transmit stations, it can be simpler to vary the location of the single receive station.

Another key question is whether terrain and land use data should be included in the analysis. They can have a significant impact on the size and shape of the interference zone, as can be seen in Figure 6.12.

If the analysis is of a specific site, it is preferable to use as much location-related information as possible and so terrain and land use data should be included where available. If neither terrain nor land use data is included in the analysis, so that path profiles are just the smooth Earth, then this could be considered to be a worst-case assumption. It can be used in generic studies to derive the maximum size of an interference zone. More realistic analysis could take account of one or more obstacles, for example, clutter close to the transmitter or receiver.

There are two standard methods to create an AA:

1. A grid of equal-sized square pixels, with the calculation of link metric X undertaken at the centre of each pixel. This approach is better at handling cases where the parameter under study does not monotonically increase or decrease with distance – for example, when there

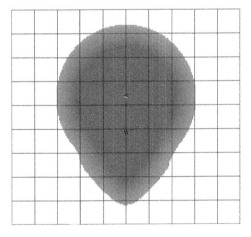

Figure 6.12 Example area analysis block plot without (left) and with (right) terrain data. *Screenshot credit: Visualyse Professional. Terrain credit: SRTM*

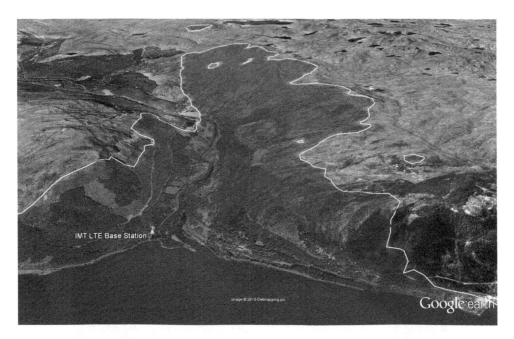

Figure 6.13 3D visualisation of coverage of IMT LTE macro base station. *Screenshot credit: Google Earth, image (c) 2015 Getmapping Plc*

   is terrain data. In addition, a number of analysis methods and propagation models such as location variability assume a pixel-based approach. It does, however, typically have larger computational requirements
2. Parameter X calculated on points along a set of lines radiating from a central location. This can be computationally efficient as path profiles can be reused for all points on that line, and for some scenarios, such as creation of coordination contours, it is the radial distance that is the required output. It is less able to handle variation between radial lines, particularly where terrain data is involved, and can lead to inaccuracies if data is merged onto a rectangular grid.

   The grid approach is more widely used as it is flexible, generic and accurate and maps onto pixel-based propagation models unlike the radial method. However, the radial method is appropriate for the generation of coordination contours and so is included in the Appendix 7 algorithm described in Section 7.4.

   An AA can also be viewed in 3D visualisation tools overlayered on terrain to show coverage of a wanted system (such as in Fig. 6.13) or hanging in space, showing the intersection of interference with a satellite's orbit (as in Fig. 6.14). Multiple AAs created at different heights can be used to define 3D volumes.

6.5.2   Boundary Analysis

Boundary analysis considers the edge of a polygon rather than its internal area. This is often an important concept for coordination between countries, where a *PFD* or field strength has been

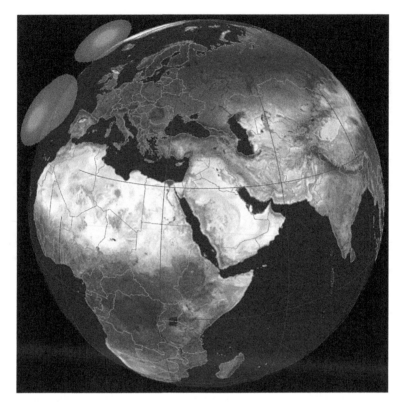

Figure 6.14 AA at height of a satellite's orbit. *Screenshot credit: Visualyse Professional. Overlay credit: NASA Visible Earth*

defined on the border or coastline. In this case, it is necessary to test whether (say) a new transmitter will meet or exceed the threshold and hence have to be coordinated with the neighbouring administration. To do this, it is necessary to define a set of test points along the line sufficiently close together to capture the variations in terrain.

Example 6.6

The static analysis in Example 6.2 was extended to undertake a boundary analysis using a set of test points deployed every 200 m along a border. At each point, the *PFD* from the IMT LTE BS was calculated and then compared against the threshold of $T(PFD) = -154.5$ dBW/m^2/4 kHz from 5.430A. Figure 6.15 shows the boundary line and how the *PFD* varies along this line.

It can be seen that there is a large variation in *PFD* due to the different terrain path profiles between the boundary test points and the base station so that, that in some circumstances, the *T(PFD)* is exceeded by a considerable margin. While in many cases the use of terrain can reduce interference, in some cases, it can make it worse if it raises the height of the transmit antenna over a flat plain.

Boundary analysis can be extended to define a border zone, usually specified via a distance from the border line that must be protected.

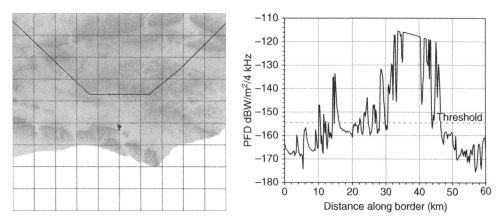

Figure 6.15 Example border polygon (left) and PFD along border (right)

Combinations of boundary line and AA can be used to study more complex deployments, such as:

- Communication to/from ships and aircraft, defined by sea and air routes, respectively
- Communication to/from vehicles, defined by roads
- Communication to/from pedestrians, focussing on traffic hot spots.

This increase in detail in the definition of service locations should lead to more accurate results – but they will become ever more specific to that scenario.

The AA approach can be extended to more detailed analysis at each point – such as the area Monte Carlo methodology described in Section 6.10.

6.6   Minimum Coupling Loss and Required Separation Distance

The static analysis methodology in Section 6.3 used a set of inputs, including well-defined locations, to derive various link metrics, which could then be compared against thresholds. This was extended in Section 6.4 by varying one or more parameters, in particular geographic or frequency separation, and identifying the impact on interference. It was noted that this is a particularly valuable technique, as separation is a useful metric to assess the difficulty of two services or systems sharing the radio spectrum.

The MCL methodology reverses the standard link budget in order to calculate the loss and hence separation required to meet a specified threshold. From Equation 5.63, the generic equation to calculate the I/N is

$$\frac{I}{N} = P'_{tx} + G'_{tx} - L'_p + G'_{rx} - L_f + A_{MI}(tx_I, rx_W) - N \tag{6.5}$$

This I/N is then compared against a threshold, with the requirement for compatibility:

$$T\left(\frac{I}{N}\right) \geq \frac{I}{N} \tag{6.6}$$

The MCL approach reverses this, using the threshold condition to calculate the require separation, so from

$$T\left(\frac{I}{N}\right) \geq P'_{tx} + G'_{tx} - L'_p + G'_{rx} - L_f + A_{MI}(tx_I, rx_W) - N \tag{6.7}$$

It follows that

$$L'_p - A_{MI}(tx_I, rx_W) \geq \left[P'_{tx} + G'_{tx} + G'_{rx} - L_f\right] - N - T\left(\frac{I}{N}\right) \tag{6.8}$$

While this approach is based on I/N, other link metrics could be used. For example, ERC Report 101 (ERC, 1999) describes a MCL method that calculates the maximum interference based on the wanted system operating at a receive level 3 dB above the reference sensitivity. In this case, the interferer must be limited to the noise floor to protect against harmful interference. This approach is making a number of assumptions about how the victim system link has been engineered that will not be appropriate for all scenarios. In particular, it should be noted that the Report was developed based on an example scenario involving mobile networks.

The key to MCL is to work from the maximum permitted level of interference (whether using a I, I/N or C/I metric) back to the necessary total loss due to geographic or geographic plus frequency separation between the interfering transmitter and victim receiver. This will, however, require some assumptions about geometry to calculate the antenna gains, as these will be considered as inputs into the MCL.

For co-frequency scenarios, which use the bandwidth adjustment factor A_{BW} rather than mask integration factor A_{MI}, it is possible to use an alternative form of Equation 6.8 in which the output is simply the required path loss:

$$L'_p \geq \left[P'_{tx} + G'_{tx} + G'_{rx} - L_f + A_{BW}\right] - N - T\left(\frac{I}{N}\right) \tag{6.9}$$

Example 6.7
The parameters from the static analysis in Example 6.1 were used as inputs into a co-frequency MCL calculation. The input parameters and derived values are shown in Table 6.14.

The result from the MCL calculation is the minimum loss between interfering transmitter and victim receiver required to ensure the scenario just meets the specified threshold. This is often developed further to get more useful information, such as:

- For co-frequency scenarios, the required separation distance given a propagation model and its associated parameters
- For non-co-frequency scenarios, the required separation distance for a specified frequency offset given a propagation model and its associated parameters
- For non-co-frequency scenarios, the required frequency offset for a specified separation distance given a propagation model and its associated parameters.

Table 6.14 Example MCL calculation

Field	Symbol(s)	Value	Calculation
Interferer power (dBW)	P'_{tx}	13.0	—
Interferer peak gain (dBi)	$G'_{max}(tx)$	18.0	—
Interferer relative gain (dB)	$G'_{rel}(tx)$	−18.4	Using F.1336-4
Receive peak gain (dBi)	$G_{max}(rx)$	36.9	—
Receive relative gain (dB)	$G_{rel}(rx)$	−29.9	Using S.465
Receive feed loss (dB)	L_f	0.0	—
Bandwidth adjustment factor	A_{BW}	0.0	Using Equation 5.5
Receiver temperature (K)	T	100.0	—
Victim bandwidth (MHz)	BW_V	30.0	—
Noise in victim bandwidth (dBW)	N	−133.8	Using Equation 3.43
Threshold I/N (dB)	$T(I/N)$	−13.0	—
Required path loss (dB)	L'_p	166.4	Using Equation 6.9

For some propagation models, it is relatively straightforward to go from required propagation loss to distance. An example would be free space path loss for which the calculation is

$$d_{km} = 10^{[L_{fs}-32.45-20\log_{10}(f_{MHz})]/20}$$ (6.10)

Propagation models such as Hata/COST 231 have a general format:

$$L_p = A + B\log_{10}(d_{km})$$ (6.11)

where A and B depend upon antenna heights, frequencies and environment (urban/suburban, etc.). This can be reversed simply using

$$d_{km} = 10^{[L_p-A]/B}$$ (6.12)

Other propagation models, such as P.452, P.1546, P.1812 and P.2001, are too complex to be reversed analytically, and so alternatives must be used. An efficient approach is to use a binary search algorithm, starting with distances known to be below and above the target. Alternatively a set of input variation analysis runs can be undertaken, such as those in Examples 6.3 and 6.4.

Example 6.8
The MCL methodology in Example 6.7 was extended to study non-co-frequency scenarios by using the mask integration adjustment factor A_{MI}. The coupling loss for a range of differences in centre frequencies was calculated and then the separation distance required derived using a binary chop search algorithm and the smooth Earth propagation model in P.452 for 20% of time. The results are shown in Figure 6.16.

In this example, the minimum separation distance required for large frequency offsets was a little over 5 km. This is likely to be conservative and the reality a lot better: at these frequency offsets, the TX and RX spectrum masks are defined by their spurious domain values, which, as

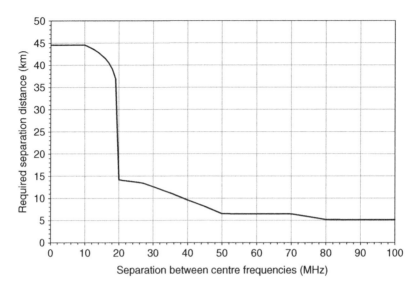

Figure 6.16 Required separation distances for given differences in centre frequencies assuming P.452 propagation model

discussed in Section 5.3.6, are conservative and relevant for occasional spikes rather than continuous values.

Sets of geographic plus frequency separations can be used as part of, or the basis of, frequency assignment processes. Examples of methodologies based upon these concepts include Rec. ITU-R SM.337 (ITU-R, 2008b) and the planning of private land mobile systems in Australia, documented in LM 8 (ACMA, 2000). In the latter, there is a multistage process that involves the culling (i.e. removal) of frequency channels that do not meet a set of (frequency, distance) criteria. These parameters would have been developed using MCL style techniques as described in this section.

The MCL methodology can be extended by using analytic methods (described as the 'Enhanced MCL' method in ERC Report 101) as covered by the following section.

The separation distance is a useful output as it gives an idea of the difficulty of sharing and indication of regulatory options. It can be combined with the input variation method to see how sensitive the required separation distance is to changes in input parameters.

Example 6.9
The MCL methodology in Example 6.7 was extended to analyse how the required separation distance calculated using the P.452 propagation model varied according to ES elevation angle. The results are shown in Figure 6.17.

This approach can help understand some of the unmanageably large number of potential static analyses that could be considered. However, there are limitations in all studies that use methodologies based on static analysis, including input variation, area and MCL analysis. In particular, they are unable to model the dynamic behaviour of stations to derive metrics such as event lengths or likelihoods of interference. For this it is necessary to consider dynamic analysis (in Section 6.8) and Monte Carlo analysis (in Section 6.9).

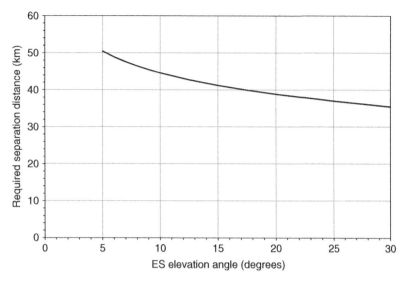

Figure 6.17   Separation distance required as a function of ES elevation angle

6.7   Analytic Analysis

Analytic methods can be used to directly calculate the answers to some interference questions without the need for simulation. These problems tend to be highly specific and controlled, but this approach can have benefits, in particular speed and accuracy, if a solution is available. The downside is that often the scenario is too complex to be solved via a simple set of equations, or in order to be solvable, it must be oversimplified. This approach is becoming less prevalent since the widespread availability of sophisticated simulation tools. However analytic methods continue to have a role in interference analysis as they can be combined with other methods, such as pre-processing inputs or post-processing the results of simulations.

Example 6.10
The MCL methodology in Example 6.7 suggested that a propagation loss of 166.4 dB is required between the IMT LTE BS and the ES to ensure the I/N threshold is met. Assuming the Hata/COST 231 propagation with parameters $A = 100$ dB and $B = 35.7$ dB, what would be the area of the interference zone?
From Equation 6.12,

$$d_{km} = 10^{[166.4 - 100]/35.7} = 72.4 \text{ km}$$

Hence the area of the interference zone is

$$A = \pi(72.4)^2 = 16,483 \text{ km}^2$$

This is unlikely to be correct because:

- The distance calculated is greater than the maximum recommended for the Hata/COST propagation model as specified (though within the range for the version in ERC Report 68)
- The interference area is assumed to be circular, but as was seen in Section 6.5, the zone is likely to have a keyhole shape
- It was assumed that the geometry was flat and there are likely to be differences compared to using a more realistic spherical Earth
- Hata/COST 231 gives the median or 50% of time loss, but the ES threshold is for 20% of the time.

It would be preferable to use an AA methodology on a spherical Earth with a more sophisticated propagation model such as Rec. ITU-R P.452, as in Section 6.5.

A more useful example of the application of analytic methods involves the calculation of probabilities for non-GSO satellite scenarios. Consider a satellite in a circular orbit with radius R and inclination angle i. If the period of the orbit is not a fraction of the Earth's rotation, its ground track (as described in Section 3.8.4.5) will gradually drift until all points on the truncated sphere with radius R called the **orbit shell** (as shown in Fig. 6.18) will be populated.

This orbit shell is defined in terms of (latitude, longitude) as it takes into account the rotation of the Earth. Satellite orbits remain either fixed in inertial space (assuming a point mass model)

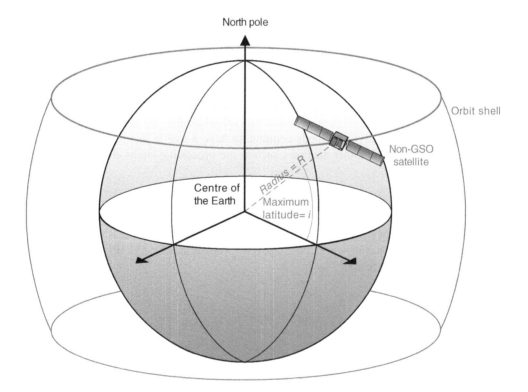

Figure 6.18   Orbit shell for non-GSO satellite

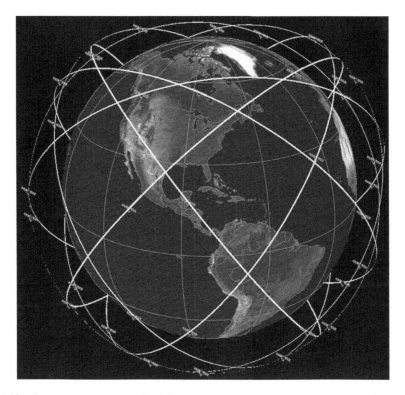

Figure 6.19 Space tracks of network LEO-D in Rec. ITU-R M.1184. *Screenshot credit: Visualyse Professional. Overlay credit: NASA Visible Earth*

or drift slowly (assuming other orbit propagators such as J_2). The orbit shell is the result of these **space tracks** (as in Fig. 6.19) being smeared out due to the rotation of the Earth.

Due to symmetry around the polar axis from the Earth's rotation, all longitudes in the orbit shell will be equally likely, but the probability of a satellite being at a specified latitude varies and depends upon the inclination angle. A simplified method to define the position vector of a circular orbit satellite is

$$r = R \begin{pmatrix} \cos(2\pi\tau) \\ \sin(2\pi\tau)\cos(i) \\ \sin(2\pi\tau)\sin(i) \end{pmatrix} \tag{6.13}$$

Here τ is in the range [0, 1] with equal likelihood and hence can be used as the probability metric. Therefore,

$$\sin(lat) = \frac{z}{R} = \sin(2\pi\tau)\sin(i) \tag{6.14}$$

So

$$\tau = \frac{1}{2\pi}\sin^{-1}\left[\frac{\sin(lat)}{\sin(i)}\right] \tag{6.15}$$

Note there are two solutions, one for the satellite latitude increasing and the other decreasing. Hence, by multiplying by two, the probability that a satellite will be in a given latitude range $[lat_1, lat_2]$ where $lat_2 < lat_1$ can be calculated as

$$p(lat_1, lat_2) = \frac{1}{\pi}\left\{ \sin^{-1}\left[\frac{\sin(lat_2)}{\sin(i)}\right] - \sin^{-1}\left[\frac{\sin(lat_1)}{\sin(i)}\right]\right\} \tag{6.16}$$

Example 6.11
For a satellite in circular orbit with inclination angle $i = 45°$, how much more likely is it to be within 5° of its maximum/minimum latitude than within 5° of the equator?

The probability density function (PDF) of this satellite is shown in Figure 6.20 using 1° bins. The satellite will be above 40°N for 13.7% of the time and a similar percentage below 50°S. It will be within 5° of the equator for 7.9% of the time. Hence it is 3.5 times more likely to be at latitudes greater than 40° than within 5° of the equator.

Intuitively this can be understood by noting the satellite has the most rapid change in latitude as it passes through the equatorial plane and least change when at its highest latitude where it is moving due east (or, very rarely, west).

This theoretical approach was extended in Rec. ITU-R S.1257 (ITU-R, 2002b) to derive formula for a number of useful metrics, including:

- The probability that a satellite is in a certain range of locations in its orbit, in particular, in the main beam of a terrestrial or Earth station's antenna
- The time taken for a non-GSO satellite to traverse the main beam of a terrestrial or Earth station's antenna
- The azimuth that would give the highest likelihood of there being a satellite within the main beam of a terrestrial or Earth station's antenna.

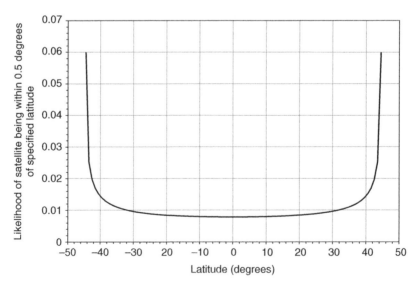

Figure 6.20   Example PDF of satellite latitude

The Recommendation includes theory, derivation, worked examples and comparison against simulation.

In most cases, analysis using simulation tools is simpler and more flexible in generating statistics such as the probability of interference and event lengths. However, it can be very helpful to have an indication of which azimuth is most susceptible to interference, in particular to:

- Configure a simulation for the worst-case scenario
- Analyse a database of receive stations (e.g. point-to-point fixed link assignments) to identify which might be most susceptible to interference.

The worst-case azimuth is calculated at a given latitude and for a given terrestrial station antenna's elevation angle and is based on assumptions that the non-GSO satellite constellation is non-repeating and in a circular orbit. If there are multiple non-GSO satellites in a constellation, it is assumed they all have the same orbit radius and inclination angle.

The inputs are:

i = non-GSO satellite orbital inclination angle
ε = elevation angle of terrestrial station
L_0 = latitude of terrestrial station
R = radius of non-GSO orbit
R_e = radius of the Earth.

It is first necessary to calculate the geocentric angle θ_ε corresponding to the elevation angle using

$$\alpha = \sin^{-1}\frac{R_e}{R}\sin\left(\varepsilon + \frac{\pi}{2}\right) \tag{6.17}$$

$$\theta_e = \pi - \left(\varepsilon + \frac{\pi}{2}\right) - \alpha \tag{6.18}$$

Then the worst-case azimuth from true north can be calculated using

$$\Lambda_1 = \cos^{-1}\left[\frac{\pm \sin i - \cos\theta_\varepsilon \sin L_0}{\sin\theta_\varepsilon \cos L_0}\right] \tag{6.19}$$

Note that only one of the ± signs will give a valid azimuth and that due to geometric symmetry there is another worst-case azimuth at

$$\Lambda_2 - 2\pi \ \Lambda_1 \tag{6.20}$$

These are calculated assuming a hypothetical beamwidth of zero: in reality the worst-case azimuth will be slightly offset, usually by between half and twice the 3 dB beamwidth.

Example 6.12
A non-GSO constellation is proposing to use the parameters of LEO-F from Rec. ITU-R M.1184 (ITU-R, 2003a) with height $h = 10,355$ km, implying orbit radius $R = 16,733.1$ km,

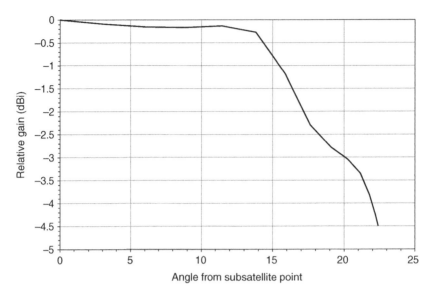

Figure 6.21   Satellite gain pattern for mask of spot beams

and inclination angle $i = 45°$. An operator of terrestrial fixed links at a latitude of 51°N is concerned about interference from the satellite's service downlinks: what is the worst-case azimuth for an elevation angle $= 0°$? The satellite's emissions have been simplified to a peak *EIRP* density $= 46.1$ dBW/MHz and satellite antenna gain mask (i.e. combination of all the spot beams) as in Figure 6.21.

From the equations earlier, converting the angles into radians:

$$\alpha = \sin^{-1}\frac{6378.1}{16,733.1}\sin\left(0+\frac{\pi}{2}\right) = 0.391\,radians$$

$$\theta_e = \pi - \left(0+\frac{\pi}{2}\right) - 0.391 = 1.180\,radians$$

So

$$\Lambda_1 = \cos^{-1}\left[\frac{\sin 0.785 - \cos 1.190\sin 0.890}{\sin 1.180\cos 0.890}\right] = 0.787\,radians = 45.1°N$$

However with a fixed link antenna with beamwidth of 5° and peak gain $= 30$ dBi, the actual worst-case azimuth is slightly offset, as can be seen by the plot of average *I/N* (*FDP*) by azimuth in Figure 6.22. This plot can be generated through simulation using dynamic, Monte Carlo or probabilistic methodologies as described in the following sections.

Example 6.12 shows how analytic methods can be used to prepare for a more detailed simulation by identifying sensitive cases and geometries. Another example of

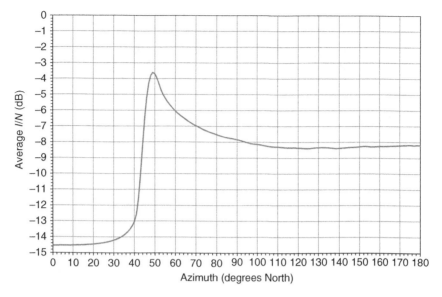

Figure 6.22   Average I/N by FS receiver azimuth

this is the worst-case geometry (WCG) for analysing non-GSO FSS networks against *EPFD* limits in RR Article 22 as described in Section 7.6, though this is sufficiently complex to be described as an algorithm rather than an analytic methodology. The example also showed how analytic methods often require simplification, in this case that the antenna beamwidth was zero and that using actual non-zero beamwidths led to different results.

The theory defining the probability that a satellite might be in a specified latitude range can be extended further to develop a probabilistic simulator that generates interference statistics by considering all possible states, as described in Section 6.10.

6.8   Dynamic Analysis

The previous sections have described methodologies that give snapshots of interference, sometimes with geographic variation, but without providing information of how it changes in the time domain. In reality radio systems, both wanted and interfering, are continually changing due to:

- Stations and their platforms move, including pedestrians, vehicles, ships, aircraft and satellites, as in Figure 6.23
- Communication traffic varies, as services from low-rate data to streaming video are switched on or off
- Propagation conditions vary, so that reflections, rain, sleet, snow or sand can cause additional attenuation or enhancements.

These variabilities can have a host of secondary effects, so that traffic can change transmit power, station movement alters antenna gains, movement and propagation effects could require

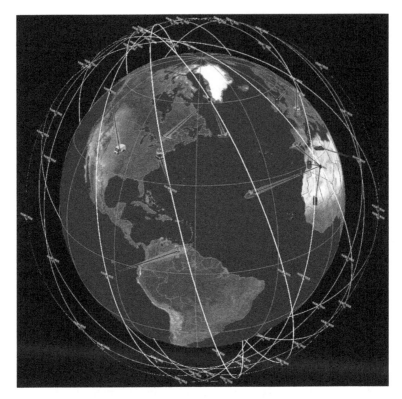

Figure 6.23 Dynamic analysis example with moving pedestrians, vehicles, ships, aircraft and satellites. *Screenshot credit: Visualyse Professional. Overlay credit: NASA Visible Earth*

a response by power control and rain can increase the noise on a link. A static analysis can only go so far to analyse phenomena that vary in the time domain, and so it is necessary to use alternative methodologies such as **dynamic analysis**, which can answer:

- What is the likelihood of interference?
- What is the CDF of interference?
- How often will there be interference events?
- What is the longest interference event that can be expected?

Dynamic analysis requires methods to specify how input parameters vary in the simulation's time domain. For example:

- Terrestrial stations, such as ships and aircraft, could move on great circle paths as described in Section 3.8.3.3. Multiple great circle paths could be combined to create shipping and aircraft routes
- The position of satellites, both GSO and non-GSO, can be predicted using the methods described in Section 3.8.4.4

- Propagation models tend not to specify explicitly how the loss varies in time and so it is necessary to use Monte Carlo methods and sample at random as described in Chapter 4.

It is also necessary to define two key parameters for a dynamic simulation: the time step size and number of time steps (or equivalently the run duration). The simulation then iterates over all the time steps and for each one updates the dynamics and propagation models before undertaking the interference calculation, which could involve any of the metrics, whether $\{C, I, C/I, I/N, C/(N+I), PFD, EPFD\}$. With multiple time steps, it will be possible to derive statistics such as average and worst value or show the full CDF, as in Figures 5.46 and 5.47. If a threshold is given for each of these metrics, then it is possible to calculate interference event statistics, such as number and duration. Note that with N time steps there will be N+1 samples including the zero-th step i.e. initial configuration.

The selection of the time step size and number of time steps can require significant consideration to ensure results are valid and accurate. Some of the issues and potential resolutions are demonstrated in the following examples. While the examples are based on satellite dynamics, they show generic issues that are applicable to other scenarios.

Example 6.13

For the non-GSO constellation with parameters in Example 6.12, what are the characteristics of an interference event when a satellite is in the main beam of a terrestrial point-to-point FS link with (Az, El) = (180°N, 0°E) located at (latitude, longitude) = (51°N, 0°E) as in Figure 6.24?

The plot of interference at the FS receiver against time is shown in Figure 6.25 for two time steps, 10 and 500 seconds.

This example shows one of the potential issues in the selection of the time step: it must be small enough to capture significant changes in the interference level. In this case, using a time step of 500 seconds led to an underestimation by nearly 15 dB of the worst interference level. The fine time step also shows two characteristics of this scenario:

- The time profile of the highest interference event is closely related to the victim antenna main lobe pattern, which in this case is parabolic
- Only half of the main lobe is visible, as when the satellite is in its 'lower half' it is below the horizon, that is, elevation angle is less than zero. If the victim had been pointing at a higher elevation angle, for example, if it was a satellite Earth station, then the whole of the main lobe would be visible in the interference plot.

On the other hand, it is important not to have too fine a time step in order to ensure run times are manageable. A suitable time step size can be calculated from the victim antenna beamwidth, the rate at which the interferer moves through that beam and the required number of samples or hits, N_{hits}, of an interference event.

The accuracy will be higher with a smaller time step and hence larger N_{hit}. A range of values have been suggested, such as $N_{hit} = 5$ in Rec. ITU-R S.1325 (ITU-R, 2003f) for studies to $N_{hit} = 16$ in Rec. ITU-R S.1503 (ITU-R, 2013p) for regulatory approval analysis. With a parabolic gain pattern, this results in errors within the main beam that vary as in Table 6.15.

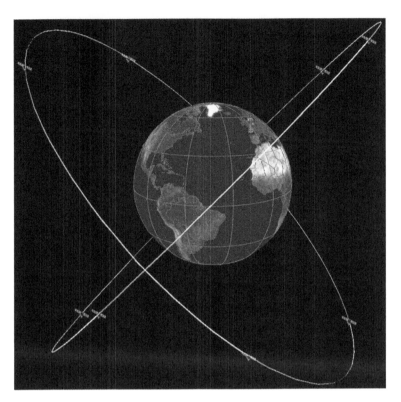

Figure 6.24 Example non-GSO MSS constellation interferer with victim FS receiver. *Screenshot credit: Visualyse Professional. Overlay credit: NASA Visible Earth*

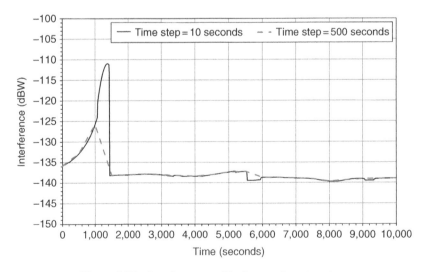

Figure 6.25 Interference profile for two time step sizes

Table 6.15 Maximum errors for given values of N_{hit}

N_{hit}	5	16
Maximum error at centre beam (dB)	0.12	0.01
Maximum error at 3 dB beam edge (dB)	1.08	0.36

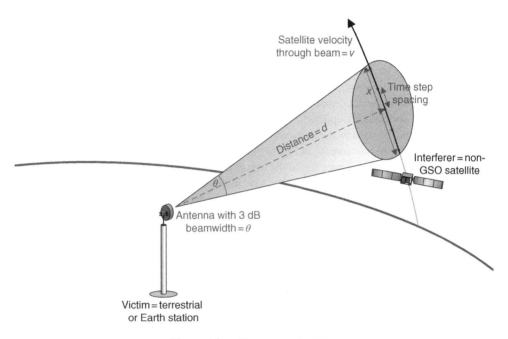

Figure 6.26   Time step calculation

Using the geometry as shown in Figure 6.26, a simplified set of equations to calculate the time step are

$$x = 2d \sin \frac{\theta}{2} \tag{6.21}$$

$$t_s = \frac{x}{v N_{hit}} \tag{6.22}$$

Other more detailed methodologies are given in S.1325 and S.1503 that take account of the rotation of the Earth.

Example 6.14

For the scenario in Example 6.13, what would be a suitable time step assuming $N_{hit} = 5$ or 16?

Calculating the velocity of a non-GSO satellite using Equation 3.147, the derived time step sizes are as shown in Table 6.16.

It can be seen that a time step size of 500 seconds would be too large, but 10 seconds would be sufficiently accurate for most purposes.

Table 6.16 Calculated time steps for Example 6.14

N_{hit}	5	16
d (km)	15,469.9	15,469.9
x (km)	1,349.6	1,349.6
v (km/s)	4.9	4.9
Time step (s)	55.3	17.3

This calculation gives the time step that captures a main beam geometry interference event: it is also necessary to consider the run duration and number of samples needed to achieve the required resolution in the statistics.

The run duration is the total time from beginning to end of the dynamic simulation during which the station positions are updated at each time step. The run duration is usually determined by the geometry and the need to achieve a uniform unbiased sampling of all possible station positions. The probability of interference in Figure 6.26 would be biased if the satellite went through the main beam twice but only once round the rest of its orbit. The objective is usually to run the simulation in the time domain until all stations have looped round all possible positions an integer value equal or greater than one. However, for some scenarios, the length can be either long or difficult to determine, and this leads to further questions about what statistics should be measured.

Example 6.15

The scenario from Example 6.12 involving calculating the FDP for a range of FS azimuths from a non-GSO MSS constellation was repeated using dynamic analysis. The run duration was set at 365 days with time step 10 seconds, requiring 3,153,600 steps. The resulting *FDP* vs. azimuth were compared against the result from a similar number of Monte Carlo samples as in Figure 6.27.

The following resources are available:

Resource 6.6   Visualyse Professional simulation file 'MSS into FS worst azimuth.sim' was used to generate the results in Figure 6.27.

Resource 6.7   Visualyse Professional simulation file 'MSS into FS worst azimuth – new height.sim' was used to check the impact of changing the satellite orbit height.

It can be seen that there is a very significant difference between the two plots of *FDP* vs. azimuth. The dynamic simulation *FDP* differs by about 3.5 dB from the Monte Carlo result, which raises the following questions: what is causing the delta, and which curve is correct?

The key to understanding the difference is to plot the (lat, long) points, which represent the position of one of the non-GSO constellation satellites over the 3.1 million time steps for both methodologies. These identify which of the points on the orbit shell described in Figure 6.18 will be sampled during the run, as shown in Figures 6.28 and 6.29.

The dynamic simulation was updating the position of the satellite using the orbit mechanics described in Section 3.8.4.4, while the Monte Carlo approach was selecting the satellite position at random using the method described in Section 6.9, which will ensure that the orbit shell is uniformly sampled.

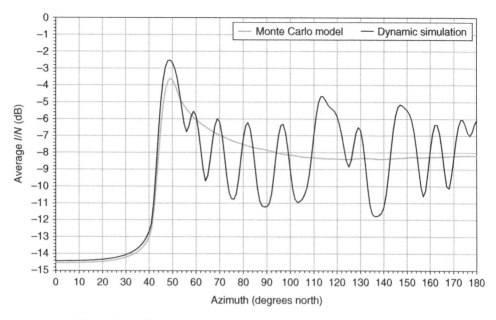

Figure 6.27 Comparison of dynamic and Monte Carlo simulation results

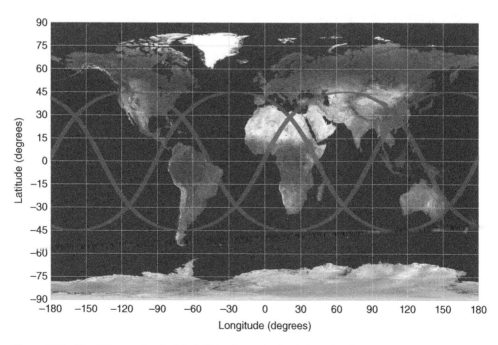

Figure 6.28 Partially populated orbit shell for dynamic simulation. *Overlay credit: NASA Visible Earth*

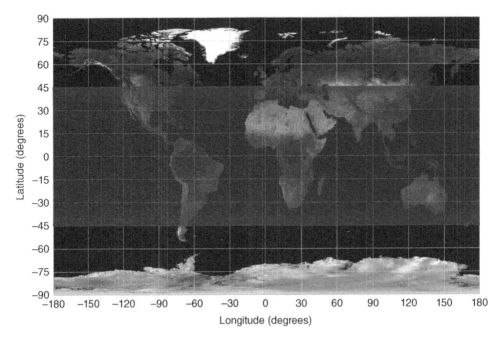

Figure 6.29 Fully populated orbit shell for Monte Carlo simulation. *Overlay credit: NASA Visible Earth*

For the dynamic simulation, the time taken for the satellite to complete one orbit assuming a point mass gravitational model can be calculated using Equation 3.142 as

$$P = 2\pi\sqrt{\frac{(16,733.1)^3}{398,601.2}} = 21,541.5 \text{ seconds} = 5.98 \text{ hours}$$

Note that the arguments are very similar though more complex for other orbit models such as J_2.

After this time, the satellite will have returned to the same location in ECI coordinates, while the Earth will have rotated an angle that can be calculated using Equation 3.128:

$$\theta = \omega_e \nabla t = 0.004178074°/s \times 21,541.5\, s = 90.0022°$$

Hence in four orbits, the Earth will rotate

$$4\theta = 360.008° = 0.0088°$$

So the LEO-F constellation is very close to be synchronous. A year is about 366 sets of 4 orbits, so the total drift of the constellation over 365 days is only

$$366.4\theta = 366 \times 0.0088° = 3.2°$$

Therefore the satellites will not populate the entire orbit shell within 1 year, and Figure 6.27 is an accurate prediction of what point-to-point fixed links would actually observe over 1 year. To fully sample the orbit shell, it would be necessary to run for $90°/3.2° \sim 28$ years.

This represents a significant computational requirement, and for this reason, Rec. ITU-R M.1319 (ITU-R, 2010a) suggested the use of artificial precession. This adds an orbit precession factor so that over the required time period, each satellite will return to its starting point, completing the population of the orbit shell. Adding an artificial precession rate of 0.995°/day to the longitude of the ascending node to each satellite is enough to ensure that after 365 days, the constellation has completed one full rotation around the equator.

However, it is worth noting what this mechanism is actually doing: it is averaging the statistics over 28 years and yet most interference thresholds are defined for average annual or worst month. Satellites do not actually use either artificial precession or Monte Carlo methods: they follow the orbit prescribed by gravity and perturbations such as the atmosphere. Using these methodologies can, therefore, hide a potential interference issue due to the variation in interference from a non-GSO constellation that is close to a repeating ground track.

While dynamic simulation methodologies require that these types of issues be considered, unlike the simpler Monte Carlo approach, it is worth noting that this is a useful and informative process, which identified a real interference issue. The analysis also can be used to identify a solution, in that the problem came about because of the close synchronisation between the Earth's rotation period and that of the orbit. By changing the satellite's height by only a small amount, it is possible to break this connection and ensure that the orbit drifts at a rate sufficient to fill the orbit shell in a year. In this case raising the height by 7.35 to 10,362.3 km is sufficient to ensure a smoother distribution of satellite location likelihoods, and the FDP vs. azimuth plot is then the same as for the Monte Carlo analysis.

This change in orbit height can be (and has been) considered as an interference mitigation technique: by increasing the orbit drift, interference is averaged quicker, leading to lower variation ranges in interference across the Earth's surface. This mitigation technique would not have been discovered by Monte Carlo or probabilistic methodologies.

However, some systems do use repeating tracks for good reasons, such as:

- Satellite networks with ground tracks that are designed to repeat, using station keeping fuel to maintain a repeating track (e.g. the proposed SkyBridge constellation to facilitate resource management or Earth exploration satellites that aim to maintain the same angle with the sun when over selected locations)
- Aircraft on routes between airports will always use the same start and end points
- Ships sailing towards ports will not deviate from the navigational part of the channel.

In none of these cases will there be an exact repeating of the track as there will be variations due to perturbations, weather, tide and pilot choice. A dynamic simulation can also capture this variability, so that the track is not a simple line but a ribbon, where the width represents the scope for cross-track motion.

It is also necessary to ensure there are enough tracks through the main beam in a similar way to ensuring the time step is sufficiently fine, as can be seen in Figure 6.30.

The final consideration is the need for sufficient samples to gain statistical significance at the required percentage of time.

Example 6.16

A simulation is considering interference from an Earth Station on a Vessel (ESV) into fixed links pointing perpendicular to the shoreline. The ship is sailing parallel to land at a speed of 20 knots (nautical miles/hour) at a distance of 20 km offshore. Using Equations 6.21 and

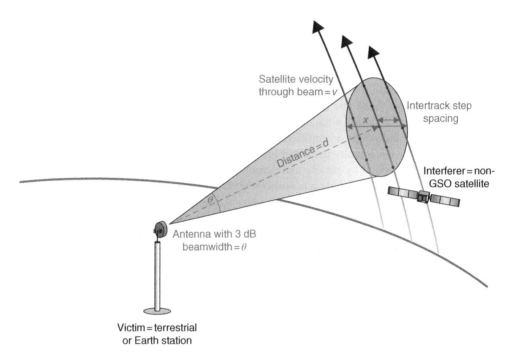

Figure 6.30   Ensuring sufficient intertrack resolution

Table 6.17   Calculation of time step and number of time steps for ship example

Total ship track length	200 NM
Distance to ship's track = d	20 km
Ship speed	20 knots
$N_{hit} = N_{track}$	5
Time step	33.9 s
Total number of time steps	5305

6.22 and N_{hits} = 5, the time step size can be calculated as in Table 6.17. With a similar number of N_{tracks} = 5 through the FS receive beam and a simulation range of 100 nautical miles (NM) either side of the beam centre (i.e. 200 NM in total), the number of time steps is then 5305. The lowest probability that can be calculated is 0.01885%.

However, this is likely to be insufficient to calculate short-term interference statistics. While it is best to use statistical tests of significance, a rule of thumb is that at least 20 samples are needed below the threshold probability. Hence the minimum number of samples required to resolve a threshold percentage p_{min} is

$$N_{min} = 20\frac{100}{p_{min}} \tag{6.23}$$

For a threshold of $p = 0.01\%$ of the time, it is therefore necessary to have at least 200,000 time steps.

One approach is to specify the total run time due to the dynamics involved and then adjust the time step size, if needed, to create the required number of samples. If working with histograms and CDFs, it will also be necessary to select a bin size fine enough to capture details of their distributions. For example, the algorithm in Rec. ITU-R S.1503 (ITU-R, 2013p) specifies a bin resolution of 0.1 dB.

As well as station motion, dynamic analysis can involve variations in the propagation conditions. Most of the models defined in ITU-R Recommendations are probabilistic in nature, meaning that they give likelihoods of propagation loss, fade or enhancement, not predictions of how these vary in the time domain. When modelling propagation and station dynamics, it is therefore necessary to use a hybrid approach, part Monte Carlo (for the propagation) and part-time sequence (for the station positions). An alternative would be to use a pure Monte Carlo methodology as described in the following section. However that would lose the benefits of a dynamic simulation, such as the ability to answer questions relating to interference event lengths and likelihoods.

Multiple dynamic analysis runs can be used to undertake parametric or input variation analysis, identifying the impact of changes in assumptions on output values or statistics.

A significant benefit of dynamic analysis is its ability to assist in building a mental picture of a complex scenario and discuss results with others. Dynamic analysis is intuitive and widely implemented in simulation tools, which usually provide graphic feedback and visualisation features that improve understanding of the scenario being studied. This can often be the most important factor, as it is through understanding a problem that it can be analysed and preferably solved without requiring an exhaustive search over the whole range of input parameter permutations. In addition, it allows proposals and the results of studies to be explained to others within an organisation or at a meeting with others (e.g. at the ITU-R or in a coordination). This can be enhanced by providing snapshots – static analysis at key moments, such as when the interfering station is directly in the main beam of the victim.

6.9 Monte Carlo Analysis

6.9.1 Methodology

The static analysis method described in Section 6.3 showed how link or interference metrics such as $X = \{C, I, C/I, I/N, C/(N+I), PFD, EPFD...\}$ could be calculated from a set of inputs that completely defined a scenario at one instance. It would be necessary to specify one value of all of the required inputs in each of the categories of:

- Station locations plus antenna gain patterns and pointing angles
- Link budget parameters, including transmit powers, bandwidths, polarisations, frequencies, transmit spectrum mask and gain patterns
- Receiver characteristics including, where necessary, receive spectrum mask, noise temperature and feed loss
- Propagation environment including models and associated parameters, most importantly the associated percentage of time (if required).

The calculation would require a geometric framework to derive distances and angles and possibly access to geoclimatic databases including terrain and land use data.

The calculation of the derived link or interference metric X can be parameterised by the associated set of input variables $V_1...V_N$ as in

$$X = X(V_1, V_2...V_N) \qquad (6.24)$$

However static analysis only gives one snapshot, and in many cases, these inputs cannot be defined simply via a single value. For example:

- Stations move in time, as described in the previous section
- Antennas can change their pointing in time, such as a radar antenna rotating in azimuth or a remote sensing satellite scanning its target
- Traffic can appear at different locations across a service area – such as mobile users within a network's coverage
- Users can also require a range of communication links with different parameters such as bandwidth and target RSL
- Users can change the frequency they are operating on due to a system's multiple access method
- Propagation varies throughout the day and year as the weather changes.

From these variations, secondary effects can occur – for example, changes in location can lead to alterations in transmit power if APC is used. There can also be uncertainty: some of these inputs could only be known to a certain accuracy or via a distribution of likelihood.

Repeatedly undertaking static analysis changing one of the inputs (e.g. V_1) to be one of a number m of different values will result in a set of Xs as output:

$$X[1...m] = X(V_1[1...m], V_2...V_N) \qquad (6.25)$$

This approach was used in the input variation analysis to plot, for example, how a link metric like I/N varied with distance. The set of output values can also be used to generate statistics including mean, standard deviation and the cumulative distribution function (CDF).

By choosing the set of inputs $V_i[1...m]$ at random from a distribution such that each value selected has an equal likelihood, it is possible to generate a series of outputs $X[1...m]$, which also have equal likelihood. Additionally, if the inputs are selected from the full range of feasible values, then the outputs will correspondingly cover the full range of possibilities. From the resulting distribution $X[1...m]$, it is then possible to derive statistics relating to the output variable X, in particular probabilities.

This approach is described as **Monte Carlo analysis** and can include variations in any number of parameters so that the resulting statistics $S[X]$ are a convolution of all the input variable distributions, taking into account the underlying geometry, propagation environment and link budgets. These statistics could include the probability that a threshold is met (or exceeded) and an output variable's time average.

Figure 6.31 shows how the Monte Carlo method builds on static analysis: at the core is the deterministic calculation of metric X from one complete set of input parameters. Each static analysis is described as a **trial**.

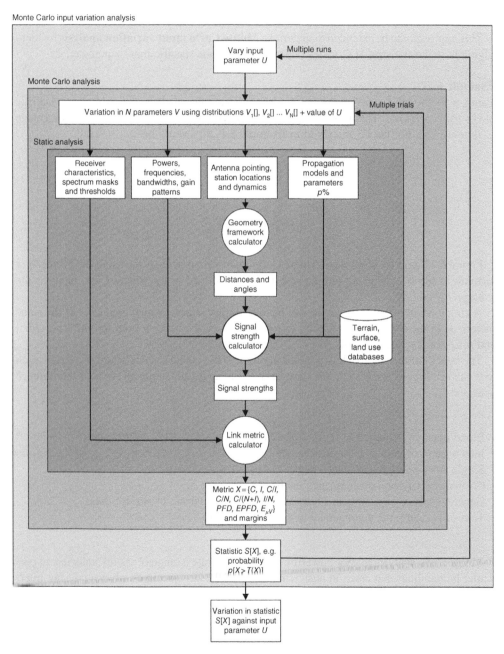

Figure 6.31 Flow diagram for static analysis, Monte Carlo methodology and Monte Carlo input variation analysis

If at each trial input variables are selected at random from their distributions so that samples have equal likelihood, then the result will be a distribution of output variable X where each value also has equal likelihood. By taking sufficient samples it is then possible to derive probabilities that X meets or exceeds a threshold $T(X)$, typically as a probability or percentage $p(X)$.

This approach can be extended to undertake **Monte Carlo input variation analysis** by identifying how the statistics $S[X]$ varies depending upon one specific input parameter, U.

Example 6.17

Using a metric of $X = I/N$, a Monte Carlo analysis can calculate the probability statistic $S[X] = p[I/N > T(I/N)]$. This probability can then be compared against a threshold percentage of time $p_T(I/N)$. For the ES threshold in Example 6.1, this would be

$$T\left(\frac{I}{N}\right) = -13\,\mathrm{dB}$$

$$p_T\left(\frac{I}{N}\right) = 20\%$$

Figure 6.22 gave an example of Monte Carlo input variation analysis, using the FS receiver azimuth as the U parameter, and the statistic being derived in the Monte Carlo analysis the average I/N or FDP.

A single Monte Carlo simulation can have multiple outputs, which could be all the same metric, as in these $FDPs$ by azimuth, or different, such as deriving both the $p[I/N > T(I/N)]$ and $p[C/(N+I) < T(C/(N+I))]$.

The benefit of this approach is it builds upon the static analysis calculation, and hence each trial can be checked and tested. It is very flexible in that it can handle all types of interference analysis:

- Based upon any metric $X = \{C,\ I,\ C/I,\ C/N,\ C/(N+I),\ I/N,\ PFD,\ EPFD,\ \text{etc.}\}$
- Inter-system or intra-system interference
- Single-entry or aggregate interference
- Co-frequency or non-co-frequency analysis
- Generic analysis (e.g. smooth Earth) or specific (with terrain or surface data)
- Terrestrial, maritime, aeronautical or satellite networks.

Monte Carlo analysis is very powerful, as it can generate a range of useful statistics, in particular probabilities that assist in answering many of the key questions when analysing interference. An example of the flexibility of the Monte Carlo approach is described in Section 7.9 'Generic Radio Modelling Tool' (GRMT), which derived statistics relating to interference using the following equation:

$$I = P'_{tx} + G'_{tx} - L'_p + G'_{rx} - L_f + A_{MI}(tx_I, rx_W)$$

Other implementations of Monte Carlo methodologies to analyse radio systems can be found in SEAMCAT, documented in ERC Report 68 (CEPT ERC, 2002)/ITU-R Report SM.2028 (ITU-R, 2002a) and Visualyse Professional.

Monte Carlo analysis also assists in reducing the computational problem described in Section 6.3, which identified that the total number of possible static analysis combinations for n variables with N_i combinations each was

$$N_{total} = N_1 N_2 N_3 \ldots N_n \tag{6.26}$$

Instead the Monte Carlo approach takes m samples of each input variable to create a histogram and hence CDF that gives likelihood of link metric X occurring given the distribution of these n input variables.

There are, however, various drawbacks and issues that must be considered relating to the Monte Carlo methodology:

- It needs the distributions of the input variables to be specified. In some circumstances, the distribution is clear, but in many, it requires additional assumptions – see Section 6.9.2
- It is necessary to consider how many trials should be calculated to ensure the required statistical significance of the outputs – or alternatively, given a set of trials, how significant are the resulting statistics – as discussed in Section 6.9.6
- A large number of trials might be needed to achieve the required level of statistical significance, with implications on the computational requirements
- Monte Carlo analysis does not provide information about event lengths or time between events, which would require dynamic analysis. Similarly the lack of state information means that statistics relating to successful call completion over multiple time steps cannot be generated directly
- Monte Carlo analysis would not, on its own, identify subtle synchronisation effects in the dynamics of stations, as highlighted in Example 6.15
- It is necessary that the variation in inputs is in the same dimension: in particular care must be taken when considering variations in likelihoods of fixed deployments and time variations, as these are not the same, as discussed in Section 6.10
- The distribution of values for each variable needs to be specified for the full range of probabilities [0, 1], but there can be restrictions (e.g. propagation models that are limited in their applicable percentages of time)
- It is necessary to be clear what output is being calculated and relate that to the required statistic, for example, interference threshold
- It is necessary to be clear which timescales are involved, whether microsecond, second, day, month or year (see discussion on radars in the succeeding text)
- There can be dependencies between variables that makes the assumption of independence invalid
- The analysis can become increasingly complicated, hence hard to understand and explain.

When undertaking Monte Carlo analysis, it is often helpful to start by identifying which question you are trying to answer (i.e. what outputs are required) rather than simply considering which inputs could be varied. It can also be useful to consider which questions would be asked by varying some of the main inputs.

A key requirement for Monte Carlo simulation is a good pseudorandom number generator, which has a long repeat cycle, uniform PDF and independence of samples. The generator should also be based upon a specifiable seed to allow runs to be repeated and to permit the comparison of results that come from the use of different input seeds.

6.9.2   Variation of Inputs

The most common form of Monte Carlo interference analysis is to study likelihoods or statistics in the time dimension. Hence a typical question to answer would be:

For the given scenario, what is the probability that the *C/I* would be below the threshold *T(C/I)*?

Therefore, the inputs that could be considered for variation would be those that change over time, such as propagation conditions, locations, frequencies, antenna pointing and activity. Power could be either varied directly or indirectly via locations and a power control algorithm.

Different types of input variable could be modified, for example:

- Booleans: is a link state = on or off, where there is a probability defined for one of the states
- Floating point number: a value selected from a distribution as described in Section 3.10, such as transmit power or antenna pointing angle
- From a set of equally likely possible values, such as a frequency from a channel plan as described in Section 5.1
- As a location, using a method that allows a position to be defined in three-dimensional space according to rules appropriate for that type of station.

How a location is varied will depend on the station and service type. For example, for mobile stations, the location could be defined as uniform across an area specified via a circle or hexagon as in Figure 6.32.

The coverage of PMR systems can be modelled as a circle centred on the base station, though more detailed analysis can consider locations where the wanted signal strength is above a threshold, as in Section 7.2. For IMT LTE systems, it is usually more appropriate to use a hexagonal deployment. The simplest assumption is that the mobile is equally likely to be located at any point within that area, and this is appropriate for generic studies or where precise traffic data

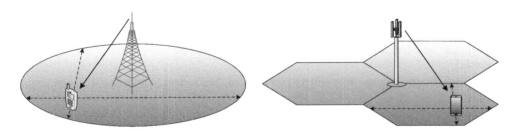

Figure 6.32   Variation in mobile location with defined area

is not available. If additional information is available for specific sites, then, for example, hot spots or individual streets could be modelled. Multiple, overlapping, areas could be specified, such as having a traffic load for a background area that contains a smaller hot spot at a specific location.

This variation in mobile locations could apply to either or both of the wanted and interfering systems. When used for the victim system, there are implications on the statistics generated between:

- Randomising the location of the mobile at each trial
- Fixing the location of the mobile for all trials, for example, at edge of coverage.

This has implications on the statistics generated and so is discussed further in the following section.

Aircraft and ships could be modelled using an area within which they would be equally likely to occur or alternatively on a line connecting waypoints, with the width of the line depending on how much variation there would be across that track.

A non-GSO satellite that does not use repeating ground tracks can be Monte Carloed by selecting at random the longitude of the ascending node and the mean anomaly:

$$\Omega = \Omega[0, 2\pi] \qquad (6.27)$$

$$M = M[0, 2\pi] \qquad (6.28)$$

For circular orbit systems, the mean anomaly can be replaced by the true anomaly ν. A repeating satellite constellation would require a model that selects one of the tracks at random with an appropriate amount of cross-track variation (e.g. due to station keeping range).

If the satellite is part of a constellation, then one satellite should be located at random using (Ω, M) and then the other satellite's locations defined in such a way as to maintain the necessary relative positions.

Some systems have antenna pointing that varies in time, such as a radar that scans around in azimuth with a fixed elevation angle. How this is modelled depends upon whether the radar is an interferer or a victim:

- Radar as interferer: the azimuth can be selected at random equally likely in the specified range, for example, [000°, 360°]
- Radar as victim: the threshold, such as I/N, must be met for each azimuth, and hence a set of fixed azimuths must be used, gathering statistics in each direction.

Another issue relates to the time domain of the simulation and the pulsed nature of the radar signal. If the interferer is a radar rotating in azimuth, then the interference it generates will vary in time due to the antenna pointing angle changing (relatively slowly) and due to the pulses (much more rapidly). It is necessary to identify the best way to handle these two scales, and this will directly impact the statistics generated, via one of the following:

1. Use the radar's transmit power averaged over the time between pulses: the statistics generated will accordingly be the (say) I/N where I is averaged over that time frame

2. Use the radar's peak power with an associated activity factor defined by the radar's duty cycle: the statistics generated will then be (say) the I/N averaged over the much shorter pulse duration
3. Use the radar's peak power without an activity factor: the statistics generated then will be (say) the peak I/N during a time frame, which should be greater than that between pulses.

In addition there can be variation in propagation at both the short- and the long-term timescales.

It is important to note that signal strength is always time averaged to a degree though usually over a period much shorter than the timescale of variation in the simulation.

The pointing angles of tracking antennas of non-GSO satellite Earth stations are dependent upon the satellite's dynamics and rules such as minimum elevation angles or the need to avoid certain directions (e.g. towards other satellites or the GSO arc). The azimuth and elevation of a GSO satellite ES's antenna will only vary if the satellite has a slight inclination or eccentricity.

A key variation in the time domain is propagation. One propagation model designed for Monte Carlo analysis is P.2001 as described in Section 4.3.7 as it accepts as input a random percentage in the range [0, 100%]. Other propagation models either have restricted range, such as P.452, which only covers $0.001\% \le p \le 50\%$, while others give the loss for a single percentage of time, typically the mean value for which $p = 50\%$. Neither of these are ideal: the former will bias the results as P.452 will not include situations when the interfering signal is faded, while the latter will not model time variation of radio wave propagation.

For short-range paths (e.g. mobile links in urban environments) while there would be less variation in propagation loss due to effects such as troposcatter, there will be changes in the multi-path and clutter loss as positions alter. An alternative in these circumstances could be to use an additional variation over the median loss via a distribution and standard deviation, as discussed in Section 4.3.5.

Propagation models can also have statistics that are valid over either worst month or average annual, with selection depending upon what output statistics are required, which will be determined by the relevant thresholds. When generating statistics over this time frame, it is worth checking that this has been taken into account by other variables. In particular, it would be preferable if the transmit power and traffic/activity factors should be defined for the average day rather than busy hour. It can be acceptable (or necessary, given availability of input data) to use busy hour traffic levels, but this should be noted as a conservative assumption during modelling.

There is the potential for subtle correlations between traffic and propagation in that radio wave propagation can be different at night time compared to during the day, and there is a similar variation in traffic levels between these two periods. However, given the lack of information about these factors, it is acceptable to assume they are independent.

6.9.3   Output Statistics and U Parameter Variation

Monte Carlo analysis can generate statistics for any of the link or interference metrics $X = \{C, I, C/I, C/N, C/(N+I), I/N, PFD, EPFD, E_{\mu V}\}$. These statistics $S[X]$ include:

- Probability that a threshold $T(X)$ is met or exceeded
- Value of X at the required probability threshold $p_T(X)$
- Average and standard deviation of X.

In addition to these link or interference metrics, Monte Carlo analysis can be used to analyse any derived value, such as

- Transmit power, possibly including aggregation. An example would be to use a full Monte Carlo simulator of traffic within an IMT LTE cell to calculate average transmit power, which could be used in another simulation
- Antenna pointing angles, for example, to calculate likelihood of a non-GSO Earth station pointing in a certain direction
- Antenna gain, in particular towards the horizon. An example would be to use Monte Carlo analysis to calculate average gains of a non-GSO ES towards the horizon to use in another simulation
- Propagation loss. An example would be to use Monte Carlo analysis to calculate the average diffraction loss in an urban area using a high-resolution surface database, which could then be compared against values generated using Hata/COST together with location variability.

The most common statistics generated by Monte Carlo analysis are probability related, such as:

- What is the probability that the $T(I/N)$ is exceeded?
- What is the probability that the $T(C/I)$ is exceeded?
- What is the probability that the $T(C/(N + I))$ is exceeded?

These can be reversed, for example:

- What is the I/N level at the threshold percentage of time?

This requires the generation of a CDF, which can be used to identify the bin that has an associated percentage of time nearest to the threshold, which means the output accuracy is limited by the bin resolution. However this is a useful metric as it shows how far in dB the interference has to be reduced to meet the threshold.

In general, there is a non-linear relationship between input variable V and output X, except that:

- Modifications in the interferer transmit power directly leads to equivalent changes to interfering metrics such as I and I/N
- Modifications in the interfering link activity factors directly lead to changes in the percentages of time for a specific I or I/N.

More complex analysis could provide additional outputs, allowing statistics to be generated on metrics such as capacity, and hence identify the capacity reduction due to interference.

By selecting and modifying the input parameter U, additional information can be gathered. For example, by locating a set of mobile test points at distance $d = \{d_1, d_2...d_m\}$ from the base station, it is be possible to derive information about how the coverage (defined as area over which the $T(C/(N + I))$ is met for a given percentage of time) is degraded due to interference.

One question is the interpretation of the statistics when applied to groups of users as raised in CEPT ERC Report 101 (CEPT ERC, 1999). If the availability probability for a mobile system is 95% when randomising the location of the handset over a cell, it could be due to:

- 5% of locations suffer interference 100% of the time
- 100% of locations suffer interference 5% of the time
- Or some combination of locations and time.

If the mobile position is randomised between samples, then there is no way of identifying from the output probability the breakdown between locations and time. This aggregate probability might be the metric that system operators are interested in and so this would be acceptable. Alternatively, a combination of Monte Carlo and AA could be used, as discussed in Section 6.10.

The U parameter is often used as a way to identify methods that could mitigate interference and hence facilitate sharing, such as:

- Power: what is the maximum transmit power that would be permitted while ensuring that $p[X \leq T(X)] \leq p_T(X)$?
- Distance: what is the separation distance required to ensure that $p[X \leq T(X)] \leq p_T(X)$?
- Frequency: what frequency offset or guard band would be required to ensure that $p[X \leq T(X)] \leq p_T(X)$?
- Spectrum mask: what transmit or receive spectrum mask would be required to ensure that $p[X \leq T(X)] \leq p_T(X)$?
- What combination of mitigation methods (power reduction, antenna downtilt, etc.) would be required to ensure that $p[X \leq T(X)] \leq p_T(X)$?

6.9.4   Example Monte Carlo Analysis

Example 6.18
The static analysis in Example 6.1 was extended using Monte Carlo methods by modelling the base station's transmissions on a link-by-link basis rather than aggregate power per antenna. Each sector was assumed to have up to 10 mobiles receiving a 2 Mbps data carrier.
The following resource is available:

Resource 6.8   Visualyse Professional simulation file 'Monte Carlo analysis example.sim' was used in this example.

The following parameters were varied at each Monte Carlo trial:

1. Location: each mobile was located at random within its sector using uniform distribution
2. Hour of day: selected at random and used with the time of day traffic model
3. Link activity: to take account of burstiness of traffic
4. Location of mobile: whether indoors or outdoors, and if indoors the amount of consequential additional path loss
5. Location variability: to take account of multipath and clutter losses
6. Percentage of time: associated with the interfering propagation model, namely, P.2001.

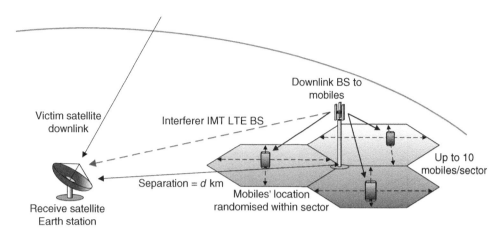

Figure 6.33 Monte Carlo IMT LTE downlink example

The scenario is shown in Figure 6.33 and is based upon the IMT LTE link budgets derived in Section 6.9.5.

Each downlink used APC with a target $RSL = -131.4$ dBW $= -101.4$ dBm. For each trial, the total interference at the ES was dependent upon how many links were active and the power required in each case to close the link (if feasible). Each wanted propagation path was assumed to be fully de-correlated, while each interfering path was assumed to be fully correlated (given that all started and ended at exactly the same station).

The U parameter varied was distance, initially in the range $d = [30, 35]$ km and the key output statistics $S[X]$ was the probability $p[I/N \leq -13$ dB$]$, with results as shown in Table 6.18 together with the I/N calculated for 20% of time using static analysis.

Table 6.18 Monte Carlo results compared to static analysis

Methodology	Monte Carlo	Static
Measure	$p[I/N < -13$ dB$]$	I/N for $p = 20\%$
$d = 30$ km	41.2%	9.9 dB
$d = 31$ km	34.2%	8.5 dB
$d = 32$ km	27.5%	7.2 dB
$d = 33$ km	21.8%	6.3 dB
$d = 34$ km	17.0%	5.8 dB
$d = 35$ km	13.0%	4.8 dB

Note that the static analysis values extended the approach of Example 6.3 to aggregate interference from all three of the base station's sectors and across the entire 30 MHz of ES receiver bandwidth, resulting in an aggregation factor of 5.3 dB.

It can be seen that the Monte Carlo analysis suggested that the ES interference threshold would be met with a separation distance of between 33 and 34 km, with the CDF for the latter distance shown in Figure 6.34. Binary search could be used to calculate a more exact required separation distance, as discussed in Section 6.9.6. Average link availability for the mobiles was 99.5%. At this distance, the static analysis suggests the threshold would be exceeded by about 19 dB.

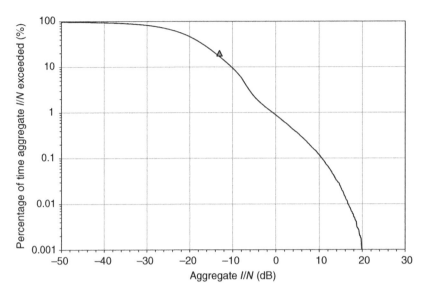

Figure 6.34   CDF of aggregate I/N at the ES after 500,000 Monte Carlo trials for a distance d = 34 km

In a more detailed analysis, it would also be necessary to consider the short-term threshold as the peak *I/N* was around +20 dB, greater than the threshold used in JTG 4-5-6-7 studies.

6.9.5   LTE Downlink Link Budget

In order to undertake the analysis in the previous section, it was necessary to create a link budget for the IMT LTE downlink. The parameters given in Table 6.4 identified the scenario as involving a suburban macro base station with sector radius 600 m, but that leaves a large scope for variation in terms of thresholds and data rate. The service type was assumed to be 2 Mbps with requirement for indoor operation, similar to the Ofcom 4G coverage obligations (Ofcom, 2012b).

With a data rate of 2 Mbps within a bandwidth of 10 MHz, the allocation of 180 kHz resource blocks to users can be derived using the assumptions in Table 6.19.

Link budgets were then developed as in Table 6.20 for a user at the edge of coverage assuming:

- The user is indoors, with loss derived by extrapolating from Ofcom figures
- Hata/COST path loss with location variability standard deviation = 5.5 dB
- 3 dB margin was included for fast fading
- 3 dB margin for interference (mostly intra-system interference from other cells and sectors).

The table gives two cases:

1. Average case, using mean location variability and averaged power
2. Peak case, using 95% location variability and calculating the power required to close the link.

Table 6.19 Calculation of number of users in 10 MHz

Data rate (Mbps)	2
Channel width (MHz)	10
Resource blocks	50
Resource block bandwidth (kHz)	180
Total occupied bandwidth (MHz)	9
Assumed modulation	16QAM
Data rate/bandwidth (b/Hz)	4
Code rate	0.80
Signalling overhead (%)	25
Usable data rate/resource block (kbps)	432
Resource blocks/Mbps	4.63
Rounded resource blocks/Mbps	4.70
Total number of users	10

Both cases resulted in closed links (i.e. achieving or bettering their target *RSL*), which suggests that this link budget is suitable for the scenario under consideration, and so it was used to derive the simulation parameters given in Table 6.21.

It can be seen that the link budgets were not sufficient on their own and additional assumptions had to be made, such as the number of users indoors or outdoors. The analysis also took account of the variation in link activity due to:

- Burstiness: traffic is often bursty, as even during peak traffic in the busy hour, a base station will not be transmitting 100% of the time
- Time of day variation: the traffic was assumed to vary during the day according to Figure 6.35, using the profile given in 'How much energy is needed to run a wireless network?' (Auer et al., 2012).

The time of day profile was sampled in the Monte Carlo analysis by taking an hour at random in the range [0..23] and using the associated traffic level for all links during that trial.

Note that assuming the average activity over the day (around 59%) would give a different result to modelling the full variation in traffic. The key difference is that at busy hour, the probability of any particular user being active taking only burstiness into account is 0.5; hence the likelihood of all 10 in a sector being active is $p = 9.77 \times 10^{-4}$. However if the activity was kept at the single averaged figure of $0.59 \times 0.5 = 0.297$, then the likelihood of all 10 being active drops to $p = 5.4 \times 10^{-6}$. In over 50,000 trials, the difference in the distributions of number of mobiles that are simultaneously active over the three sectors (30 in total) is shown in Figure 6.36.

In general, including more input variables in a Monte Carlo simulation increases the range between output variables minimum and maximum, in this case from [1–20] to [0–24] active users.

6.9.6 Statistical Significance

The random nature of the Monte Carlo methodology introduces uncertainty in the output, so how is it possible to determine if the results are statistically significant?

Table 6.20 Example LTE link budget

Typical link budget	Average power and location	Peak power and location	Notes
Data rate (Mbps)	2	2	Reference[b]
Frequency (MHz)	3 600	3 600	Scenario
Total base station power (dBW)	13	13	46 dBm minus 3 dB cable loss
Channel overhead (dB)	1	1	Reference[a]
Number of users	10	10	From Table 6.19
Power per user (dBW)	2.0	9.1	Calculated
Bandwidth (MHz)	9	9	From Table 6.19
Peak gain (dBi)	18	18	Reference[c]
Relative gain edge of coverage (dB)	−5	−5	Estimated worst case
EIRP (dBW)	15.0	22.1	Calculated
Cell radius (km)	0.6	0.6	Reference[c]
Environment	Suburban	Suburban	Reference[c]
Hata/COST 231 path loss (dB)	121.8	121.8	Calculated
Location variability sigma (dB)	5.5	5.5	Reference[c]
Percentage of locations (%)	50	95	Average or with variation
Location variability loss (dB)	0.0	9.0	Calculated
Indoor–outdoor loss (dB)	19.7	19.7	Reference[b]
Fast fading margin (dB)	3	3	Reference[a]
Total path loss (dB)	144.5	153.5	Calculated
Receive gain (dB)	0	0	Reference[a]
Receive signal (dBW)	−129.5	−131.4	Calculated
Receive noise figure (dB)	7	7	Reference[a]
Receiver noise (dBW)	−127.4	−127.4	Calculated
C/N (dB)	−2.0	−4.0	Calculated
Interference margin (dB)	3	3	Reference[a]
Usable SINR (dB)	−5.0	−7.0	Calculated
Threshold SINR (dB)	−7.0	−7.0	From Reference[a] scaled by data rate
Target RSL (dBW)	−131.4	−131.4	Calculated
Margin (dB)	2.0	0.0	Calculated

[a] WCDMA for UMTS – HSPA Evolution and LTE (Holma and Toskala, 2010)
[b] 4G Coverage Obligation Notice of Compliance Verification Methodology: LTE (Ofcom, 2012b)
[c] Draft New Report ITU-R [FSS-IMT C-BAND DOWNLINK] (ITU-R, 2014b).

The first consideration is to look at the CDF as in Figure 6.37 that shows the result for 9000 and 10,000 samples for the case in Example 6.22 where $d = 30$ km. The two CDFs are very closely aligned and the Kolmogorov–Smirnov two-sample test would likely reject the hypothesis that the distributions are different.

Table 6.21 Parameters used in simulation

Transmit frequency (MHz)	3 600
Transmit bandwidth (MHz)	9
Number of users/sector	10
Maximum power/user (dBW)	3
Power control range (dB)	40
Target *RSL* (dBW)	−131.4
Receive noise temperature (K)	1453.4
Wanted signal propagation model	Hata/COST
Correlation between wanted paths	None
Environment	Suburban
Location variability (dB)	5.5
Percentage of users indoor (%)	50
Percentage of users requiring maximum indoor loss (%)	10
Maximum indoor loss (dB)	20
Time of day profile	See Figure 6.35
Burstiness of traffic (%)	50
Propagation model for interfering path	P.2001
Correlation between interfering paths	Full

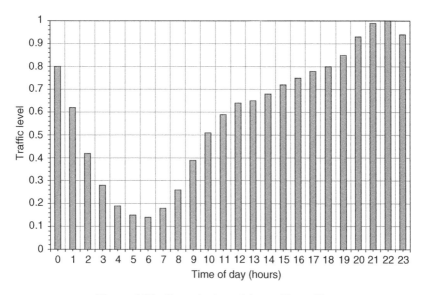

Figure 6.35 Example time of day traffic profile

However there can still be significant variation in statistics relating to the percentage of time that a threshold is met, that is, $p[X < T(X)]$, even after 10,000 trials, as can be seen in Figure 6.38. This figure also shows the variation in $p[X < T(X)]$ using a different random number seed.

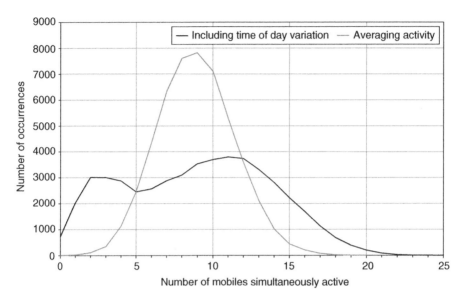

Figure 6.36   Impact of modelling time of day variation in activity

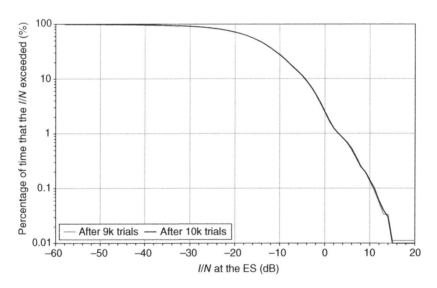

Figure 6.37   CDF of I/N after 9000 and 10,000 Monte Carlo trials

It can be seen that there is still significant variation in the output statistics $S[X]$ even after thousands of samples. In addition, the value after 10,000 samples was slightly different between the two runs that used different random number seeds.

If a suitably strong random number generator is used, then running K times for N steps with different seeds should give the same statistical results as running for $N.K$ steps and dividing the results into K subsets. These K subsets can then be used to create statistics, which can be the basis of significance tests.

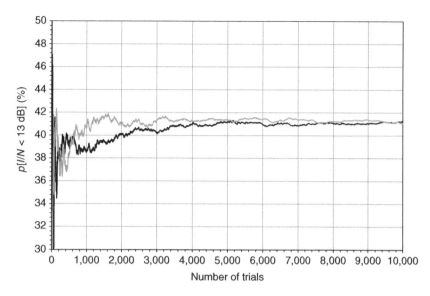

Figure 6.38   Variation in p[I/N < 13 dB] against the number of Monte Carlo trials for two seeds

Example 6.19
The Monte Carlo analysis in Example 6.18 using a separation distance of $d = 30$ km was run for 3,500,000 trials. The trials were split into 3500 groups of 1000 trials, and a histogram of the range of values of the $p[I/N < 13$ dB] across the 3500 groups was generated, as shown in Figure 6.39. For these groups, the mean and standard deviation were also calculated and used to plot a normal distribution: it can be seen that there is a reasonably good match with the histogram.

This is an example of the **central limit theorem**, which states that the distribution of the sample mean is approximately normal if the sample size is large enough. It can be seen that a very large number of trials can be required – 3.5 million in this case.

The standard deviation of this distribution depends upon the number of trials used to calculate the output statistics $S[X]$: the larger the sample size, the smaller the standard deviation, with the decrease in standard deviation relating to the square root of the number of samples.

As discussed in Section 3.10, the normal distribution can be used to create a range $[X_{min}, X_{max}]$ using the mean, standard deviation and confidence interval.

Example 6.20
The Monte Carlo analysis in Example 6.18 using a separation distance of $d = 30$ km was run for 500,000 trials. In every 10,000 trials (i.e. after $N = \{10,000, 20,000, 30,000\ldots500,000\}$ trials), the results were split into 10 groups each comprising $N/10$ trials. The mean and standard deviations over all groups were then derived from the statistic $S[X] = p[I/N < -13$ dB] for each group. This was used to generate a 95% confidence interval as shown in Figure 6.40.

This approach can be used to:

- Identify the confidence in the results of Monte Carlo analysis after N trials
- Identify whether further trials are required to achieve a specified level of confidence.

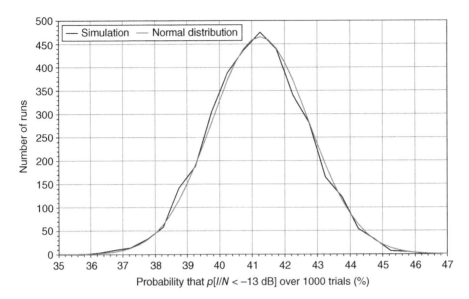

Figure 6.39 Histogram of p[I/N <−13 dB] averaged over 1000 trials calculated in 3500 runs compared against normal distribution

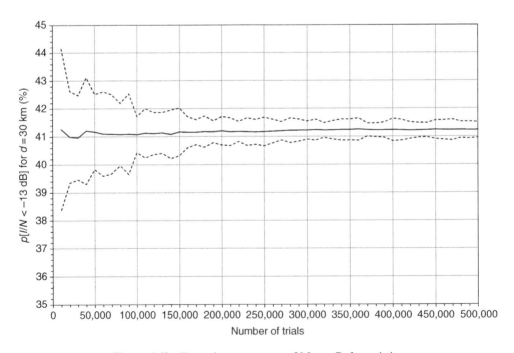

Figure 6.40 Example convergence of Monte Carlo statistics

In the example earlier, the final 95% confidence level after 500,000 trials was $p[I/N < -13$ dB$] = 41.24 \pm 0.28\%$.

The analysis in Example 6.18 varied the distance d in the range $d = [30, 35]$ km to identify the minimum that would just meet the threshold:

$$p[I/N < -13 \text{ dB}] \text{ for no more than } p_T(I/N) = 20\%$$

This requires sufficient samples to ensure that the test has been met or that there is sufficient confidence in the hypothesis that

$$H0 : p[I/N < -13\text{dB}] \leq p_T(I/N) = 20\%$$

Example 6.21
The Monte Carlo analysis in Example 6.18 suggested that the minimum separation distance required to meet the ES's interference threshold was between 33 and 34 km. Finer resolution of the required distance could be achieved using a binary chop algorithm, starting with a value of $d = (33 + 34)/2 = 33.5$ km. How many Monte Carlo trials would be needed to have 95% confidence as to whether the result for $d = 33.5$ km was above or below the threshold?

Figure 6.41 shows the convergence of the statistics as the number of trials increases compared to the threshold of $p_T(I/N) = 20\%$. It can be seen that there can be 95% confidence that the threshold is met after 100,000 trials. If the binary chop procedure were to continue, then the next distance to evaluate would be $(33.0 + 33.5)/2 = 33.25$ km.

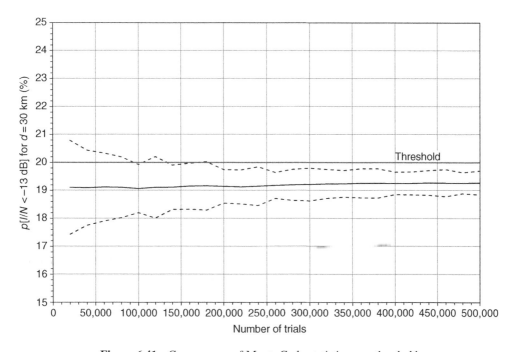

Figure 6.41   Convergence of Monte Carlo statistics near threshold

Near the threshold, the number of samples required to achieve a specified confidence in the results will increase. When testing whether a threshold $p_T(X)$ is met, checking the significance of the results of Monte Carlo analysis can ensure:

- For cases that clearly meet the threshold, that only the relatively small number of trials required are undertaken to avoid excessive computational overhead
- For cases that are close to the threshold, that sufficient trials are undertaken to give confidence in the results.

6.9.7 Deployment Analysis

The most common form of Monte Carlo analysis is of stochastic (not sequential) events in the time dimension and hence probabilities of occurrence. The positions of any stations at fixed locations are therefore input assumptions, but another approach is to consider the likelihood of deployments and hence derive the probability of the type of geometry that could lead to interference.

This can be useful information for a regulator to assess whether to accept a new entrant into a band. It might be that there is a significant interference problem for a given deployment of stations, but how likely is that arrangement?

An example would be point-to-point fixed links: once deployed their locations and antenna pointing angles do not change, unlike mobile users. Hence it is not generally appropriate to convolve deployment likelihood with time variation, as for specific configurations, the geometry will be fixed and hence those links have a deployment probability $p = 1$.

A difficulty of this approach is the lack of thresholds relating to what would be an acceptable percentage of deployments that are permitted to suffer interference. In many scenarios, the answer is none: the existing service must be protected at all locations and for all pointing directions. For example, in Article 22 of the RR, there are EPFD limits to protect GSO satellite systems from non-GSO FSS systems. These apply '*at any point on the Earth's surface visible from the geostationary-satellite orbit*' and '*for all pointing directions towards the geostationary satellite orbit*' (ITU, 2012a).

However, for highly directional antennas, there can be very low likelihoods of alignment, and this could be one approach to facilitate sharing.

One problem about Monte Carlo analysis of deployments is that often actual station locations and antenna orientations are not random. For example, the capacity of bands used for point-to-point fixed links is often driven by critical routes where demand is greatest. Outside these trunk routes, with geometries determined by the locations of major cities, deployment could be sparse.

To a certain extent, existing deployments can be used as a resource and can be used to create the distribution that can be sampled to generate antenna pointing angles in Monte Carlo trials. However, the distribution of these angles can vary across a geographic area due to biases from the underlying major trunk routes. A database of existing links could be used as a source of locations and antenna pointing angles, selecting one at random for each trial.

There are further dangers in making assumptions about deployments for scenarios involving multiple interferers, as demonstrated in the following example.

Example 6.22

The static analysis in Example 6.1 was extended using Monte Carlo methods by varying the deployment parameters as per Table 6.22. The resulting CDF of I/N is given in Figure 6.42 and shows that for the majority (98.8%) of deployments, the I/N calculated was less than the static analysis. However for 100% of samples, it exceeded the threshold $T(I/N) = -13$ dB.

While any input parameter could be randomised to create a Monte Carlo scenario, it is not always useful or meaningful to do so. In particular, in this example, the ES azimuth was varied from pointing directly at the interfering IMT LTE base station to point up to 10° away. But at a separation distance of 15 km, this angle would cover a geographic distance of 2.6 km, within which it is plausible to suggest there would be other base stations. Hence the ES could still be pointing towards an IMT LTE base station, as shown in Figure 6.43.

One approach to analysing deployments where there is also a time variation component is using the two stage Monte Carlo methodology described in Section 6.10.

When there are very large numbers of fixed location transmitters, the deployments can be modelled statistically. In this case, the deployment distribution could be used as one of the input variations of a time-based Monte Carlo analysis, though it would be more statistically vigorous

Table 6.22 Example MC deployment parameters

Input variable	Value in Example **6.1**	MC value distribution
ES elevation angle	10°	Uniform [10°–20°]
ES azimuth angle to BS direction	0°	Uniform [0°–10°]
IMT BS downtilt	6°	Uniform [3°–9°]
IMT BS azimuth to ES direction	0°	Uniform [0°–360°]

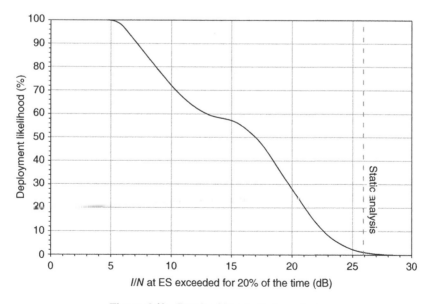

Figure 6.42 Result of Monte Carlo analysis

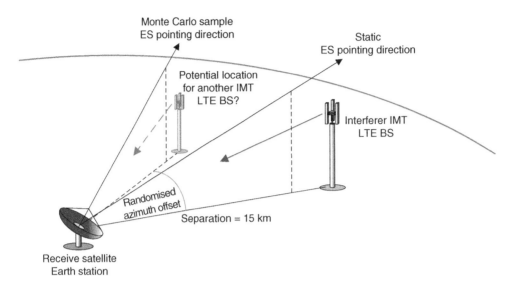

Figure 6.43   Impact of IMT LTE BS deployment on randomisation of ES azimuth

to run with one randomly generated deployment and compare the results against a different randomly generated deployment.

A specialised form of the deployment Monte Carlo model is the *N*-systems algorithm whereby the number of randomly deployed transmitters in an area is increased until the interference-limited maximum density is reached, as described in Section 7.8.

6.9.8   Conclusions

The Monte Carlo methodology can be used to analyse a wide range of interference scenarios in detail. In many cases, the analysis should provide results that are more accurate and lead to reduced constraints, facilitating sharing. This can be achieved by specifying inputs via ranges, which tends to avoid the aggregation of pessimistic assumptions that can be necessary when undertaking static analysis or MCL (as in the development of 2.6 GHz block edge masks discussed in Section 5.3.8).

Some scenarios include inherently variable geometry and probabilistic link and interference metrics – for example, non-co-frequency analysis between systems that operate at random locations across an area. Monte Carlo methodologies can analyse these systems in more detail and with more accuracy than, say, static analysis. It can also be a helpful way to derive input parameters used by other simulations, such as calculating the average *EIRP*.

However, the methodology is more complex and can require significant computational resources and additional modelling assumptions. This can make it hard to understand or explain the simulations created and analysis work undertaken.

As Monte Carlo methods introduce uncertainty via the random element, it is necessary to consider and assess the statistical significance of results.

There can also be limits on the information available regarding key inputs such as modelling mobile locations as having uniform density across an area. However this is common to most interference analysis: the best approach is to use the most accurate information available and then document the assumptions clearly.

6.10   Area and Two-Stage Monte Carlo

The previous section described how Monte Carlo methods can be used during interference analysis to derive statistics $S[X]$ by undertaking multiple trials in which inputs to a static analysis were selected at random from defined distributions. In most cases, the analysis was undertaken in the time domain though, alternatively, fixed station deployments could be varied. The Monte Carlo approach was built upon to determine how the statistics $S[X]$ depended upon the U parameter.

Just as the static analysis was extended to analyse how interference varies geographically using the AA (as in Section 6.5), so is it also possible to employ a **Monte Carlo AA** method. In this case, a scenario is defined for a single location and then one of the deployments varied across an area. At each location, a full Monte Carlo run is undertaken to derive the statistic $S[X]$ associated with that pixel, which can then be displayed either as a coloured block or using contours.

Example 6.23

The static analysis in Example 6.18 was extended to use a Monte Carlo AA methodology by moving the ES across an area as shown in Figure 6.44. At each point, a full Monte Carlo run was used to calculate the statistic $p[I/N \leq -13 \, \text{dB}]$ after 50,000 trials. The resulting contour plots in Figure 6.45 show:

- Static analysis contour: where the I/N for 20% of time $= -13 \, \text{dB}$
- Monte Carlo analysis contour: where the $p[I/N \leq -13 \, \text{dB}] = 20\%$.

Note that in the static AA the ES was moved rather than the IMT BS; hence compared to Figure 6.10, the shape is inverted in the north–south direction. In addition the I/N is an aggregate of all sectors at the base station and across the whole 30 MHz to allow comparison with the Monte Carlo results. Grid lines are 10 km apart.

The AA approach provides additional information and so is very helpful to understand the variation of interference across a geographic region. It can also be combined with terrain, as was shown in Example 6.5. However, it can require significantly greater computational resources than static AA. The example here used:

- Grid = 21 × 21 locations – 441 test points
- Monte Carlo run at each location with 50,000 trials
- Total trials = 22.05 million.

The computational requirements are one reason why the broadcast coverage calculations include methods to analytically derive the statistical power summation of wanted and interfering signals at each pixel, as described in Section 7.3.3.

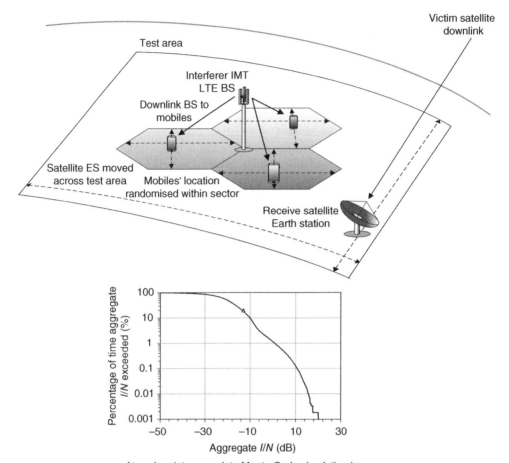

Figure 6.44   Monte Carlo area analysis in Example 6.23

Section 6.9 described how there are two separate dimensions in which Monte Carlo analysis could be undertaken, namely, deployment and time, and it was recommended that these two are not convolved together. However, these two dimensions can be kept separate to analyse deployment and time variability using a **two-stage Monte Carlo** methodology.

Consider the scenario in Figure 6.46, which involves:

- Victim: receive satellite ES
- Interferer: fixed wireless access (FWA) network's uplink from the terminal station (TS) to the central station (CS).

The FWA network is proposing to provide high-speed data services to customers at fixed locations (e.g. home or office). Therefore the interference into the ES will depend upon which customers are signed up as that will define the geometry – in particular how many are in

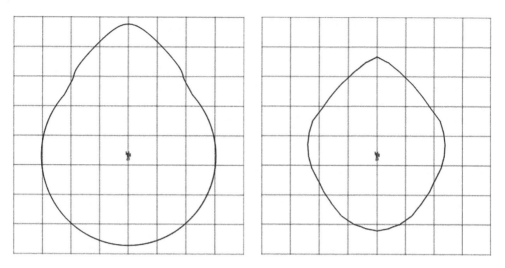

Figure 6.45 Comparison of static AA (left) with Monte Carlo AA (right)

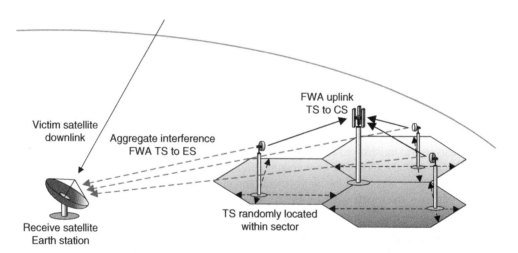

Figure 6.46 FWA uplink interference into satellite ES scenario

locations that would cause the TS to point directly at the ES. Once the customer has been signed up and is operating, the TS uplink will represent a continuous source of interference as this would be a fixed deployment: unlike the mobile that is only at a specified location for a small percentage of the time.

The interference management approach will depend upon the licensing regime in the country and band involved. This could include regulatory tools such as:

- Individual TS licensing: as each customer is signed up, the regulator undertakes an interference check to ensure that operating at the specified location would not cause harmful

interference into the satellite ES. This would add a regulatory burden on the FWA operator that might not be acceptable, though it could be eased by automated or e-Licensing processes.
- Definition of an interference zone: a worst-case analysis is undertaken to identify the area around the ES where there could be harmful interference if all TS within the sector were pointing directly at the satellite ES. The FWA operator can deploy their CS anywhere but within this zone. As this is a conservative deployment assumption, it is most likely to be a larger zone than necessary.

An alternative is to use a two-stage Monte Carlo analysis:

S1. Undertake an analysis of deployment probabilities to identify the configuration of stations that results in an acceptable level of risk using a metric such as $S[X] = p[Deployment\ causes\ I/N\ within\ Y\ dB\ of\ peak]$
S2. Undertake a Monte Carlo analysis to determine $p[I/N < T(I/N)]$ assuming the deployment selected in Stage 1.

Stage 2 could be extended into a Monte Carlo AA.

Example 6.24
An FWA system is operating with the parameters in Table 6.23. The location of each TS was randomised within its sector to generate the CDF of I/N at the ES against probability of deployment as in Figure 6.47. A value of $p = 20\%$ of time was used for both the wanted and interfering propagation models. There was a difference of 6.4 dB in the I/N between the worst and 95% likelihood not to be exceeded deployment.

This analysis was undertaken assuming a single TS per sector: including more in the analysis is likely to smooth the CDF due to averaging effects.

The second stage was to fix the deployments and undertake a time-based Monte Carlo analysis. In this case, the inputs varied were:

Table 6.23 FWA sharing scenario parameters

Parameter	Static analysis value
ES parameters	As per Table 6.9
ES–FWA CS separation distance	15 km
FWA CS parameters	As per Table 6.10
FWA TS antenna pattern	Rec. ITU-R F.1245
FWA TS dish size	30 cm
FWA bandwidth	30 MHz
FWA target RSL	−107 dBW
FWA TS maximum power	0 dBW
FWA APC range	40 dB
FWA wanted signal propagation model	Rec. ITU-R P.530
FWA interfering signal propagation model	Rec. ITU-R P.2001

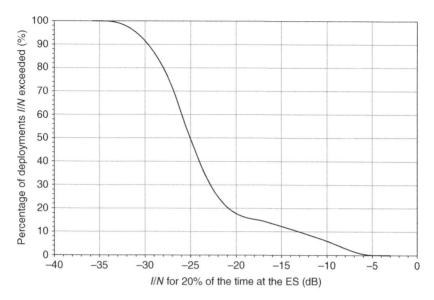

Figure 6.47   CDF of percentage of deployments for which the *I/N* at 20% of time is exceeded

- Traffic: each FWA TS to CS link was given a 50% activity ratio
- FWA TS to CS link propagation model: percentage of time selected at random, each independent
- Interfering path FWA TS to ES propagation model: percentage of time selected at random, all using the same percentage.

The two deployment configurations are shown in Figure 6.48 with a 1 km grid. For the worst-case deployment, two of the TS were pointing directly at the ES, while for the 95% not exceeded case, there was only one TS pointing at the ES and there was no direct alignment.

The differences in resulting *I/N* CDFs are shown in Figure 6.49, where it can be seen that over all percentages of time, there is a consistent reduction in interference from making non-worst-case deployment assumptions.

This technique can be used to estimate the risk in going from worst-case assumptions to more realistic deployment likelihoods. It is similar to the concept of the auxiliary contours that can be used when generating coordination contours via the algorithm in Appendix 7, as described in Section 4.3.9. However, it gives a foundation on which to quantify what would be the appropriate figure to use, in particular for complex scenarios involving multiple interferers

Another approach to analyse complex scenarios via a two-stage Monte Carlo process is given in Recs. ITU-R F.1760 (ITU-R, 2006c) and F.1766 (ITU-R, 2006d), which describes a methodology based upon the following stages:

1. Undertake a Monte Carlo model in detail of one FWA cell including all transmitters, locations, gain patterns and power control to derive as output the CDF of the aggregate EIRP (AEIRP) towards the horizon

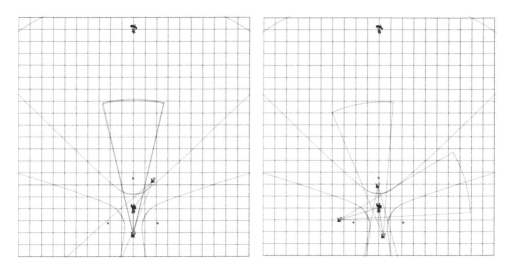

Figure 6.48 Worst-case deployment (left) with the 95% likelihood not exceeded deployment (right)

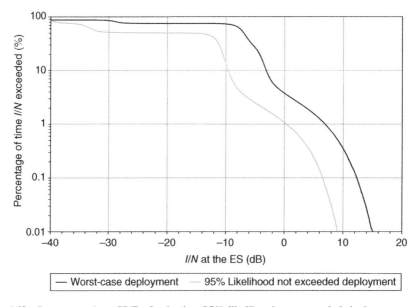

Figure 6.49 Impact on time CDF of selecting 95% likelihood not exceeded deployment vs. worst deployment

2. Undertake a wide area simulation involving hundreds or thousands of FWA cells using the AEIRP CDF as an input to calculate the aggregate interference into a radio astronomy observatory.

Technically this approach does imply the use of a deployment distribution in a time-based Monte Carlo analysis via the AEIRP CDF, which as previously been noted is not recommended. However, with a large number of FWA cells, the deployment should be statistical with distribution that follows the AEIRP CDF.

6.11 Probabilistic Analysis

The Monte Carlo methodology has, as inputs, the probability distributions of key variables, which are sampled at random via a convolution to generate the required output statistics S $[X]$. This random nature introduced uncertainty, requiring tests on the statistical significance of the results.

An alternative is to identify the sampling density required of each probability distribution and directly in sequence check all possible values to derive a quasi-analytic calculation of the $S[X]$, described here as **probabilistic analysis**.

This methodology is the basis of Rec. ITU-R S.1529 (ITU-R, 2001e), which can be used to analyse interference between non-GSO FSS systems and GSO or other non-GSO systems.

The key concept is to identify the orbit shell, which can be divided into cells as in Figure 6.50. Each cell should be sufficiently small that the link or interference metric X would not vary significantly across it, using the same principles as was used to calculate the time step size in Section 6.8. Then the probability that the satellite is in the cell can be calculated.

For example, for non-repeating circular orbits, the probability that the satellite is within a cell of (latitude, longitude) can be calculated from the:

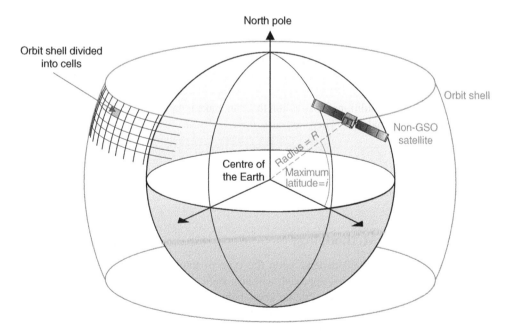

Figure 6.50 Dividing orbit shell into cells for probabilistic analysis

- PDF in latitude: Equation 6.16
- PDF in longitude:

$$p(long_1, long_2) = \frac{1}{2\pi}[long_2 - long_1] \qquad (6.29)$$

Rec. ITU-R S.1529 also includes equations to handle repeating ground track systems and satellites in elliptical orbit.

The satellite is then located at the centre of the cell and the rest of the constellation configured around it so that a static analysis can be calculated. The result is then, for that configuration, the link or interference metric X and the associated probability. By iterating over all cells, it is possible to derive a full CDF of $S[X]$.

This approach has the advantage that the accuracy can be controlled by the size of the cell and the resulting CDF should cover all possible permutations of positions. This is an advantage for algorithms that are being used for regulatory purposes where go/no go decisions should avoid uncertainty.

There are, however, a number of drawbacks:

- There is no value in intermediate statistics, unlike in Monte Carlo analysis where if the threshold is met or not met by a significant margin, it can become clear very quickly that no further trials are required. However, probabilistic analysis requires that all samples are calculated before any conclusions can be made
- Some run times can be extremely long if there are a large number of feasible permutations. This comes back to the issue discussed in Section 6.3 where the possible number of combinations of input values was seen to quickly reach levels that were computationally infeasible. As noted in the previous bullet, a probabilistic analysis run requires that all samples be calculated before the statistics can be assessed
- Unlike the dynamic analysis methodology, there is no information on event lengths or time between of events. Synchronisation effects such as those in Section 6.8 would also not be directly observable
- The method involves additional abstraction that can make it harder to understand and explain the scenario being analysed. There are also fewer software tools designed to employ this approach.

Probabilistic analysis methods tend, therefore, to have restricted usage compared to static, area, dynamic or Monte Carlo analysis.

6.12 Selection of Methodology

The methodology to use will depend upon the type of interference analysis and scenario involved. In many cases, the first approach will be a static analysis to check that link budgets are complete and are giving reasonable numbers before other methodologies are considered. Complex scenarios could require a range of methodologies, and Table 6.24 describes some of the advantages and disadvantages of each.

As noted at the beginning of this chapter, a good starting point is what questions must be answered, and Table 6.1 showed which methodologies could be used to address these example

Table 6.24 Advantages and disadvantages of interference methodologies

Methodology	Advantages	Disadvantages
Static	Simple and quick to implement Easy to understand and explain Good for checking scenarios	Unable to model time or geographic variations Large number of combinations of inputs to consider
Input variation	Relatively straightforward and clear approach Helps understanding by giving direct answers to questions about how metrics vary with an input	Lots of inputs could vary Can be limits on number of input parameters that can be varied in the same run
Area and boundary	Relatively straightforward Shows interference problems graphically, hence a good way to gain understanding Links into regulatory solutions such as interference zones and boundaries	Difficult to combine with other variations In particular, the time domain tends to be fixed
MCL	Simple to calculate Generated value of required total loss between systems a useful metric	Limited in ability to model input variations Other approaches can generate the same results in a more transparent manner Often requires conservative assumptions
Analytic	Direct calculation of result	Applicable to limited range of scenarios
Dynamic	Simple to understand and explain Can identify subtle dynamic behaviour of real systems Able to model behaviour of moving stations Able to generate statistics $S[X]$ Able to generate statistics about event durations	Can lead to long run times Need to calculate the time step and run time correctly Propagation models typically are modelled statistically rather than time sequence, so dynamic analysis can require Monte Carlo components
Monte Carlo	Powerful: able to model variation of any number of inputs Generates statistics $S[X]$ that can be used to answer key study questions	Complicated and can be hard to understand Needs careful selection of input distributions Have to check for statistical significance of results Unable to calculate statistics about event lengths
Area Monte Carlo	Powerful: able to model time variation of inputs and generate statistics $S[X]$ across an area	Can involve very long run times Similar issues as Monte Carlo analysis
Probabilistic	Can be accurate and efficient	Complicated Can lead to very long run times Need to calculate cell size Unable to calculate statistics about event lengths

questions. Another useful starting point are the forms of the interference metrics and the type of analysis required, in particular whether a regulatory check or a general study.

General principles for the selection of methodology that could be considered include:

1. It should be able to answer the questions posed in a clear form
2. It is helpful if it can, if necessary, be modified or extended to answer additional questions
3. The aim should be to use the best methodology and data that is available, accepting that there are likely to be limitations on the accuracy of both
4. A significant factor in the methodology to use will be what input parameters are available and what outputs are required
5. The accuracy of the results is likely to be driven by the least reliable inputs and methodologies: reliable outputs require reliable inputs
6. The methodology should produce the required results within the time frame available and so should be consistent with computational resources
7. It is important that the methodology is understandable both by the user and also others in the community
8. It is important that the methodology can be agreed upon during interactions with other organisations.

In some cases, it is not necessary to choose a methodology as regulations identify a specific algorithm. But it is worth noting that these algorithms were developed within the industry using principles such as those listed earlier, in order to achieve a regulatory goal, whether:

- Assign a frequency that would ensure the system would not suffer or cause harmful interference (e.g. fixed link in Section 7.1 or PMR in Section 7.2)
- Check whether the proposed system should be subject to a coordination process (e.g. Appendix 7 in Section 7.4 or Appendix 8 in Section 7.5)
- Check whether the proposed system meets hard limits that must not be exceeded (e.g. the EPFD limits in Article 22, as discussed in Section 7.6).

In addition, the GRMT generic interference engine in Section 7.9 selects which is the most appropriate methodology from {Static, Area, Monte Carlo, Area Monte Carlo} using a decision tree that takes account of the input parameters, in particular the deployment types.

These specific algorithms can also be a useful source of ideas about possible approaches to studying interference analysis.

It is often useful to consider multiple methodologies in the same study, such as to start with static analysis and then to move to AA when the link budgets are agreed or to use a Monte Carlo analysis to calculate statistics of EIRP, which are then used in an AA.

6.13   Study Projects and Working Methods

Whereas there are specific algorithms to follow when undertaking regulatory checks, studies have much more flexibility in methodologies, tools and working methods. The process of an idealised study is shown in Figure 6.51. Here a study smoothly goes from start to data collection, analysis, reporting and then project end.

Figure 6.51   Flowchart of idealised study project

The reality is likely to be more complicated, with an iterative approach that has multiple layers of feedback, both internally and externally, as shown in Figure 6.52 with stages:

1. The starting point for the first cycle is the question to be addressed, such as what would be the impact of a proposed new system or service?
2. Interference analysis will require input parameters, so there will be a data collection phase, which could include terrain, land use or surface databases
3. From the range of methodologies described in this chapter, it will be necessary to select one or more and identify a software tool that can be used to implement these methodologies
4. The analysis would then be undertaken, often with iterations either as new information is discovered or in order to identify how the required metric varies due to changes in inputs. A further data collection phase could be required if key parameters in an iteration are not available
5. These results would then be documented in reports and technical notes. These can differ between an internal audience (which could discuss the implications for that organisation) and an external audience (that could use a specific document template and exclude organisation-specific details)

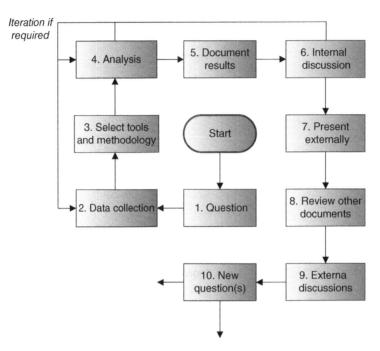

Figure 6.52   Iterative study project

6. The results could then be discussed internally, possibly with a requirement to undertake additional studies
7. The external format of the study report could then be forwarded to external organisations, such as the ITU-R, national regulators, regional regulators or other companies for coordinations
8. Often it is necessary to also review the work undertaken by other organisations, checking their approach and conclusions. Key questions will be what assumptions did they use and were the methodologies selected appropriate?
9. The various contributions can then be discussed at external organisations.
10. This can raise additional questions so that the cycle begins again.

As can be seen, a key attribute is that there is frequent feedback between the various parties, both internally and externally. This ensures that studies do not head off at a tangent and keep the project focussed on the key objective, namely, how to use interference analysis to answer the sharing question.

Analysis can be undertaken in two stages:

1. Calculate the regulatory constraint that would permit sharing between services, such as the derivation of the PFD on boundary threshold in Example 5.48.
2. Calculate whether a specific system meets the regulatory constraint, such as the PFD on boundary calculation in Example 6.6.

Some general rules that can be helpful are:

- It is best to start with a simple scenario and introduce additional complexity only when there is confidence it has been constructed correctly
- The input data might not be 100% accurate, but usually this is acceptable if it is the best available. In some cases, it can be worth asking whether it is possible to get improved data (e.g. station positions, propagation models, gain patterns or terrain databases)
- It is generally easier to share like systems with like, to see if there are options to reduce differences (e.g. heights, powers, densities, etc.)
- It only takes one wrong input or assumption for the output to be wrong, so wherever possible, double-check inputs and intermediate values.

6.14   Further Reading and Next Steps

This chapter has built on the interference calculation in Chapter 5 to identify the various methodologies that can be used to analyse interference. It was noted that a good starting point is to consider what questions the interference analysis must address as that can suggest which methodology would be most appropriate.

A range of methodologies were described including:

- Static analysis
- Input variation analysis
- Area and boundary analysis
- MCL
- Analytic analysis
- Dynamic analysis
- Monte Carlo analysis
- Monte Carlo AA
- Probabilistic analysis.

In each case, examples were given of analysis using either a sharing scenario involving IMT LTE base stations sharing with satellite ESs or non-GSO MSS sharing with fixed links.

The advantages and disadvantages of each methodology were described along with working practices that could be used during study work. It was noted that whereas for study work there is considerable flexibility in the selection of methodology, for many regulatory checks, the algorithm will be specified. The next chapter looks at these specific algorithms and services, and it can be considered as supplementary information to understand the methodologies described in this chapter and also as a source of ideas for new algorithms.

7

Specific Algorithms and Services

The previous chapters described the interference calculation, building on fundamental concepts such as geometry and propagation models, and then showed how it can be used in a number of methodologies to analyse sharing scenarios. While some examples were service and system specific, the concepts were generic and could be applied to a wide range of studies.

This chapter, however, considers specific algorithms and services using the ideas and methodologies described earlier in this book. In many cases the algorithm is part of a regulatory process, whether national or international, and has been chosen and designed to be suitable for the scenario in question. The following algorithms and services are covered:

1. Planning of fixed links including interference thresholds, high/low sites and frequency/ channel selection
2. Coverage and frequency assignment of private mobile radio (PMR) systems including compatibility analysis for exclusive or shared systems
3. Broadcasting services, including threshold calculations, coverage predictions, statistical power summation and single-frequency networks
4. Earth station coordination
5. GSO satellite coordination including coordination triggers, detailed coordination and coordination constraints
6. Non-GSO EPFD analysis using the algorithm in Rec. ITU-R S.1503 to check against the limits in Article 22
7. The radar equation
8. The N-Systems methodology to derive interference-limited deployment densities and associated ratios
9. The Generic Radio Modelling Tool (GRMT) and parameter-based flexible licensing
10. White space device (WSD) algorithms.

Interference Analysis: Modelling Radio Systems for Spectrum Management, First Edition. John Pahl.
© 2016 John Wiley & Sons, Ltd. Published 2016 by John Wiley & Sons, Ltd.
Companion website: www.wiley.com/go/pahl1015

Issues addressed, where applicable, include background history to the algorithm, its regulatory status and whether it relates to a coordination trigger, coordination analysis, the licensing process or a hard limit. The methodology used is described and whether it involves static, dynamic, area (grid or radial approach) or Monte Carlo analysis. Technical parameters are identified including thresholds, metrics, station deployments, antenna pointing methods, gain patterns and link budget parameters, together with a worked example.

The following resource is available:

Resource 7.1   The spreadsheet 'Chapter 7 Examples.xlsx' contains example fixed service link budgets, broadcasting RSL and k-LNM calculations, and GSO *DT/T* and *PFD* calculations.

In addition, a number of simulation files are available as resources, as described in the following sections.

7.1   Fixed Service Planning

7.1.1   Overview

Fixed service (FS) links are used to provide connectivity for a wide range of services and are used globally in a multitude of frequency bands. Applications include providing core trunking for fixed telephony and data networks, backhaul for mobile services, multimedia distribution, telemetry gathering, thin route and local links and short-term connections for outside broadcasts. FS has a number of advantages over wired alternatives such as optic fibre, in particular speed of deployment and relatively low capital expenditure (CAPEX), which is of importance for lower-capacity routes where laying fibre would not be cost effective, such as to remote islands.

A number of architectures could be considered, including:

- **Point-to-point** (PtP): a link from a transmit to receive station, with directional antennas at either end, typically with the return on another frequency (duplex paired band operation), though there can be simplex operation using time division between directions. Simplex operation reduces the number of channels required and can be spectrum efficient if there is significant asymmetry in the traffic between directions, though it can introduce a delay to access the carrier
- **Point-to-multipoint** (PtMP): in this case there is a central station (CS) communicating with a number of terminal stations (TS) as in the fixed wireless access (FWA) system in Figure 6.46. This type of system can provide services such as video and multimedia distribution or high-speed Internet access
- **Multipoint-to-multipoint** (MPtMP): this approach combines aspects of both PtP and PtMP to create a mesh that connects a number of fixed stations together. An example application would be the provision of backhaul to small cell base stations in urban environments where each BS would have antennas that are pointed (either physically or electronically) at other stations.

This section is primarily focussed at planning of PtP fixed links, though the considerations are similar for the other architectures. In PtP links there will be, for each hop or direction, a transmit station with a directional antenna and another one at the receive end. For line-of-sight FS links these antennas point directly at each other, but for some long oversea paths, where it is not feasible to use repeaters, the signal can be reflected off the troposphere and so antennas must be pointed at the common volume.

The characteristics and applications of PtP fixed links can vary by band, with typically:

- Lower bands (e.g. under around 10 GHz) being used for longer hops with lower deployment densities, such as trunk routes or across bodies of water
- Upper bands (e.g. at and above around 18 GHz) being used for shorter hops with higher deployment densities, such as backhaul for mobile base stations.

For a PtP link with given receiver characteristics (in particular, the noise figure), the radio performance will be primarily driven by two factors:

1. The degradation of the wanted signal due to fading
2. The presence of unwanted or interfering signals.

Increasing the transmit power would reduce the susceptibility of the link to both of these factors but at the cost of causing higher levels of interference into other links plus requiring greater energy. Operating at higher transmit powers could lead to the radio spectrum being used in an inefficient manner whereby other potential users are denied access because one system is not operating at the minimum power level required.

The solution is to design or plan the link using the general principles in Sections 5.8 and 5.9.

7.1.2 Link Planning

The inputs to the link planning process are:

N = noise, calculated from noise figure and bandwidth using Equation 3.60
$T(C/(N+I))$ threshold for the selected carrier, using manufacturer's data or the values in Table 3.6
G_{tx} and G_{rx} = the transmit and receive antenna's peak gains
M_i = the assumed interference margin, typically 1 dB
M_s = the system margin, which should be as low as possible, potentially zero.

The first step is to calculate the receiver sensitivity level (RSL), which was given in Equation 5.126 as

$$RSL = N + M_i + M_s + T\left(\frac{C}{(N+I)}\right) \qquad (7.1)$$

This methodology is described in ETSI TR 101 854 (ETSI, 2005), which gives typical values of noise figure and system margin (which it calls industry margin) depending upon frequency band.

Example 7.1

A PtP system is operating with the characteristics in Table 7.1. The required RSL can be calculated from the noise (Equation 3.60) and $T(C/(N+I))$ (from Table 3.6):

$$N = 10\log_{10}(290) + 7 + 10\log_{10}(14) + 60 - 228.6 = -125.5 \text{ dBW}$$

$$RSL = -125.5 + 1 + 1 + 20.5 = -103 \text{ dBW}$$

This RSL can be precalculated as shown in the Ofcom Technical Frequency Assignment Criteria (TFAC) OfW 446 (Ofcom, 2013).

From the RSL it is possible to calculate the transmit power required using the following:

$$P_{tx} = RSL + M_{fade} - G_{tx} + L_p - G_{rx} + L_f \tag{7.2}$$

The fade margin M_{fade} can be calculated using a propagation model such as Rec. ITU-R P.530 described in Section 4.3.3. Hence the fixed link can be assigned a transmit power that should be sufficient to achieve the required bit error rate (BER) with 1 dB of margin for interference.

Example 7.2

Two PtP systems, A and B, are operating with characteristics as in Tables 7.1 and 7.2. What transmit power should they use?

The fade margin in each case can be calculated using Rec. ITU-R P.530 (see Section 4.3.3) with an unavailability of 0.01%. Rec. ITU-R P.530 includes two sub-models – multipath and rain fade – and for the frequencies involved the loss will be dominated by rain fade. The free space path loss, fade margin and gaseous attenuation were then used to calculate the required transmit power as shown in Table 7.3.

Table 7.1 Example FS parameters

Noise figure (dB)	7
Bandwidth (MHz)	14
Interference margin (dB)	1
System margin (dB)	1
Modulation	16QAM

Table 7.2 Link design parameters

Link design	A	B
Frequency (GHz)	28.1	28.1
Availability (%)	99.99	99.99
Length (km)	8.4	5.5
Transmit antenna gain (dBi)	35	35
Receive antenna gain (dBi)	35	35
Receive feed loss (dB)	1	1

Table 7.3 Link power calculation

Link design	A	B
RSL (dBW)	−103.0	−103.0
Fade margin (dB)	38.1	27.9
RSL + fade margin (dB)	−64.9	−75.1
Free space path loss (dB)	139.9	136.2
Gaseous attenuation (dB)	1.1	0.7
Transmit gain (dBi)	35.0	35.0
Receive gain (dBi)	35.0	35.0
Feed loss (dB)	1.0	1.0
Total path loss	71.9	67.8
Required transmit power (dBW)	7.0	−7.2

These two links are both proposing to operate at 28.1 GHz, and so it is also necessary to check that they won't cause interference into each other. This task could be undertaken by the national regulator as part of its responsibility for spectrum management and issuing a licence or the spectrum block could be privately managed.

7.1.3 Interference Thresholds

The links were designed with an interference margin of 1 dB, and hence it is possible to calculate the aggregate interference threshold. However, it would be difficult to calculate the total interference into each receiver as:

- There could be a large number of interferers in densely populated bands leading to excessive computational requirements
- It would be necessary to have good propagation models able to handle the degree of correlation between each of these paths.

In practice it is simpler to apportion the aggregate interference into a set of single-entry thresholds as discussed in Section 5.9. The apportionment ratio will depend upon the expected density levels, but in Ofcom OfW 446 it is assumed that the worst-case interference will be equivalent to the aggregate interference being split between four single entries.

Example 7.3
The PtP system described in Example 7.1 used an interference margin of 1 dB and had receiver noise = −125.5 dBW. With an assumed aggregation factor $n = 4$, the long-term interference thresholds are as given in Table 7.4.

Short- and long-term thresholds can be derived from the following two static cases:

1. When the wanted signal is faded and the interference is at its median (i.e. 50% of time) level
2. When the wanted signal is at its median level and the interfering signal is enhanced.

This approach is based on the assumption that the wanted and interfering paths are not correlated and that the unavailability figure is sufficiently small that it is very unlikely that there

Table 7.4 Calculation of long-term interference threshold

I_{agg}/N (dB)	−5.9
n	4.0
N (dBW)	−125.5
I (dBW)	−137.4

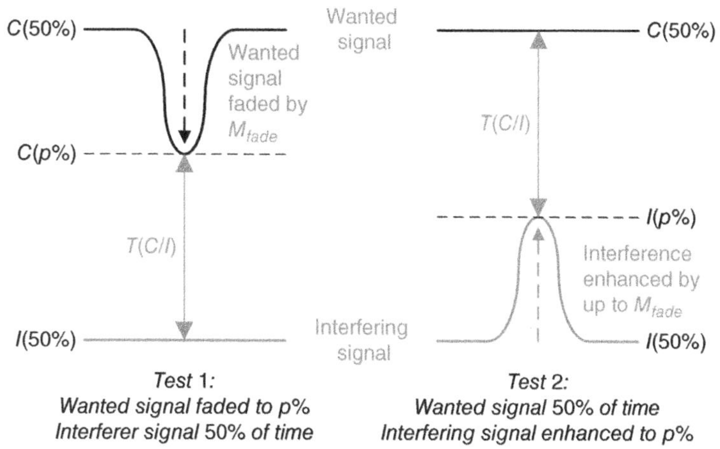

Figure 7.1 Two interference calculations for FS-FS analysis

would be a deep fade of the wanted signal at the same time as a large enhancement of the interfering signal. Therefore the fade margin is available to counter enhancements in the interfering signal.

Graphically this can be displayed using power line diagrams as in Figure 7.1.

Hence the two interference thresholds are as given in Table 7.5.

Table 7.5 Thresholds for PtP interference calculation

Threshold	Long-term	Short-term
Interference level	$T_{lt}(I)$	$T_{lt}(I) + M_{fade}$
Percentage of time	50%	$p\%$

The interference threshold $T(X)$ could be defined in the form of $T(I)$, $T(I/N)$ or $T(C/I)$ as the static nature of the scenario means they are interchangeable. This approach to the definition of interference thresholds could be applied to other services, in particular when analysing interference from satellite Earth stations, as discussed in Section 7.4.

Whereas the wanted signal used free space and Rec. ITU-R P.530 as the propagation model, the interfering signal can be calculated using Rec. ITU-R P.452 together with a terrain database. Typically PtP fixed links are located above clutter to ensure there is line-of-sight to the receive antenna and hence it is not necessary to include a land use database. However it could be helpful to check there isn't any terrain obstructing the path or within the Fresnel zone, described in

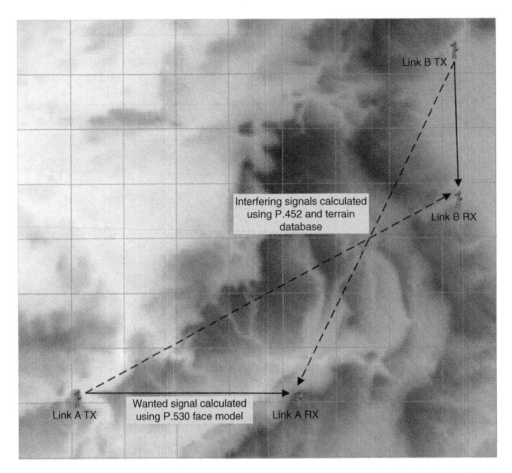

Figure 7.2 P2P FS systems interference analysis. *Screenshot credit: Visualyse Professional. Terrain credit: SRTM*

Section 4.3.2 (also see boxed text after Section 7.1.5). It is also a conservative approach to neglect clutter for interfering paths.

Example 7.4
The locations of the two FS systems from the previous example are shown in Figure 7.2. The interference levels were calculated in both directions for the two cases of long- and short-term using the P.452 propagation model together with the SRTM terrain database. The results of the interference analysis are given in Table 7.6: it can be seen that the margins are significantly greater than zero.

Resource 7.2   The Visualyse Professional simulation file 'FS to FS example.sim' was used to generate the results in this example. The associated terrain file 'Resource 7-2 terrain.gen' is also available.

The antenna patterns used were Rec. ITU-R F.699, which is a general-purpose antenna pattern for modelling PtP fixed links. In the case that there are large numbers of interferers (as in

Table 7.6 PtP FS interference link budgets

Interfering link budgets Threshold type	B into A		A into B	
	Long-term	Short-term	Long-term	Short-term
TX power (dBW)	−7.2	−7.2	7.0	7.0
TX peak gain (dBi)	35.0	35.0	35.0	35.0
TX relative gain (dB)	−32.2	−32.2	−32.4	−32.4
Propagation model	P.452	P.452	P.452	P.452
Percentage of time (%)	50.0	0.01	50.0	0.01
Path loss (dB)	188.6	178.7	195.4	187.0
Receive peak gain (dBi)	35.0	35.0	35.0	35.0
Receive relative gain (dB)	−38.7	−38.7	−38.7	−38.7
Receive feed loss (dB)	1.0	1.0	1.0	1.0
Interference (dBW)	−197.7	−187.8	−190.5	−182.1
Threshold $T(I)$ (dBW)	−137.4	−99.3	−137.4	−109.5
Margin $M(I)$ (dB)	60.3	88.5	53.1	72.7

the FS/MSS example in Section 6.2.2), this can lead to an overestimation of the aggregate interference, and so it can be more appropriate to use the antenna pattern in Rec. ITU-R F.1245.

PtP fixed links can be deployed with high densities without causing harmful interference due to the use of highly directional antennas, which:

- Reduces the area that is 'lit up' by the transmitting signal
- Reduces the likelihood of there being coupling between an interfering transmit antenna's location plus azimuth and the victim receive antenna's location plus azimuth
- Reduces the transmit power required to close the link.

In addition, there is the potential for terrain shielding to reduce the interfering signal to be below that calculated using free space path loss.

The maximum density that can be achieved is discussed further in Section 7.8 on the N-Systems methodology.

The regulatory status of this type of algorithm is either as part of a licensing/assignment process or during detailed coordination. Hence if a country notifies they intend to deploy a new PtP fixed link (either bilaterally or via the ITU-R), this type of analysis can identify if it would cause a problem into any neighbouring country's links.

7.1.4 High versus Low Site

Often PtP links are connected together in multiple hops to provide connectivity over a greater area. In addition, resources such as towers can be used by multiple operators at the same time. In these situations it is necessary to consider the potential for interference between different links operating from the same tower and whether it is identified as a **low or high site**.

In Figure 7.3 there are two bidirectional links between:

1. Station A and Station B using frequencies f_1 and f_2
2. Station B and Station C using frequencies f_3 and f_4.

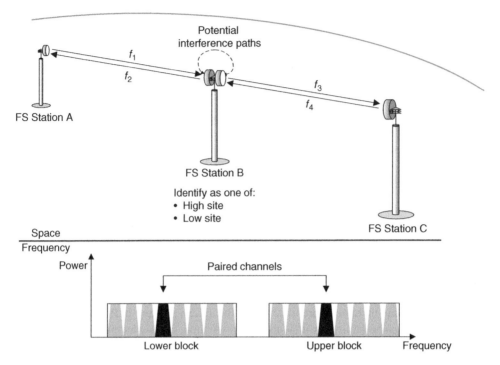

Figure 7.3 Multiple PtP links and a high/low site

At Station B, the extremely short distances between transmit and receive antennas mean there is the potential for interference between:

1. The antenna transmitting on frequency f_3 and the one receiving on f_1
2. The antenna transmitting on frequency f_2 and the one receiving on f_4.

These two pairs of frequencies will be selected from a channel plan, shown in the lower half of the figure, in which there are upper and lower blocks. The potential for interference will be reduced if the two transmit frequencies are taken from either the low or high block and the two receive frequencies are taken from the other, as this will maximise the frequency separation.

A high site is where all the transmit frequencies are selected from the upper block, and a low site is where all the transmit frequencies are selected from the lower block. There can also be dirty sites where a mixture of blocks are used, which could require careful planning to avoid interference (e.g. select frequencies from extreme ends of the blocks).

7.1.5   Channel Selection

A number of methods can be used to select the next channel to try from a set (e.g. channel plan) including:

- Start from one end and use the first channel for which the interference thresholds are met
- Start at different ends depending upon the modulation order, for example, so that low-order modulations are at one end of the block and high at the other

- Test a range of channels and select the one that would lead to the largest positive margin over the thresholds
- Test a range of channels and select the one that would lead to the lowest positive margin over the thresholds.

The spectrum efficiency of each of these methods has been the subject of some research (Flood, 2013), and alternative planning methods have been suggested, such as considering a joint probability model (Flood and Bacon, 2006).

Fresnel Zones and Scottish Whisky

I was once in a pub on the Scottish island of Islay and got chatting with a local who I discovered worked for the telecommunications company responsible for the local fixed links. We had a rather surreal discussion about Fresnel zones over our drinks, and I asked about the analysis they did of the oversea paths connecting the island to the mainland, given tidal variations in the height of the reflection plane.

While there are indeed algorithms that can be used to support the frequency planning task, he admitted he took a rather more pragmatic approach. 'We just give it a go and if we're not meeting our availability targets we get a bigger antenna'.

In these beautiful but remote parts of the United Kingdom, the supply of spectrum is much greater than the demand, and so the pressures to use the most efficient technology are less important than getting a link that definitely works. With the wild storms that can blow off the Atlantic bringing snow and sleet with them, it's best to have a bit of margin spare, especially as these links are the islanders' principal connection to the outside world.

7.2   Private Mobile Radio

7.2.1   Overview

A licence class that is as ubiquitous as the fixed link is private mobile radio (PMR), sometimes called business radio (BR), part of the land mobile service (LMS). It is used for a wide range of applications including:

- Construction workers
- Distribution systems
- Emergency services such as health and police
- Security guards
- Taxi despatch
- Transport systems
- Etc.

Traditionally the service was predominantly voice using analogue modulation, but, with the transition to digital, additional services can be offered, in particular tracking the location of vehicles by polling for a position determined by GPS. The 12.5 kHz bandwidth used was determined by the need to provide analogue voice communications, though legacy systems can use 25 kHz and digital systems can split a 12.5 kHz channel into two 6.25 kHz carriers. A range of frequencies are used for PMR in parts of VHF and UHF: for example, in the United Kingdom the bands are around 55.75–87.5, 138–207.5, 425–449.5 and 453–466 MHz.

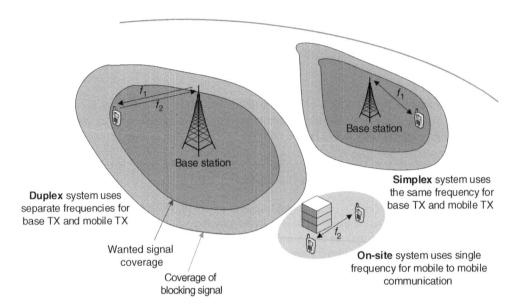

Figure 7.4   Types of PMR system

One key characteristic of PMR systems is that users are active for the duration of a call and then cease transmissions. There can also be multiple mobiles within an area, which can lead to contention when one user wishes to transmit but another is already using the channel. Most systems include mechanisms to avoid this, such as listen before transmit. An existing transmission will block other users if the signal detected is strong enough.

The architecture of the main three types of PMR system is shown in Figure 7.4, namely:

1. **Simplex**: a base station provides coverage to one or more mobile users within a service area, with the same frequency used for communications in both directions
2. **Duplex**: a base station provides coverage to one or more mobile users within a service area, with different frequencies used for communications in the downlink (base transmit) and uplink (mobile transmit) directions
3. **On-site**: mobile users communicate directly with each other on a single frequency.

There can be additional stations in a deployment, such as repeaters and fill-in stations, but this section will be based upon these three service types. Different terminology could be used – for example, the ACMA refers to these as single- or two-frequency systems in LM 8 (ACMA, 2000). Services can also be shared between multiple users, using what are described as common base stations.

7.2.2   Coverage Calculation

Typically the base station transmits with a constant power defined in a licence or other assignment process. Using its location, transmit power, antenna pattern and a propagation model, it is possible to predict the locations covered by the signal S in terms of:

- Sufficient signal to provide the wanted service when $S \geq T(C)$
- Insufficient signal to provide the wanted service but with the potential to block other users from also using the channel at the same time when $T(C) > S \geq T_{block}(I)$
- Insufficient signal to either provide the wanted signal or block other users but potential to degrade their service $T_{block}(I) > S \geq T(I)$.

Example 7.5

A duplex PMR system is operating with the parameters in Table 7.7. The resulting predicted coverage is shown in Figure 7.5 assuming the propagation model in ITU-R Rec. P.1546 using $p = 50\%$ of time and $q = 50\%$ of locations with both a terrain and land use database. Grid lines are spaced every 10 km.

The transmit power chosen (either by the operator or regulator) should be sufficient to cover most of the required area, but there are likely to be gaps due to variations in terrain and clutter loss. There can be a range of powers between PMR systems depending upon the coverage requirements: a taxi service might need to communicate with drivers 50 km from the base station, while a crane operator talks to a controller just 100 m away. As noted previously, it is good spectrum management practice to use the lowest power necessary to provide the requested service, and this can be incentivised using a licensing regime that charges more for high-power transmitters. Another factor that influences the range is the antenna height, and typically wide area systems will require antennas to be located higher than systems that only need to cover one specific site.

The appropriateness of the power selected can be measured using ratios of the area covered by the transmitted signal to the range requested. Consider Figure 7.6, which shows:

- The area the PMR operator has requested for their coverage, namely, 10 km from their base station
- The predicted coverage taking into account terrain and land use data.

In an ideal world these two would precisely overlap, but in reality there will be gaps in coverage within the wanted area and spillage outside it. Two factors can be defined:

Coverage ratio F_c:

$$F_c = \frac{N_{PSA}}{N_{RSA}} \tag{7.3}$$

Table 7.7 Example PMR coverage parameters

Base station latitude (°N)	51.50797
Base station longitude (°E)	−0.09521
Base station transmit frequency (MHz)	165.03
Mobile station transmit frequency (MHz)	169.85
Base station ERP (W)	15
Mobile station ERP (W)	5
Transmit antenna height (m)	20
Receive antenna height (m)	1.5
Bandwidth (kHz)	12.5
Wanted signal threshold $T(C)$ (dBm)	−104
Blocking signal threshold $T_{block}(I)$ (dBm)	−116

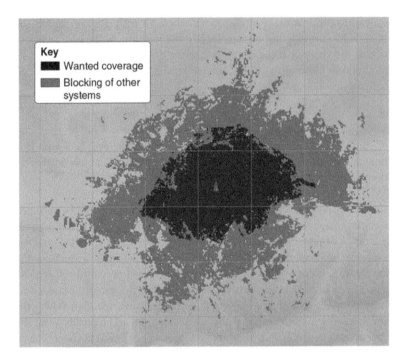

Figure 7.5 Example PMR coverage. *Screenshot credit: Visualyse Professional. Terrain and land use credit: Ofcom & OS*

Pollution ratio F_p:

$$F_p = \frac{N_{PCA}}{N_{PSA}} \tag{7.4}$$

where

N_{PSA} = number of pixels in the protected service area (PSA), which are those that are within both the requested service area (RSA) and predicted coverage area (PCA)
N_{RSA} = total number of pixels in the RSA
N_{PCA} = total number of pixels in the PCA.

A perfect coverage would be when $F_c = F_p = 1$, but this is unlikely to occur and instead

$$F_c \leq 1$$

$$F_p \geq 1$$

The nearer both ratios are to one, the more appropriate the transmit power.

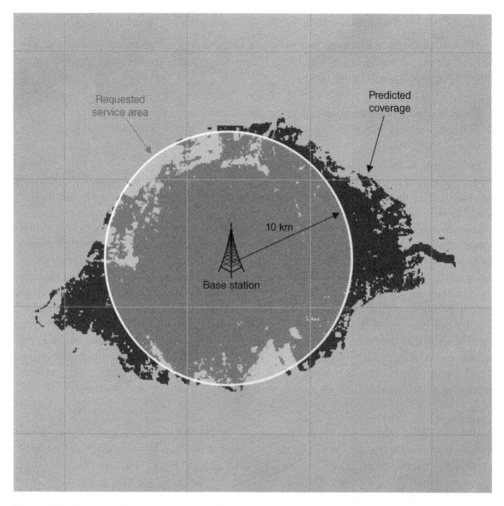

Figure 7.6   Requested service area and predicted coverage. *Screenshot credit: Visualyse Professional. Terrain and land use credit: Ofcom & OS*

Example 7.6
For the PMR system described in Example 7.5, the ratios are

$$F_c = 0.855$$

$$F_p = 1.211$$

These values are both close to 1, indicating an appropriate power level.

Factors that can be used to manage the coverage and pollution ratios include the base station antenna's gain pattern and pointing angles. For example, a directional antenna could be pointed with higher gain towards areas that otherwise would lack coverage and reduced gain towards areas where coverage would spill outside the RSA. A number of gain patterns suitable for PMR systems are described in Section 3.7.8.

7.2.3 PSA and Uplink Calculations

The PSA is important in interference calculations as that is where mobiles are assumed to operate, with all pixels having equal likelihood. The starting point is the coverage area that can, if required, be processed in two stages:

1. Pixels outside the requested coverage area are excluded
2. Isolated pixels are removed and holes filled to make a more contiguous shape.

An example of this process is shown in Figure 7.7.

More advanced filtering could take account of specific locations – for example, identifying locations that are more likely or can be excluded, in particular depending upon whether the user is in a vehicle or on foot.

The coverage prediction in this example was generated using the propagation model in P.1546 using the median signal values $p = 50\%$ of time and $q = 50\%$ of locations. These are the values in the Ofcom planning methodology for PMR systems in Ofw 164 (Ofcom, 2008b), but other propagation models could be used, such as P.1812 or P.2001 and other percentages of time or location. For example, the ACMA's LM 8 describes a service model based upon a 90% of locations and 1% of time to meet the enhanced availability requirements of their emergency services.

It is worth noting that for most PMR services there is also a requirement for the return link from the mobile to the base station. This uplink link budget must also be closed, and this can become the limiting factor on range as there can be restrictions on the power available or permitted for the mobile, especially handheld. To a degree a lower handset power can be balanced by a lower noise figure of the base station receiver, but this is likely to be only a few dB less.

Example 7.7
The PMR system operating with the parameters in Table 7.7 was designed for vehicular mobiles with similar transmit power as the base station. If handsets were used with an ERP = 500 mW, then the uplink coverage would be reduced as in Figure 7.8.

7.2.4 Thresholds and Propagation Model

The coverage area in Example 7.5 was predicted using the propagation model in P.1546 and the two thresholds of $C_{min} = -104$ dBm and $I_{block} = -116$ dBm. But what would be the impact of

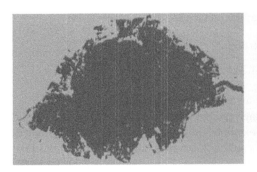

Figure 7.7 Raw coverage area (left) and service area after filtering (right). *Screenshot credit: Visualyse PMR. Terrain and land use credit: Ofcom & OS*

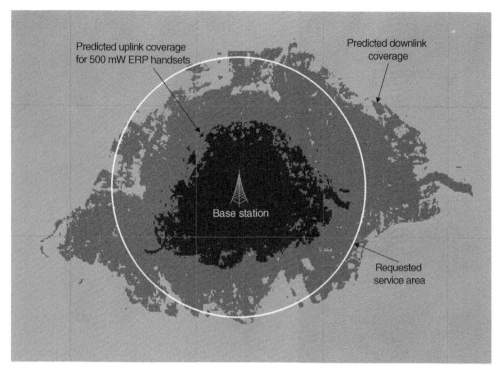

Figure 7.8 Example uplink versus downlink coverage. *Screenshot credit: Visualyse Professional. Terrain and land use credit: Ofcom & OS*

using other propagation models or alternative thresholds? The propagation model in particular can make a significant difference, as can be seen in the following example.

Example 7.8

The coverage of the PMR system from Example 7.5 predicted using Rec. ITU-R P.1546-5 and Rec. ITU-R P.1812-3 is compared in Figures 7.9 and 7.10. In both cases $p = 50\%$ and $q = 50\%$. It can be seen that there is a significant difference, which is reflected in the coverage and pollution ratios given in Table 7.8.

But which model is right? This is where a regulator could consult with users or undertake its own measurement campaign. Each propagation model is designed for specific scenarios and tested against measurement data that will have its own biases in it. Furthermore, each model can be considered work in progress, with a series of updates as more information becomes available.

The use of more modern propagation models such as P.1812 (described in Section 4.3.6) should be more accurate as the standard deviation error against measurement data is lower than that for P.1546 (described in Section 4.3.5). It also takes into account the full path profile rather than just horizon elevation angles and average terrain heights. In addition, it is able to use surface databases to model urban areas in high resolution, unlike P.1546. However such accuracy often has requirements on data capture, as locations must be precisely known to avoid prediction errors.

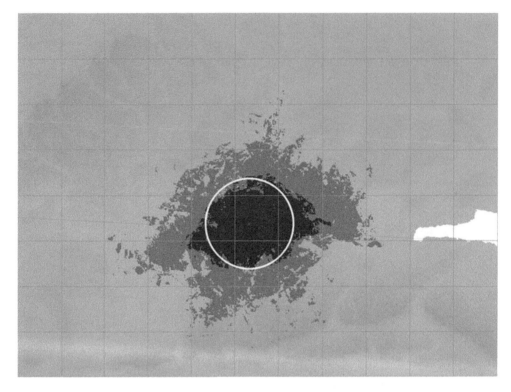

Figure 7.9   Example PMR coverage predicted using P.1546

Other models could be considered, in particular Rec. ITU-R P.2001 (see Section 4.3.7), which has the advantage that it can also consider fading of the wanted signal, predicting coverage areas met for (say) 99% of the time.

However changing the propagation model could require replanning and lead to situations where assignments that were previously considered acceptable exceed interference thresholds.

A second issue is whether the thresholds are set at appropriate values. As technology improves there is generally an ability to operate with a lower RSL, and hence reduced transmit power, improving spectrum efficiency. However in a band in which there are large numbers of legacy systems operating with older, higher thresholds, the benefit would be lessened. In addition, the new systems with lower thresholds would be more vulnerable to interference, leading to only minor changes in separation distances. Furthermore, manufacturers can design their systems for specific markets, and hence the equipment thresholds could be set at the existing values.

As is often the case in interference analysis, it is usually beneficial for like to share with like so that the victim and interferer have similar characteristics. Therefore it could be appropriate to identify that equipment with lower RSLs be used in new bands where there are no legacy systems to consider.

7.2.5   Compatibility Checks

A number of licensing models can be used for PMR systems, ranging from undertaking a technical analysis for each site to allowing users to deploy base stations anywhere within a

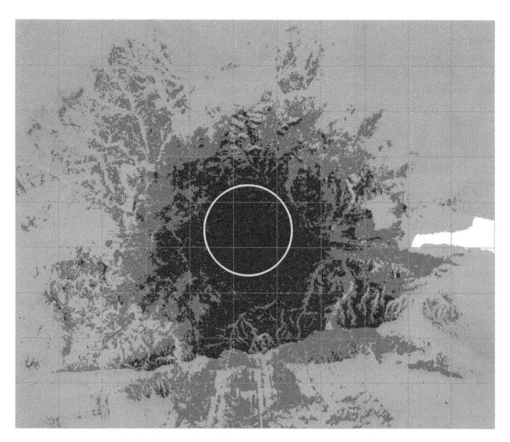

Figure 7.10 Example PMR coverage predicted using P.1812. *Screenshot credit: Visualyse Professional. Terrain and land use credit: Ofcom & OS*

Table 7.8 Impact of propagation model on example PMR F_c and F_p

Propagation model	P.1546	P.1812
Coverage ratio F_c	0.855	0.938
Pollution ratio F_p	1.211	4.257

licensed area. Low-power systems, in particular those providing on-site services, could be permitted to operate on a licence-exempt basis, selecting from a predefined set of channels the one that minimises interference or blocking.

When licensed on a site-by-site basis, it is necessary to undertake an assessment of the potential for interference to or from other PMR systems, and this can be done at a number of levels, as described in the following. It can also be useful to undertake checks on intermodulation products as described in Section 5.3.7.

7.2.5.1 Frequency/Distance Checks

The simplest check is on the frequency and distance separation between the interfering and victim base stations. A MCL methodology can be used to identify the minimum distance required for a number of frequency offsets using the methodology in Rec. ITU-R SM.337.

An example of this can be found in the ACMA description of frequency assignment requirements for the LMS (ACMA, 2000). The distance excluded varies by frequency offset, so that, for example, in the mid VHF band the separation varies from 140 km for offsets less than 1.25 kHz to 200 m when the frequency offset exceeds 1.29 MHz.

These can be used for both co-frequency and non-co-frequency analyses, unlike the following methods that are focussed on co-frequency cases.

7.2.5.2   Interfering Signal on or within Boundary

Generic approaches such as frequency/distance separations derived via MCL methodologies do not take account of site-specific information, in particular terrain and land use data. The necessary separation between co-frequency base stations will vary depending upon the geography: the distance for hilly locations could be significantly less than that for flatter terrain. What is important for radio systems is the interfering power, and hence an alternative approach that should result in improved spectrum efficiency is to check for the interfering signal at the boundary of each victim's service area.

This is the approach taken in Ofcom's TFAC Ofw 164 for non-shared assignments, which identifies the threshold:

Interference level: $T(I) = -116\,\mathrm{dBm} = -146\,\mathrm{dBW}$ in 12.5 kHz
Percentage of time: $p = 50\%$

This constraint can be applied either at the edge of an area licence or within the protected coverage area for site licensing.

These types of checks represent examples of area or boundary analysis methodologies.

7.2.5.3   Channel Occupancy Checks

For many PMR users, one of the key advantages of this licence type is that they have exclusive access to the frequencies they use at the required locations. This exclusivity ensures that they can operate with very high reliability in terms of access to the channel, unlike a shared service such as public mobile networks that can suffer from congestion. As the signal strength that leads to blocking is lower than that required to provide a service, the area that must be kept clear is larger than the protected coverage area. A large sterilised area is costly in terms of spectrum denied to other users and hence likely, in a pricing regime that matches technical spectrum opportunity cost, to be financially expensive.

In areas where there is high congestion, it can be useful to offer PMR users the option to operate on channels shared with other licensed systems. Sharing leads to the potential for blocking of one licensee by another, but this can be acceptable if:

- Some PMR users can accept a delay in communication – for example, taxi despatch messages or an industrial site giving non-urgent updates
- The delay can be short and/or have low probability of occurring. Example 5.27 described how with six PMR users sharing a channel for short voice communications the probability of blocking was $p = 0.107$

- There can be expected to be a consequential reduction in price. So if a channel was shared with another user, then each could be expected to pay half the licence fee of an exclusive channel
- In heavily congested areas this might be the only way to gain access to the required PMR services.

This shared usage implies that there will be overlap in coverage areas between PMR systems, and hence it is not possible to simply check the PFD on boundary or interfering signal within an area. Instead an algorithm is required that calculates the degree of channel occupancy, as described in the following section.

7.2.6 Channel Sharing Ratio

When PMR channels can be shared, the key metric is the probability that another user will be active such that there is blocking. This can be calculated via the channel sharing ratio R_{cs}.

Consider the scenario in Figure 7.11 where two PMR systems are overlapping and using the same frequency for their downlinks. If all locations in the victim's PSA are equally likely, then the probability that it is in a location that could be blocked is the ratio of the overlap area to the total PSA. This can be described mathematically as

$$R_{cs} = \frac{PSA_V \bigcap PCA_I(I > I_{block})}{PSA_V} \tag{7.5}$$

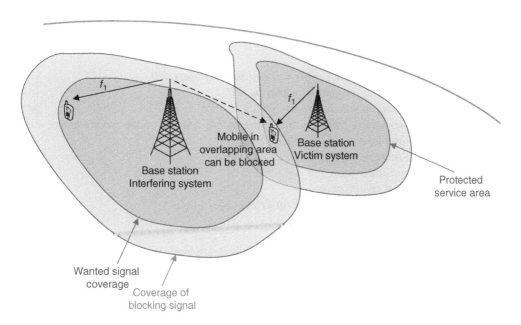

Figure 7.11 Overlap area in channel sharing ratio calculation

Table 7.9 Calculation of channel sharing ratio

Interferer	Victim	Calculations	Equation
Base station	Base station	$R_{cs,1} = \begin{cases} 1 \text{ if } I > I_{block} \\ 0 \text{ if } I \leq I_{block} \end{cases}$	7.6
Base station	Mobile	$R_{cs,2} = \frac{PSA_V \cap PCA_I(I > I_{block})}{PSA_V}$	7.7
Mobile	Base station	$R_{cs,3} = \frac{PSA_I \cap PCA_I(I' > I_{block})}{PSA_I}$	7.8
Mobile	Mobile	$R_{cs,4}$ = Monte Carlo analysis	7.9

The method to calculate the R_{CS} depends upon the geometry, namely, whether the victim or interferer is the base station or mobile. The calculations involved in each case are shown in Table 7.9.

For the base station-to-base station case, the geometry is fixed, so the sharing ratio is either 1 or 0 depending upon whether the calculated interference is above the blocking ratio.

The case of the mobile uplink interfering with another PMR's base station receiver involves a calculation very similar to the downlink case. The difference is that the calculation needs to determine the interference that would come from a mobile at that location into the base station rather than that from the base station into a mobile. However this can be quickly calculated from the PCA by adjusting for the difference in transmit powers using

$$I' = I + P_{tx,ms} - P_{tx,bs} \tag{7.10}$$

The mobile-to-mobile case requires a Monte Carlo analysis to determine the likelihood of there being sufficient interference that there would be blocking. An example algorithm would be:

1. *Initialise statistics, setting total trials = 0 and blocked trials = 0*
2. *Locate the victim mobile at random within its PSA*
3. *Locate the interfering mobile at random within its PSA*
4. *Calculate I = interference from the interfering mobile into the victim mobile taking into account terrain and land use data (where available)*
5. *If I > I_{block}, then increment the blocked trials count*
6. *Increment the total trials count*
7. *Check for statistical significance*
8. *If further trials are required, then repeat at step 2*
9. *$R_{CS,4}$ = [Blocked Trials]/[Total Trials].*

The channel sharing ratio is an aggregate, and it is typically useful to keep track of each PMR system's total R_{cs} from all existing licences for each of its channels when issuing a new one, rather than using apportionment methodologies. Including each PMR's own use of the spectrum, the total R_{CS} is then

$$R_{cs,total}[Channel] = 1 + \sum_{i=other\ PMR} R_{cs,i}[Channel] \tag{7.11}$$

The average channel sharing ratio over all the channels used by the PMR system is then

$$R_{cs,average} = \frac{1}{Channels} \sum_{Channels} R_{cs,total}[Channel] \qquad (7.12)$$

Example 7.9

It is proposed that a new duplex PMR be introduced operating on the same frequencies and with the same parameters as that in Example 7.5 but with the base station located at (lat, long) = (51.367476°N, −0.104348°E). The predicted coverage of the new potentially interfering system and degree of overlap with the existing system's coverage is shown in Figure 7.12.

The calculated channel sharing ratio for the base station transmit direction was calculated to be $R_{cs,2}[DL] = 0.984$ – that is, there was almost total overlap between the victim PSA and where the interferer's predicted signal strength was greater than I_{block}. The channel blocking ratio for the mobile transmit case was lower at $R_{cs,2}[UL] = 0.253$, and hence the average channel sharing ratio including its own usage is $R_{cs,avg} = 1.619$.

But what would be a suitable sharing ratio threshold? One way to look at the sharing ratio is to compare it to a contention ratio on a broadband link: it identifies the number of users that could access the shared resource:

$$CR = \frac{1}{R_{cs}} \qquad (7.13)$$

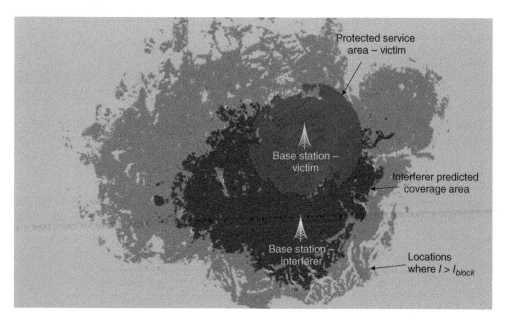

Figure 7.12 Overlap of interfering predicted coverage with victim PSA. *Screenshot credit: Visualyse PMR. Terrain and land use credit: Ofcom & OS*

Another approach would be based upon Erlang modelling. The R_{CS} that would be acceptable would depend upon the traffic levels of each individual user, and one way to quantify this traffic could be via the activity factor (AF) as described in Section 5.7.2. It is then possible to calculate the average traffic intensity per channel within the PSA of one PMR system:

$$A_{tc,i} = AF.R_{cs,i} \tag{7.14}$$

Taking into account only traffic from outside a particular PMR, the calculation is

$$A_{tc} = \frac{1}{Channels} \sum_{Channels} \sum_{i=other\ PMR} AF_i.R_{cs,i}[Channel] \tag{7.15}$$

Taking into account a system's own traffic, the calculation is

$$A_{tc} = \frac{1}{Channels} \sum_{Channels} \left[AF_{victim} + \sum_{i=other\ PMR} AF_i.R_{cs,i}[Channel] \right] \tag{7.16}$$

Example 7.10
Assuming that both of the PMRs in Example 7.9 have an AF = 0.05, then the average traffic intensity per channel within the victim system's PSA is $A_{tc} = 0.162$.

The average traffic intensity per channel can then be used to calculate the probability of blocking using the algorithms in Section 5.7.2.

In Ofcom's frequency assignment defined in Ofw 164 (Ofcom, 2008b) and the Mobile Assignment Technical System (MASTS), this methodology is used to define a quality of service (QoS) metric that should not be exceeded where

$$QoS = \sum_{Channels} \left[AF_{victim} + \sum_{i=other\ PMR} AF_i.R_{cs,i}[Channel] \right] \tag{7.17}$$

There are then thresholds based upon PMR type as in Table 7.10, which can be seen to depend upon the number of channels used.

7.2.7   Sharing with Other Services

When undertaking interference analysis with other services and sharing studies, it is rare to require this level of detail. Typically it is sufficient to make simplifying assumptions, such

Table 7.10   Ofcom Ofw 164 PMR sharing thresholds

PMR type	QoS threshold
Duplex	2
Simplex	1
On-site	1

as that the service can be modelled as a base station communicating with a mobile in a circular service area. In many cases it is sufficient to only model the base station, as interference from the mobile will be in general less.

A critical parameter in non-co-frequency modelling is the assumed density, related to the distance between co-frequency base stations. Altering this value can have significant impact on services operating in adjacent bands, as was found when the Nextel network changed from PMR to a higher-density deployment of base stations, as discussed in Section 5.3.8.

7.3   Broadcasting

The most widely used radiocommunication service is arguably audio and TV broadcasting, which has a well-developed planning algorithm that takes account of intra-service interference. A major task for regulators at the start of the twenty-first century was the transition of TV broadcasting services from analogue to digital, increasing spectrum efficiency to permit more channels and/or additional services such as mobile, and this required extensive re-planning.

At its core the broadcasting planning algorithm is an area analysis methodology, but it contains specific features that are of interest to interference analysis and general radio modelling. The following subsections describe:

- How the threshold field or signal strength is calculated for the service being planned
- How the coverage is calculated, taking into account the variation of wanted and interfering signals across a pixel
- How multiple wanted and/or interfering signals are summed using statistical power summation methods
- Specific issues relating to single-frequency networks (SFNs).

This section focusses on the DVB family of standards used in the majority of countries including Europe. Other standards are available, including those from the Advanced Television Systems Committee (ATSC) that are based upon 8VSB (vestigial sideband modulation) as used in the United States.

7.3.1   Threshold Calculation

Digital broadcasting is based on a set of standards such as DVB-T and DVB-T2, which are based on parameters that can be optimised to meet specific requirements. In particular:

- Receivers can be fixed (i.e. rooftop antenna), portable (which could be either indoor or outdoor), mobile (e.g. pedestrians or vehicles) or handheld
- A range of bandwidths are available, though typically there is channelisation to (say) 8 MHz
- A wide range of modulations are available (e.g. QPSK, 16-QAM, 64-QAM, 256-QAM)
- Code rates include 2/3 and 3/4
- Number of OFDM subcarriers (e.g. 2 k, 8 k, 32 k)
- Guard interval as fraction (e.g. 1/32, or in μS, e.g. 7, 28).

Planning therefore starts with the selection of the required service type based upon these parameters. While the modulation, code rate and guard interval/time will have an impact on the usable data rate, the starting point for the planning calculation is to derive the required wanted signal level.

An example of the approach that can be used to calculate the required receive signal (here called the RSL) is given in Rec. ITU-R BT.1368 (ITU-R, 2014d). The key inputs are the $F_N =$ noise figure of the receiver and $T(C/N) =$ required S/N at the input to the receiver, as in

$$RSL = T\left(\frac{C}{N}\right) + F_N + 10\log_{10}(kT_0B) \qquad (7.18)$$

But what is the $T(C/N)$ to use? Given the modulation and code rate, it is feasible to calculate the theoretical S/N needed to achieve the required BER, for example, those given in Table 3.6, noting that these were given for a $BER = 10^{-6}$ without coding, while DVB could accept a lower BER and include coding.

In practice other values are used, which include additional factors such as margin for fading and receiver implementation. The fade model can be either Rician for rooftop reception where a single signal is likely to dominate or Rayleigh for other receivers with potentially multiple signals (see Section 4.2.6). Examples of these additional factors can be found in:

- Frequency and Network Planning Aspects of DVB-T2 (EBU, 2014)
- Technical Parameters and Planning Algorithms (Joint Frequency Planning Project, 2012).

Example 7.11
An example calculation of the target RSL is given in Table 7.11.

Many of the references, such as those described previously, then convert this wanted signal level into a power flux density or field strength by including, where applicable:

- Receiver feed loss L_f
- Receive antenna gain, G_{dBd} here specified relative to a dipole (dBd) rather than isotropic (dBi).

The first step is to calculate the effective area of the antenna in $dB(m^2)$ using the effective area of an isotropic antenna in Equation 3.227:

$$A_e = G_{dBd} + 2.15 + A_{e,i} \qquad (7.19)$$

Table 7.11 Example broadcasting RSL calculation

Modulation	64-QAM
Error code rate	3/4
$T(C/N) = S/N$ objective (dB)	18.6
Implementation margin (dB)	3
Noise figure (dB)	7
Bandwidth (MHz)	7.9
Target C/N (dB)	21.6
Receiver noise (dBW)	−128.0
RSL (dBW)	−106.4

Then the required power flux density can be calculated from the *RSL* and line feed loss:

$$PFD = RSL - A_e + L_f \tag{7.20}$$

Finally, using Equation 3.225, the field strength is

$$E_{\mu V} = PFD + 145.8 \tag{7.21}$$

The *PFD* and field strength are useful for regulatory checks and measurement, but the actual performance of the receiver is driven by the *C*, *N* and *I*, which are therefore the key metrics used in this book. Therefore while it is noted how to calculate the field strength and *PFD*, the following sections continue to use the $\{C, I, N\}$ set of metrics.

It is worth comparing this methodology with that given in Section 5.8 on link design and in particular noting the key difference that the target RSL was calculated in Equation 7.18 without an explicit interference margin.

While there could be some interference margin included in the additional factors, the principal reason for calculating *RSL* without it is that the intra-system interference is included in the coverage prediction described in the following section. This is, therefore, an example of an alternative approach to handling interference in the design of radio systems.

An implication is that this approach is implicitly using the *C/N* threshold as the $T(C/(N + I))$.

For the methodology in Section 5.8, explicit margin for interference was included in order to derive the *RSL*. This was a useful approach for FS link design, as described in Section 7.1, in that the geometry was known and fixed; hence the required transmit power could be derived. A similar approach is available for coverage-based systems such as PMR in that the coverage requirement is an input and the result of link design is the power required (or at least to balance the F_c and F_p ratios described in Section 7.2). In both these cases there is the expectation that additional FS or PMR assignments will be introduced after deployment, and margin is therefore necessary so that interference received from these new systems is not harmful.

However broadcasting takes the reverse approach: for a given transmitter the coverage algorithm described in the following section is used to calculate the population covered taking into account interference. This difference is due to the characteristics of spectrum management of broadcasting bands, in particular:

- An integrated approach is used in the planning of all transmitters, rather than (e.g. for FS and PMR) each system potentially operated by a different organisation
- Changes tend to happen on a slower scale than for the FS and PMR. Major replans, such as for the digital switchover, are rare
- Broadcasting bands tend to be harmonised internationally so that the most common sharing scenario is with another broadcast system.

Hence rather than leaving a margin for interference (in particular, to protect against interference from systems introduced in the future), the aggregate interference is calculated for a known deployment of broadcasting transmitters. This can improve spectrum efficiency as it avoids introducing margin for interference that might not actually occur. It is an example of a general principle of interference analysis in that the more information is available, the less conservative the results tend to be, as there is no need to take worst-case (or near worst-case) assumptions for unknown parameters.

However this approach introduces difficulties when defining interference thresholds for inter-service sharing or be able to accept a reduction in coverage. The implication of interference is that it will decrease the predicted coverage, which raises the question as to what would be an acceptable degradation. As noted in Section 5.9, Rec. ITU-R BT.1895 identified that the threshold for aggregate interference from all co-primary services should be

$$ T\left(\frac{I_{agg}}{N}\right) = -10 \text{ dB} \tag{7.22} $$

This suggests that the link design should include some margin to protect against interference or be able to accept a reduction in coverage. The need for margin for interference is likely to increase as additional services including white space devices (WSDs) operate close by or within frequency bands used for broadcasting services. Indeed, for DVB-T reference networks a power margin of 3 dB was used for interference-limited planning.

7.3.2   Coverage Prediction

This section describes an example of the coverage prediction method. It uses one of the standard reference planning configurations (RPCs) that were developed as part of the transition to digital services discussed at the Regional Radiocommunication Conferences (RRC) RRC-04 and RRC-06 in Geneva. In particular, the RPC network described as RPC1 for a fixed antenna was used, as specified in Table 7.12.

The fixed receiver is assumed to be at a height of 10 m and uses the gain pattern specified by Rec. ITU-R BT.419 (ITU-R, 1992a) as shown in Figure 7.13 for Bands IV and V. A mobile receiver would likely have an isotropic antenna and be located at a lower height, typically below the clutter line or even indoors.

The fixed antenna peak gain can vary in frequency, but for these bands the range 7–12 dBd was identified in Rec. ITU-R BT.2036 (ITU-R, 2013b). Similar values were proposed in documents such as Joint Frequency Planning Project (2012) and the Chester Agreement (CEPT, 1997). These correspond to 9.15–14.15 dBi using the adjustment factor of 2.15 dB in Table 3.5.

Table 7.12   Parameters of DVB-T reference network RN1 RPC1

Network	RN1 RPC1
Receiver	Fixed
Number of transmitters	7
Geometry of transmitters	Hexagon
Distance between transmitters (km)	70
Service area diameter (km)	161
TX antenna height (m)	150
TX antenna pattern	Non-directional
ERP (dBW) less delta	39.8
Delta (dB)	3
EIRP (dBW)	44.95

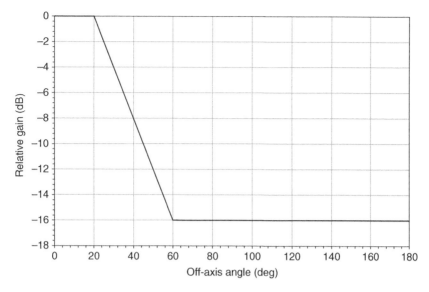

Figure 7.13   Gain pattern in Rec. ITU-R BT.419

Table 7.13   Fixed receiver characteristics from Rec. ITU-R BT.2036

Band	III	IV	V
Frequency (MHz)	174–230	470–582	582–862
Peak gain (dBd)	7	10	12
Peak gain (dBi)	9.15	12.15	14.15
Feed loss (dB)	2	3	5
Usable gain (dBi)	7.15	9.15	9.15

Sometimes the receive peak gain is quoted including the feed loss. Hence if (say) the receiver feed loss was 5 dB and the peak gain in Band V is 14.15 dBi (i.e. 12 dBd), then the overall gain would be 9.15 dBi, as shown in Table 7.13.

The standard propagation model assumed for broadcasting is Rec. ITU-R P.1546, as described in Section 4.3.5, which is one reason that this model has certain broadcasting features, such as the reference transmitter with ERP 1 kW and receiver at a height of 10 m. However it is worth noting that other propagation models are available, in particular:

- P.1812 is also a point-to-area propagation model for the frequency ranges considered, though it takes account of the full terrain path profile and can handle high-resolution surface data (see Section 4.3.6)
- P.2001 has a similar structure to P.1812 though it has the capability to handle the full range of percentages of time, including fading on the wanted path (see Section 4.3.7).

The coverage is predicted on a pixel-by-pixel basis using an area analysis methodology.

Example 7.12
The coverage for a single broadcast transmitter with the characteristics of RN1 RPC1 operating at 610 MHz with DBT-T carrier bandwidth 7.9 MHz is shown in Figure 7.14. Grid lines are spaced every 20 km and the greyscale shows locations where the coverage probabilities are {95%, 97%, 99%}.

The coverage of the wanted system is calculated using a percentage of time $p = 50\%$, while for interfering signals it is a value such as $p = 1\%$ (i.e. signal strength exceeded for no more than 1% of the time), though other values could be used. Note that the interfering signal plus a protection ratio is sometimes described as the nuisance field.

The wanted signal strength at each pixel is calculated initially for the median $q_0 = 50\%$ of locations and then a log-normal variation of signal across the pixel is assumed so that the percentage of locations actually served can be calculated using

$$z = \frac{\left.\frac{C}{N}\right|_{p=50\%,q=50\%} - T\left(\frac{C}{N}\right)}{\sigma} = \frac{M\left(\frac{C}{N}\right)\big|_{p=50\%,q=50\%}}{\sigma} \tag{7.23}$$

Typically the standard deviation $\sigma = 5.5$ dB is used for digital broadcasting, though other values could be considered, as described in Rec. ITU-R P.1546. The value z can then be used to calculate the percentages of location, q, by inverting the normal distribution as described in Section 3.10.

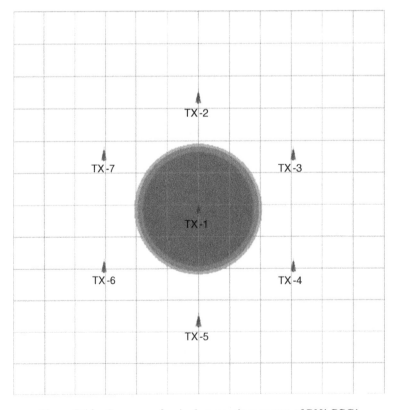

Figure 7.14   Coverage of a single transmitter as part of RN1 RPC1

Example 7.13

The C/N predicted using $p = 50\%$, $q_0 = 50\%$ at a pixel is 34.4 dB, a margin of 12.8 dB over the threshold value $T(C/N) = 21.6$ dB. With a location variability of 5.5 dB, this implies $z = 2.3$ and hence $q = 99\%$ of locations within this pixel would be covered.

Typically there is minimum percentage of locations that must be served within a pixel for it to be considered to be covered by a broadcast transmitter. In this example it was taken to be $z = 1.645$ for $T(q) = 95\%$ of locations, but other values could be used including $q = 70\%$, 90% or 99%.

From the location coverage within a pixel and a database that can identify the population in that pixel, it is therefore possible to derive the total population covered by a broadcast transmitter using

$$Population_{Total} = \sum_{i=pixel} Population_i . q_i \big|_{q_i > T(q)} \tag{7.24}$$

This example was based upon a single wanted signal and no interferers, but in practice there could be multiple wanted signals as part of a SFN as described in Section 7.3.4 and multiple interfering signals. The question is then how to calculate the availability over a pixel taking into account these multiple wanted or interfering signals, each of which is likely to vary across that pixel. This is discussed further in the following section.

7.3.3   Statistical Power Summation

The QoS at the receiver depends upon the $C/(N+I)$. Within each pixel the C and I will vary depending upon location as shown in Figure 7.15, each signal typically modelled as having a normal distribution around a mean. A **statistical power summation** calculation is required to derive the aggregate interference and hence the $C/(N+I)$.

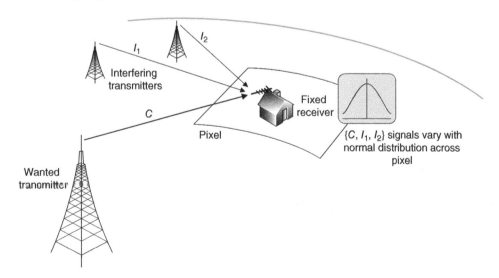

Figure 7.15   Calculation of $C/(N+I)$ from multiple normal distributed signals

One approach could be to undertake a Monte Carlo simulation to derive the $C/(N+I)$ using an algorithm similar to this for m interfering contributions:

1. Initialise number of locations served $N_S = 0$
2. Repeat for $k = 1 \ldots N_T$ trials:
   a. Generate random percentage $q = q[0, 100]$
   b. Use q to calculate $C_k(\mu, \sigma)$ assuming log-normal distribution
   c. Initialise the $(n + i)$ to:

$$(n + i) = 10^{N/10} \tag{7.25}$$

   d. Repeat for each of $j = 1 \ldots m$ interferers:
      i. Generate random percentage $q = q[0, 100]$
      ii. Use q to calculate $I_{k,j}(\mu_j, \sigma_j)$ assuming log-normal distribution
      iii. Increment the $(n + i)$ with this interfering power:

$$(n + i) += 10^{I_{k,j}(\mu_j, \sigma_j)/10} \tag{7.26}$$

   e. Calculate the noise plus interference in dBW:

$$(N + I) = 10 \log_{10}(n + i) \tag{7.27}$$

   f. Hence calculate the $C/(N+I)$ in dB:

$$\left(\frac{C}{N+I}\right) = C - (N + I) \tag{7.28}$$

   g. If $C/(N+I) > T(C/(N+I))$, then increment the number of locations served:

$$N_S += 1 \tag{7.29}$$

3. Hence calculate likelihood of availability across pixel as a percentage:

$$P = 100.\frac{N_S}{N_T} \tag{7.30}$$

Example 7.14
A Monte Carlo simulation was undertaken that used the previous methodology to combine the wanted and interfering signals given in Table 7.14 assuming the receiver from Example 7.12. After 100,000 trials the distributions of the wanted and interfering signals were as in Figure 7.16 and the resulting $C/(N+I)$ distribution in Figure 7.17.

Note that the (C, I) distributions have spikes towards their high and low ends due to the clipping algorithm in Equation 4.18 where the random number is created in the range [0, 100] but the P.1546 location variability is restricted to the range [1%, 99%].

Table 7.14   Example wanted and interfering signals

Source	Wanted	Interferer 1	Interferer 2
μ_i (dBW)	−88.9	−114.5	−120.7
σ_i (dB)	5.5	5.5	5.5

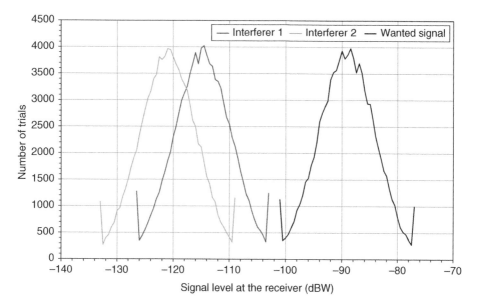

Figure 7.16   Example Monte Carlo analysis distribution of wanted and interfering signals

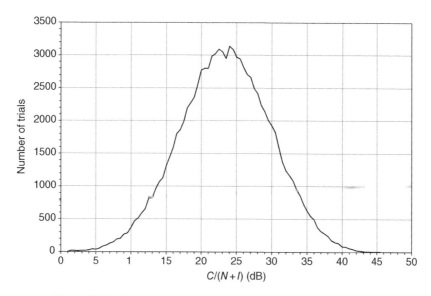

Figure 7.17   Example Monte Carlo analysis distribution of $C/(N+I)$

In this case the coverage was calculated to be $q = 62.6\%$ of locations.

This approach is flexible and generates the required metrics but is computationally intensive, in particular as it would have to be repeated for each point across a national coverage area. Alternatives are therefore usually employed that can quickly generate an approximate value, and a range of methods are available including Schwartz–Yeh, k-LNM (log-normal methodology) and t-LNM.

The k-LNM approach has a set of means and standard deviations as inputs, identified as (μ_i, σ_i) where $i = (1 \ldots n)$. They can be combined using the following methodology from:

- EBU Technical Report 24: SFN Frequency Planning and Network Implementation with regard to D-DAB and DVB-T (EBU, 2005).

First the variables are converted into the Neper scale:

$$\mu_i \leftarrow \frac{1}{10\log_{10}(e)}\mu_i \tag{7.31}$$

$$\sigma_i \leftarrow \frac{1}{10\log_{10}(e)}\sigma_i \tag{7.32}$$

Then the power distribution mean and standard deviations are calculated:

$$M_i = e^{\mu_i + \frac{\sigma_i^2}{2}} \tag{7.33}$$

$$S_i^2 = e^{2\mu_i + \sigma_i^2}\left(e^{\sigma_i^2} - 1\right) \tag{7.34}$$

These are then summed:

$$M = \sum_{i=1}^{i=n} M_i \tag{7.35}$$

$$S^2 = \sum_{i=1}^{i=n} S_i^2 \tag{7.36}$$

From these the mean and standard deviation of the resulting aggregate powers are calculated using the k factor:

$$\sigma^2 = \log_e\left(k\frac{S^2}{M^2} + 1\right) \tag{7.37}$$

$$\mu = \log_e(M) - \sigma^2 \tag{7.38}$$

These are then converted back from the Neper scale:

$$\mu \leftarrow 10\log_{10}(e)\mu \tag{7.39}$$

$$\sigma \leftarrow 10\log_{10}(e)\sigma \tag{7.40}$$

k is an adjustable parameter that depends upon the number, mean and variations of each of the contributions, but it is suggested that $k = 0.5$ is appropriate for cases where the standard deviation is between 6 and 10 dB and $k = 0.7$ if lower.

Example 7.15

The total noise plus interference of a receiver depends upon two interfering signals and the noise with values given in Table 7.15. Using the previous methodology with $k = 0.7$, the mean and standard deviation of the noise plus interference are (−112.3 dBW, 4.4 dB).

Finally the mean and standard deviation of the $C/(N + I)$ can be calculated using

$$\mu_{C/(N+I)} = \mu_{\Sigma C} - \mu_{\Sigma N + I} \tag{7.41}$$

$$\sigma_{C/(N+I)} = \sqrt{\sigma_{\Sigma C}^2 + \sigma_{\Sigma N + I}^2} \tag{7.42}$$

The location availability can then be calculated from the *z* parameter and the reverse normal distribution where

$$z = \frac{\mu_{C/(N+I)} - T\left(\frac{C}{N}\right)}{\sigma_{C/(N+I)}} \tag{7.43}$$

Example 7.16

The parameters used in the Monte Carlo analysis in Example 7.14 were used to undertake similar calculations with the k-LNM methodology, generating the results in Table 7.16. It can be seen that there is a reasonable match with the results of the Monte Carlo analysis but requiring a fraction of the computational resources.

Note that there are differences in assumptions between methodologies that imply that it is unlikely that outputs would precisely match. In particular, the Monte Carlo analysis, unlike the analytic approach, clipped the location variability to the range [1%, 99%] as specified by Rec. ITU-R P.1546.

Table 7.15 Example interference plus noise contributions

Source	Interferer 1	Interferer 2	Noise
μ_i (dBW)	−114.5	−120.7	−128.0
σ_i (dB)	5.5	5.5	0.0

Table 7.16 Results using k-LNM methodology

$C/(N + I)$ mean (dB)	23.4
$C/(N + I)$ standard deviation (dB)	7.1
$M(C/(N + I))$ (dB)	1.8
z	0.26
Location availability q (%)	60.3

7.3.4 Single-Frequency Networks

The DVB digital broadcasting standard is based upon OFDM due to its protection against multipath fading plus its ability to aggregate signals from multiple transmitters. As described in Section 3.4.4, OFDM splits a high rate data stream into a large number of lower rate subcarriers using orthogonal frequencies, which reduces self-interference. This technology mitigates against fading, which is likely to affect only a subset of carriers, and permits the combination of those multipath signals that arrive within the selected guard interval. Increasing the guard interval results in improved tolerance of multipath but at the cost of reducing the usable payload. The maximum difference in path distance that can be tolerated by the receiver's fast Fourier transform (FFT) circuits can be calculated for a given guard interval using Equation 3.173.

This ability of the receiver to combine multiple signals can be used to aggregate signals from multiple wanted transmitters all operating simultaneously on the same frequency. A SFN is planned taking into account the aggregate signal strength of all transmitters that would be received within the specified guard interval.

An example of this is shown in the space, frequency and time diagram in Figure 7.18 where:

- A fixed receiver is within the coverage area of three transmitters
- Its antenna is pointing at the primary wanted signal and is picking up the other two via its side-lobes

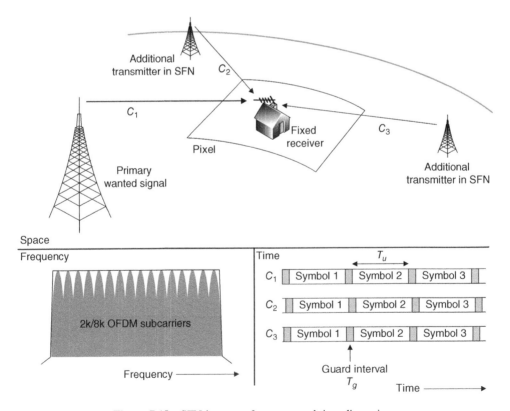

Figure 7.18 SFN in space, frequency and time dimensions

- The wanted signal in each case comprises an array of 2k or 8k OFDM subcarriers
- The distance to each transmitter is slightly different and so the symbol arrival times are slightly offset
- However the offset is less than the guard interval and so each signal could be combined in the FFT circuits.

In the planning process, these multiple wanted signals can be aggregated using the same power summation techniques described in the previous section for the interfering signal.

Example 7.17

A test point is able to receive three signals from a SFN with values as per Table 7.17 based upon the wanted signal and two interferers from the previous example. Using the k-LNM methodology described in Section 7.3.3 with $k = 0.7$, the mean and standard deviation of the total wanted signal were calculated to be (−88.2 dBW, 5 dB).

Table 7.17   Example wanted signals for SFN test point

Source	C_1	C_2	C_3
μ_i (dBW)	−114.5	−120.7	−88.9
σ_i (dB)	5.5	5.5	5.5

It can be seen that in this case the impact of the SFN is to boost the wanted signal (similar to a simple summation of the powers) and reduce the standard deviation, but neither by much. This is typically the case when using SFNs with fixed receivers: as the secondary signals are likely to be received on the far off-axis part of the antenna gain pattern, they will be at least 16 dB weaker than the primary signal (see gain pattern in Figure 7.13). The benefits of SFN are greater for low-gain receivers such as mobiles.

Another trade-off is between the size of the SFN and the guard interval required to ensure synchronisation across the network. A larger guard interval increases the number of transmitters that can be included in the wanted signal but at the cost of reduced payload.

The coverage of an example SFN is shown in Figure 7.19 based upon the results from Example 7.17 (wanted signals summed, standard deviation reduced to 5 dB). Grid lines are spaced every 20 km and the greyscale shows locations where the coverage probabilities are {95%, 97%, 99%}.

7.4   Earth Station Coordination

Satellite Earth station coordination, as defined in Appendix 7 of the RR, is an example of a regulatory algorithm, which can also be used during national frequency assignment. Satellite Earth stations, like fixed links, tend to use highly directional antennas that facilitate sharing. There remains, however, the potential for harmful interference, and hence it is necessary to undertake analysis. To avoid checking all co-frequency assignments in detail, a two-stage process is followed:

1. Identify the area within which there could be potentially affected or affecting systems using the actual parameters of the new ES and typical parameters of other existing systems

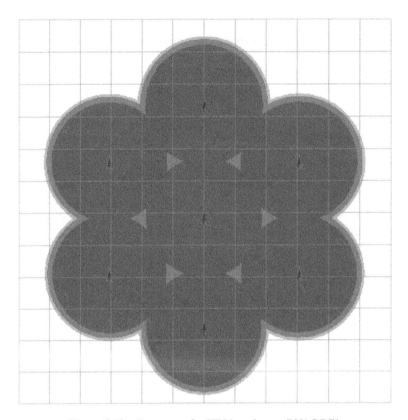

Figure 7.19   Coverage of a SFN based upon RN1 RPC1

2. Analyse those systems within this area in detail using their actual parameters to identify whether there could be cases of harmful interference.

   While this section will consider the introduction of a new satellite Earth station, it is worth noting that the process must also be followed in shared bands when a new terrestrial station is introduced. However the coordination contour is always calculated around the ES and the analysis determines whether terrestrial stations are within that contour. Note also the discussion in Section 7.5.4 on GSO satellite *PFD* and ES off-axis *EIRP* constraints.

   When a new satellite Earth station is proposed, then the first step is to identify the area around the transmit or receive station where there is the potential to cause or suffer interference. This is known as the **coordination contour** and is an example of a radial-based area analysis (see Section 6.5.1) and which uses the propagation model in Appendix 7 of the RR, as described in Section 4.3.9. This contour requires two sets of parameters:

1. The new ES's actual parameters
2. Reference or typical parameters to use for other systems in the band, most likely terrestrial fixed services but possibly another satellite Earth station where **reverse band** operation is feasible (i.e. both satellite uplinks and downlinks are permitted).

From the reference parameters an interference threshold $T(I)$ is calculated, in particular for the short-term. Given a threshold, system parameters and propagation models, it is then possible to generate a contour that defines the area within which coordination is required. This contour is calculated using smooth Earth so it in general is a worst-case assumption: this is appropriate as it is a trigger for further action, coordination, in which a more detailed assessment is made. The objective is to minimise the number of cases where there would be harmful interference but it is not identified.

Similarly the reference parameters are typically towards the more sensitive end of the range of possible values. For example, in the 17.7–18.8 and 19.3–19.7 GHz bands, Appendix 7 identifies that the analysis should assume that the FS antenna gain is 45 dBi, which corresponds to a dish size of around 1.2 m. However typical use of this band is for mobile network backhaul, which often uses smaller dishes for short hops. Furthermore, it is assumed that the antenna is pointing directly at the ES, which is likely to be the case only for a very small percentage of links.

The interference for the FS (transmit) into ES (receive) at distance = d along the line with azimuth = Az for $p\%$ percentage of time can be calculated for Mode 1 propagation using

$$I_{ST} = P_{TX} + G_{FS,\,peak} - L_{App7}(d,p\%) + G_{ES}(Az) \tag{7.44}$$

Conversely, the interference ES (transmit) into FS (receive) is

$$I_{ST} = P_{TX} + G_{ES}(Az) - L_{App7}(d,p\%) + G_{FS,\,peak} \tag{7.45}$$

In Appendix 7 these equations are reordered to calculate the required loss in a particular azimuth to just meet the threshold $T_{st}(I)$, as in

$$L_{App7}(p\%) = P_{TX} + G_{FS,\,peak} + G_{ES}(Az) - T_{st}(I,p\%) \tag{7.46}$$

The thresholds can be calculated using the equations given in Section 5.9.

Example 7.18
A GSO satellite ES is proposed to operate at Goonhilly Downs with the parameters in Table 7.18. The resulting coordination contour is shown in Figure 7.20.

Table 7.18   Example receive ES parameters

Service	FSS
Direction	Space–Earth
Frequency band (GHz)	17.7–19.7
Latitude (deg N)	50.047
Longitude (deg E)	−5.1803
Height (m)	5
GSO satellite longitude (deg E)	−30
Receiver noise temperature (K)	100
Antenna gain pattern	Rec. ITU-R 580-6
Antenna dish size (m)	2.4
Antenna efficiency	0.6

Figure 7.20   Example receive GSO Earth station coordination. *Screenshot credit: Visualyse Coordinate*

Resource 7.3   The Visualyse Coordinate file 'Goonhilly Ka band.Coord' was used to generate the results in this example.

Having generated the contour it is then necessary to identify the impact in terms of:

• Administrations to coordinate with (in this case there are none)
• Potentially affected or affecting terrestrial fixed links or reverse band satellite ESs.

The identification of potentially affected or affecting terrestrial fixed links or reverse band satellite ESs will require a search of an assignment database. These are typically managed by the relevant national regulator and may or may not be publically available. Example of countries for which information is available includes Australia (ACMA), Canada (Industry Canada) and the United States (FCC). These databases tend to focus on specific bands, such as those used for PMR, FS and satellite ES site licensing, where it would be beneficial to be aware of what constraints exist prior to submitting a licence application.

Another source of information is the ITU's database, the Terrestrial International Frequency Information Circular (IFIC). This contains stations that have been filed with the ITU and approved for entry onto their Master Register. Many administrations do not automatically file their stations with the ITU-R unless it is necessary, avoiding cases if they are:

• Limited in scope to the national administration's territory (as in the contour in Example 7.18)
• Covered under bilateral or multilateral processes with neighbouring administrations (e.g. the HCM Agreement)
• Not requiring protection from another administration's systems.

Example 7.19

A search of the Terrestrial IFIC was made for assignments that were within the contour of the ES from Example 7.18. The resulting assignments are shown in Figure 7.21. In this case the band was proposed for use by non-GSO FSS satellite networks in the 1990s, and hence there was a flurry of registration of assignments with the ITU at that time. In a real coordination it would be necessary to check for changes since that time, whether old links were shut down or new links introduced.

After the assignments have been identified, then detailed coordination can be undertaken. The process used is not mandated and can be agreed with the parties involved but typically:

- Uses actual parameters for both interferer and victim
- Uses a terrain database and a propagation model that takes account of terrain such as Rec. ITU-R P.452 (ITU-R, 2013d)
- Considers both short-term and long-term interference using the thresholds in Rec. ITU-R SF.1006 (ITU-R, 1993)
- Could consider additional mitigation methods such as site shielding (see the clutter model in Section 4.3.4, in particular Equation 4.15)
- Could consider both co-frequency and (where applicable) non-co-frequency sharing cases.

These represent a series of static analyses as described in Section 6.3.

Figure 7.21   Terrestrial fixed links extracted from Terrestrial IFIC search. *Screenshot credit: Visualyse Professional. Terrain and land use credit: SRTM*

Example 7.20

A detailed analysis was undertaken of the potential for interference into the ES from Example 7.18 from the terrestrial fixed links extracted from the Terrestrial IFIC in Example 7.19. Figure 7.22 shows the long-term I/N plotted across the 17.7–19.7 GHz band assuming a 30 MHz ES victim bandwidth using the SRTM terrain database and Rec. ITU-R P.452 propagation model using the A_{BW} and A_{MI} for the two cases of co-frequency and non-co-frequency interference analysis. It was noted that:

- The I/N interference plot was dominated by two segments, relating to a set of six 110 MHz bi-directional point-to-point links with one station located at the same Goonhilly Downs site as the ES
- The central portion of the frequency plot represents the duplex gap for the fixed links involved
- The impact on the I/N plot of the other FS links was not significant
- In cases such as this when some interferers are geographically very close it is useful to undertake non-co-frequency analysis using A_{MI} to identify how much frequency separation would be needed
- The analysis using A_{MI} produced very similar results to A_{BW} when there was frequency overlap, but there were significant differences for the non-co-frequency case.

As noted earlier, one of the reasons why the fixed-satellite service (FSS) and FS often have primary allocations in the same frequency bands is that the directivity of either system's antenna facilitates sharing. This means that if there is harmful interference at one location, it usually only requires a station to move a short distance to ensure compatibility.

Example 7.21

The satellite ES in Example 7.20 was moved around the vicinity of the Goonhilly Downs site assuming a receive frequency of 19.1 GHz. It was noted that interference was above a

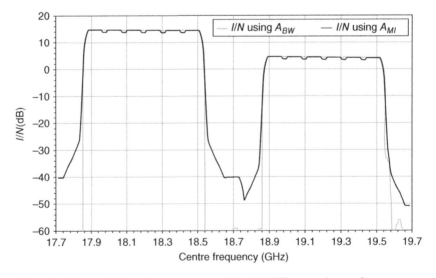

Figure 7.22 Variation in I/N for frequency range 17.7–19.7 GHz assuming co-frequency or non-co-frequency analysis

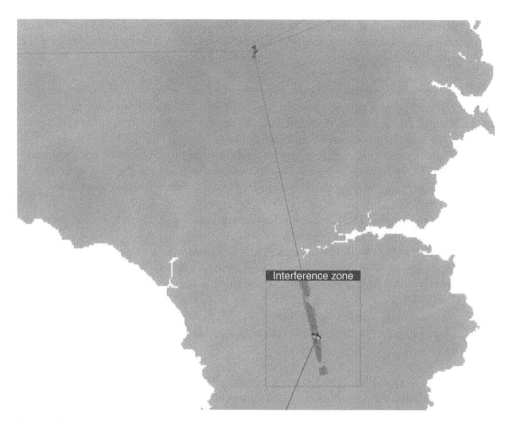

Figure 7.23 Interference zone along line between point-to-point FS stations. *Screenshot credit: Visualyse Professional. Terrain and land use credit: SRTM*

$T(I/N, 20\%) = -10$ dB for only a narrow line between the two FS stations, as can be seen in Figure 7.23. Therefore moving the ES only a short distance, potentially combined with site shielding, should be sufficient to ensure a successful coordination.

There have been proposals from the satellite industry to provide what is described as high-density fixed-satellite service (HD-FSS) in which large numbers of ES would be deployed in this and other bands. The numbers involved have made some question whether alternatives to case-by-case coordination should be considered. In discussions within CEPT (as in the draft ECC Report on enhanced access to spectrum for FSS uncoordinated Earth stations in the 17.7–19.7 GHz band), options discussed include:

- **Band segmentation**, where parts of the band could be identified for HD-FSS, in particular the duplex gap. However this gap alone might provide insufficient spectrum and operation outside it constrain development of the FS
- **Geographic segmentation**, where specific locations could be identified for either HD-FSS or FS, which would involve constraints on both services
- **Dynamic frequency selection** (DFS), in which the ES spectrum senses to detect operation of FS and selects the clear channel(s) on which to operate

- **Automated frequency assignment**, in which a database is maintained of FS and HD-FSS assignments and software automatically undertakes interference analysis to identified the preferred channels.

The last option is facilitated by the development of online services and by having an algorithm that is well defined: it would basically be following the approach used in Example 7.19 and Example 7.20. It can be considered to be applying the concepts of WSDs (as in Section 7.10) to sharing between FS and FSS systems.

The previous examples were based upon the coordination of a GSO satellite ES: the procedures for non-GSO are similar but must take account of the motion of the ES's antenna as it tracks the moving satellite. The methodology in Appendix 7 describes two approaches:

1. **Time-invariant gain (TIG)**, in which the highest gain at each azimuth is calculated and used in the calculation of the coordination contour
2. **Time-variant gain (TVG)**, in which the probability of each gain value for each azimuth is calculated and convolved with the propagation statistics. This is an example of a dynamic simulation with Monte Carlo elements, as described in Section 6.8.

The gain towards the horizon, either worst case or statistics, will depend upon the tracking strategy used by the non-GSO ES, which is likely to depend upon the operational requirements. For example:

- A customer non-GSO ES is likely to select a single satellite at a time, most likely the highest elevation, possibly taking into account GSO arc avoidance as described in Section 7.6
- A gateway or control non-GSO ES is likely to track all visible satellites down to a minimum elevation angle, also possibly taking into account GSO arc avoidance.

Example 7.22
A non-GSO ES is proposed to be located at Goonhilly Downs serving the LEO-F constellation with parameters in Section 6.2.2.2. The maximum gain towards the horizon for each azimuth is shown in Figure 7.24 assuming that there are one, two or three antenna active tracking the highest, two highest or three highest satellites, respectively. The simulation ran for a period of 1 year with a time step of 10 seconds and the altitude was adjusted to $h = 10{,}362.3$ km to ensure the satellites fully populated the orbit shell, as described in Section 6.8. The variation between azimuths for the $n = 1$ case is due to symmetry within the constellation driving the satellite selection.

It should be noted that even for cases where the gain exceeded 4 dBi, it only did so for very short periods of time, in this case under 0.1%. Therefore assuming these values were applicable for 100% of the time (using the TIG methodology) would overestimate the interference generated or experienced.

7.5 GSO Satellite Coordination

7.5.1 Regulatory Background

The geostationary satellite orbit (GSO) is unique in that a satellite located on it will appear fixed with respect to the rotating Earth. At this altitude a single satellite is able to provide

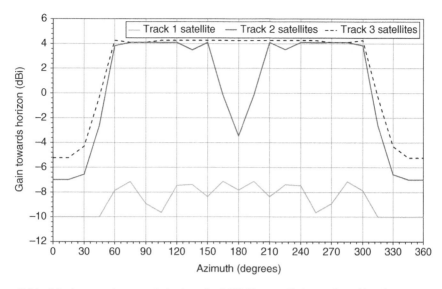

Figure 7.24 Maximum gain towards horizon for LEO-F constellation and tracking from one to three satellites

communication services over a very wide area, with approximately 40% of the Earth's surface visible. The line of this orbit in the sky is called the **GSO arc**, and positions or **slots**, defined by the longitude of the GSO satellite, are in great demand.

If satellites are located too close together, they can cause harmful interference, as shown in Example 3.32, in which it was seen that for a given ES antenna size there is an interference-driven constraint in the minimum GSO arc separation angle (i.e. the difference in GSO satellite longitudes). For this reason the GSO arc is considered a limited resource, and there are specific requirements in international regulations to ensure it is used in an efficient and equitable manner.

Two regulatory approaches are used to manage access to the GSO:

1. Coordination approach, whereby each system must undergo a technical and procedural process with existing or earlier filed systems, in the **unplanned bands**
2. Planned approach, whereby slots and channels are allocated to administrations, such as those defined in Appendix 30, 30A and 30B of the Radio Regulations, which identify the **planned bands**.

A wide range of satellite services use the GSO including to provide mobile, broadcasting and fixed services and to support operation in other services, such as navigation, intersatellite links or meteorology. The process is similar for those services for which the ES uses directional antennas, whether fixed-satellite service (FSS), broadcasting-satellite service (BSS) or control stations for other services.

Coordination is more challenging for services that are operating with low-gain antennas, such as mobile Earth stations (MES) or BSS (sound) terminals. For this reason these systems tend to require geographic and/or frequency separation to avoid harmful interference, with locations and bands agreed during bilateral or multilateral agreements between the system operators involved.

This section focusses upon coordination methodologies used to analyse sharing between GSO satellite services that have ESs with highly directional receive antennas, in particular those of the FSS and BSS. This process is based on the principle of first-come, first-served (FCFS) so that earlier networks have a higher priority over those that file at a later date.

The process is defined via the RR, in particular those in Articles 9 and 11, which includes definitions of the timetable for events, and Appendix 4, which specifies the data that must be provided. The key unit is the **filing**, which contains information that can be used to undertake an interference assessment of a GSO satellite network. The data is provided in three stages:

1. **Advanced publication information** (API): this is the initialisation trigger for the ITU's procedures and generally only gives high-level information. Within 2 years this must be followed up by the:
2. **Coordination request** (CR): this gives much more detailed information, sufficient to undertake a detailed interference assessment as described in the following sections. The date of receipt of the CR is a key milestone for the priority process. There could be modifications, additions and suppressions to the CR as more information is known during the coordination process. Within 7 years it must be followed up by the:
3. **Notification** (N-notice): this gives the final state of the proposed system with the aim of it being entered into the master international frequency register (MIFR).

The final stage is that the filing must be **brought into use** by ensuring a satellite operates at the GSO location identified in the filing on the frequency bands specified. Note that at WRC-15 the API was modified to use the basic characteristics of a CR as in Article 9.1A of the Final Acts (ITU, 2015).

The structure of the data follows the format in Appendix 4 and a high-level view of the key database tables is given in Figure 7.25, where [D] implies that there is an associated date. It is useful to have a good understanding of this structure as it will be the basis of any analysis.

The key tables are:

- **Notice**: this includes the longitude, filing administration and various ITU-R references, such as notice ID and status
- **Beam**: this relates to a transmit or receive beam of the satellite network and defines the peak gain and gain pattern, often using the shaped beam format described in Section 7.5.5
- **Group**: this is a container for the following three tables but also includes additional information such as polarisation, service area and noise temperature. Groups can have their own dates as part of multiple filings relating to the same notice. A group can be associated to a specific transponder on the satellite and hence has characteristics and constraints defined by it (e.g. bandwidth, power)
- **Emissions**: this includes the designation of emission (which defines, in particular, the bandwidth using the format described in Section 5.1) but also the maximum and minimum power and power densities
- **Frequencies**: this is simply a list of frequencies
- **Earth stations**: there are different types, such as typical (which could be deployed anywhere in a service area) or specific (which are defined at a given location), and associated

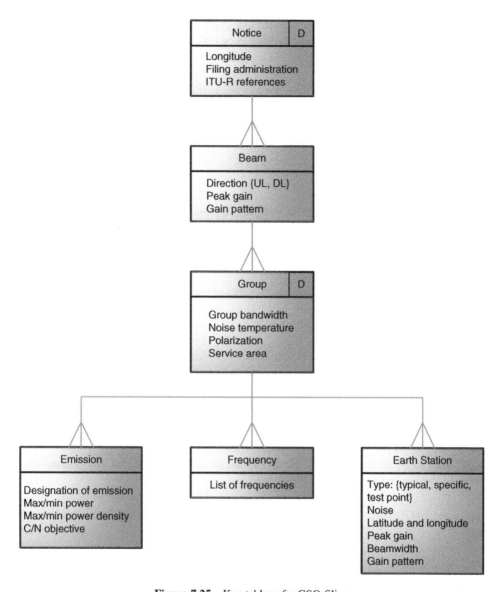

Figure 7.25   Key tables of a GSO filing

parameters such as peak gain, beamwidth and gain pattern, typically using one of those described in Section 7.5.5.

An additional table could define the connections between uplink and downlink directions (strapping), but this is generally not used. Note that non-GSO systems have additional tables to define their orbit parameters.

The data is available from the ITU-R via the Space Services **International Frequency Information Circular** (IFIC), which distributed every two weeks with the latest updates, and the **Space Radiocommunication Stations** (SRS) database, which contained the full set of GSO and non-GSO stations and associated filing information. These databases can be edited by a suite of tools provided by the ITU BR, in particular:

- **SpaceCap**: to create filing data
- **SpaceQry**: to query the databases and view the output
- **Graphical Interference Management System** (GIMS): to create, edit and view shaped beams and service areas in the format described in Section 7.5.5.

From the ITU's web site are interfaces to the space network databases:

- **Space Network Systems** (SNS) online contains Appendix 4 data of GSO filings, non-GSO filings and ES filings
- **Space Network List** (SNL) is a list of basic information concerning planned or existing space stations, Earth stations and radio astronomy stations.

Creation of a filing requires a good understanding of the service intended to be provided in terms of geographic location, frequency band, Earth station characteristics (noise temperature and, in particular, dish size) plus carrier types.

7.5.2 Coordination Triggers

A GSO system comprises two paths, an uplink (UL) and a downlink (DL). Where there are two GSO systems, there can be four potential interference paths as identified in Figure 7.26:

Case 1: from an UL into another UL or DL into another DL
Case 2: from an UL into a DL or DL into an UL due to one or other of the GSO systems operating in reverse band mode.

The standard GSO satellite coordination process involves the first of these cases and has two stages:

1. The coordination trigger, by which it is identified where further analysis is required, involves checking if there is frequency overlap (hence non-co-frequency analysis is typically not required) and one of:
    a. Difference in GSO longitude between two systems is less than the coordination arc defined for the bands involved.
    b. $T(DT/T) > 6\%$.
2. Detailed coordination, discussed further in Section 7.5.3.

The **coordination arc** is defined in the RR Appendix 5 and varies between $\pm 7°$ and $\pm 16°$ depending upon frequency band. In some bands the coordination arc is the default method unless an administration requests it be included in the coordination process based upon a DT/T calculation. Note that WRC-15 tightened the range of the coordination arc in some bands

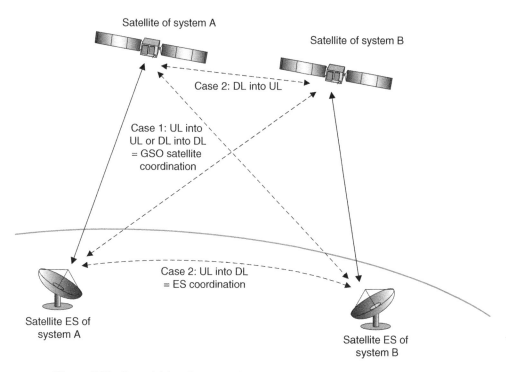

Figure 7.26 Potential interference paths to consider between GSO satellite systems

and added a PFD threshold for harmful interference outside the coordination arc in Earth–space and space–Earth directions in selected bands.

The method to calculate the DT/T is given in Appendix 8 and Rec. ITU-R S.738 (ITU-R, 1992b), which use the terms in Figure 7.27 as defined in Table 7.19. Note that θ_t is the **topo-centric angle** at the Earth station between its boresight line and the relevant off-axis direction, whereas θ_g is the **geocentric angle** at the centre of the Earth of the lines to the two satellites involved.

The DT/T for the uplink and downlink can then be calculated using (noting that all terms are in absolute and hence multiplied or divided)

$$\frac{DT_s}{T_s} = \frac{p'_e g'_1\left(\theta'_t\right) g_2(\delta_{e'})}{l_u k T_S} \tag{7.47}$$

$$\frac{DT_e}{T_e} = \frac{p'_s g'_3(\eta_e) g_4(\theta_t)}{l_u k T_e} \tag{7.48}$$

Here k is Boltzmann's constant in absolute. In most cases the DT/T can be checked against the $T(DT/T) > 6\%$ for the uplink and downlink separately. If the gain $= \gamma$ at the satellite is available (as used for the end-to-end calculation in Section 5.6), then the end-to-end DT/T can be calculated using

$$DT = \gamma DT_s + DT_e \tag{7.49}$$

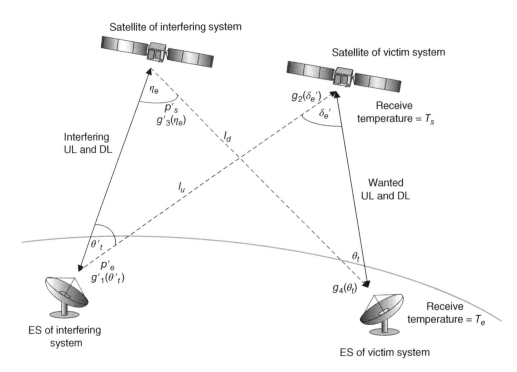

Figure 7.27   Terms in the calculation of DT/T

Table 7.19   Notation for Figure 7.27

Direction	Uplink	Downlink
Interferer's transmit power density in watts/Hz	p'_e	p'_s
Off-axis angle at interferer in direction of wanted	θ'_t	η_e
Interferer transmit gain in direction of wanted	$g'_1(\theta'_t)$	$g'_3(\eta_e)$
Free space path loss from interferer to wanted station	l_u	l_d
Off-axis angle at wanted in direction of interferer	$\delta_{e'}$	θ_t
Wanted receive gain in direction of interferer	$g_2(\delta_{e'})$	$g_4(\theta_t)$
Receive temperature in kelvin	T_s	T_e

Note that all calculations are referenced to 1 Hz (though averaged over either 4 kHz or 1 MHz) and hence there is no need to specify the bandwidth or calculate the A_{BW}. Also the propagation model is free space path loss, which is the simplest and most conservative approach: there typically will always be some additional attenuation due to atmospheric gases. Similarly there is no consideration of mitigation factors such as polarisation unless both administrations involved agree to it: this is designed to be a worst-case calculation to trigger any necessary coordination.

Often this equation is calculated in dB, so that the interference per Hz $I_0 = DT$ for the downlink, which was given in absolute in Equation 7.48, is alternately

$$I_0 = DT_e = P'_s + G'_3(\eta_e) - L_d + G_4(\theta_t) \tag{7.50}$$

While this calculation is straightforward, an example of static analysis, application is complicated by two factors:

1. There can be an extremely large number of permutations to consider taking into account the two system's sets of {beams, groups, emissions, frequencies, Earth stations}. The tendency of static analysis with many combinations of inputs to result in large numbers of cases to consider is discussed in Section 6.3
2. The DT/T calculated can be expected to vary depending upon the ES position, and hence a detailed calculation would have to scan for the worst-case location. This would involve multiple area analyses, one for each combination of parameters.

The first of these can be addressed by noting that a coordination trigger only needs to fail once to be activated and hence it is appropriate to look for the worst combination of parameters, such as highest transmit power density and lowest receive temperature.

The second can be simplified by using standardised geometry, so that the satellite beams are assumed to be pointing directly at the ES (and hence the peak gain used) that is located subsatellite, while the topocentric angle can be calculated from the geocentric angle using the approach in Rec. ITU-R S.728 (ITU-R, 1995) as

$$\theta_t = 1.1 \times \theta_g \tag{7.51}$$

Example 7.23
GSO Operator A has filed a satellite at 130°E with a spot beam peak gain of 43 dBi and average power density at 11.1 GHz of −57.5 dBW/Hz. Existing GSO Operator B has a satellite at 132.2°E with a typical ES having a noise temperature of 140 K and gain pattern defined by Rec. ITU-R S.465. Coordination is required as:

a. GSO Operator A's proposed satellite is within the coordination arc of GSO Operator B's satellite
b. GSO Operator A's downlink would exceed $T(DT/T) = 6\%$ at GSO Operator B's ES as shown in Table 7.20.

Note how the analysis at the coordination trigger stage only requires a minimal set of parameters compared to detailed coordination.

7.5.3 Detailed Coordination

The methodology used during detailed coordination is agreed between the parties involved but typically is undertaken based upon the C/I metric calculated (using the notation from Chapter 5) by

$$C = P_{tx} + G_{tx} - L_p + G_{rx} \tag{7.52}$$

$$I = P'_{tx} + G'_{tx} - L'_p - L_{pol} + G'_{rx} + A_{BW} \tag{7.53}$$

$$\frac{C}{I} = C - I \tag{7.54}$$

Table 7.20 Calculation of DT/T from satellite A into ES B

Field	Symbol(s)	Value	Calculation
Geocentric angle (deg)	θ_g	2.2	Difference in longitudes
Frequency (GHz)	f_{GHz}	11.1	Example filing data
TX power density (dBW/Hz)	P_{tx}	−57.5	Example filing data
Transmit gain (dBi)	G_{tx}	43	Example filing data
Receive noise temperature (K)	T	140	Victim parameter
Distance from satellite A to ES B	d_{km}	35,791.5	Geometry calculation
Free space path loss (dB)	L_{fs}	204.4	Using Equation 3.13
Topocentric angle (deg)	θ_t	2.4	Using Equation 7.51
Receive gain (dBi)	G_{rx}	22.4	Assuming $32 - 10\log_{10}(\theta_t)$
Interference (dBW/Hz)	I_0	−196.5	Using Equation 7.50
Noise (dBW/Hz)	N_0	−207.1	Using Equation 3.41
I_0/N_0 (dB)	I_0/N_0	10.61	$I_0 - N_0$
DT_e/T_e (%)	DT_e/T_e	1151	Using Equation 3.219

$$M\left(\frac{C}{I}\right) = \frac{C}{I} - T\left(\frac{C}{I}\right) \tag{7.55}$$

The methodology used is defined in Recs. ITU-R S.740 (ITU-R, 1992c) and S.741 (ITU-R, 1994c).

The bandwidth adjustment term has to take account of the difference in wanted and interfering bandwidths and possibly whether there are multiple interferers that overlap with the victim, as explained in Section 5.2. The degree of overlap would depend upon the groups involved, but a simplified approach would be to use the two bandwidths:

$$A_{BW} = 10\log_{10}\left(\frac{B_V}{B_I}\right) \tag{7.56}$$

The coordination process itself is initiated by the administrations involved, as they are recognised within the ITU-R system. The actual discussions, however, are usually handled by members of the relevant satellite operating companies. Most coordinations are bilaterals involving two satellite networks, but there can also be multilaterals, in particular for MSS systems. The coordination process can involve multiple satellites within each operator's network, as there can be a range of dates for each satellite's overall notice and each of its groups, resulting in a matrix of priorities. The negotiating strength of each party is related to the number of cases for which they have date priority over the other.

The inputs to the coordination process are the various filings and any previous coordination agreement(s), and the objective is to reach an agreement on how each system can operate, defined via constraints such as possible EIRP limits on satellites and Earth stations.

As with the coordination trigger stage, a difficulty is the number of permutations of {beams, groups, emissions, frequencies, Earth stations} for both systems, potentially with multiple filings to consider, with a range of date priorities. Most of the interference studies involve static methodology or area analysis based upon the C/I calculation focussing on specific scenarios, such as locations, types of Earth stations and emissions determined by the types of services that

are intended to be provided. However such analysis could also take account of future requirements to provide alternative services and at other locations.

Two additional complexities of satellite coordination are that:

1. The filing can differ from the actual satellite: there should be a degree of match between the
   two, but the regulatory and manufacturing processes occur in parallel and there can be differences (e.g. gain pattern on the satellite). In an extreme case a satellite originally developed
   for one filing might be sold to bring into use a completely different filing. Therefore one of
   the key early stages of the coordination process is an exchange of data on the real network's
   characteristics
2. The filing can define global spot beams without identifying where exactly they are pointing.
   In practice it is likely to have its geometry fixed for long periods of time to fulfil contracts
   with customers, but the operator will want to maintain as far as possible the ability to serve
   other customers at other locations in the future. Again, this is a suitable topic for discussion
   during the data collection phase.

Section 5.10.1 described how the $T(C/I)$ in Rec. ITU-R S.741 (ITU-R, 1994c) is given as

$$T\left(\frac{C}{I}\right) = T\left(\frac{C}{N}\right) + 12.2 \text{ dB} \qquad (7.57)$$

It is worth noting that the $T(C/N)$ in this equation implicitly rather than explicitly includes the
interference margin, as can be seen by considering the equivalent power line diagram given in
Figure 7.28, and hence

$$T\left(\frac{C}{N}\right) = T\left(\frac{C}{N+I}\right) + M_I \qquad (7.58)$$

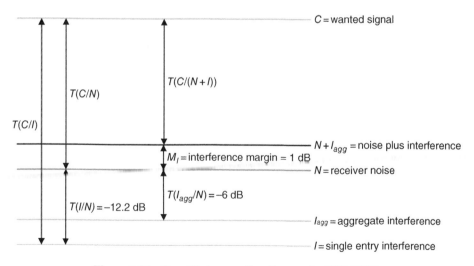

Figure 7.28   Simplified power line diagram for GSO $T(C/I)$

Example 7.24

The satellite systems from Example 7.23 are using BPSK modulation. To achieve the necessary BER, the $T(C/(N+I)) = 10.5$ dB (see Table 3.6) so with the standard 1 dB of interference margin the required $T(C/I)$ for the downlink is

$$T\left(\frac{C}{I}\right) = [10.5 + 1] + 12.2 = 23.7 \text{ dB}$$

In this downlink case it was not necessary to include in the $T(C/I)$ calculation margin for fading as propagation loss on the wanted and interfering paths can be assumed to be correlated, as described in Section 4.4.2. This was therefore calculated assuming free space propagation or 'clear air' conditions, but the wanted link budget could include additional terms for gaseous attenuation and rain fade.

Example 7.25

The GSO Operator A from Example 7.23 wishes to provide communication services to Australia using a range of ES antenna diameters = {0.6 m, 1.2 m, 1.8 m, 2.4 m, 3.6 m}. The operator is in coordination with GSO Operator B, which provides services to islands within the Indonesia and Papua New Guinea archipelagos using the shaped beam in Figure 7.29 with $G_{peak} = 36.5$ dBi. How could Operator A approach coordination?

As noted earlier, one of the greatest challenges of GSO satellite coordination is the sheer number of combinations of parameters to analyse. There is consequently a range of possible

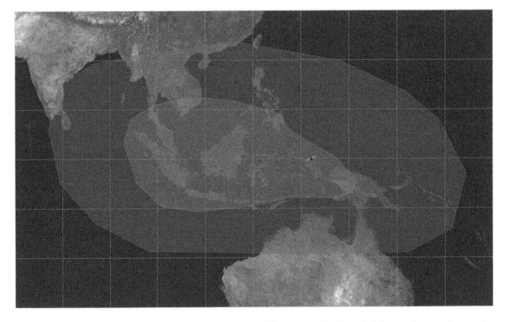

Figure 7.29   −4 and −12 dB contours of example satellite operator B downlink beam. *Screenshot credit: Visualyse Professional. Overlay credit: NASA Visible Earth*

approaches to achieve a successful coordination, each with implications on the services that can be provided.

Any satellite can achieve a successful coordination by simply agreeing to reduce their transmit power to levels that would not cause harmful interference into another network. This, however, is not a solution to satellite coordination as it is necessary to be able to:

- Transmit enough power to provide services to prospective customers
- Ensure that other satellite networks do not themselves cause harmful interference, which is made more difficult given the C/I metric if the wanted transmit power is reduced.

A starting point could be to identify a specific case of Satellite B to protect and investigate how that could be achieved while providing a useful service, for example:

Dish size = 0.9 m
Bandwidth = 24 MHz
Service area = −4 dB contour
Transmit power = 9.1 dBW.

Therefore the analysis would begin by locating the victim ES on the −4 dB contour and calculating the C/I.

Operator A will have a wide range of options for their parameters too, but a good baseline would be to assume something similar on the grounds that it is generally easier to share like with like and hence use a similar bandwidth carrier, in this case 30 MHz, with power derived from the power density of −57.5 dBW/Hz, that is, $P_{tx} = 17.3$ dBW. The resulting C/I contours are shown in Figure 7.30 with Operator A's steerable spot and Operator B's ES as close together as feasible for this scenario. At the test point the $C/I = 13.3$ dB with the wanted and interfering link budgets given in Table 7.21.

The first-cut $C/I = 13.3$ dB was 10.4 dB below the threshold $T(C/I) = 23.7$ dB, so further analysis was clearly required. A number of approaches could be considered, typically involving modifying the interferer's characteristics, in particular the transmit power, beam pointing or polarisation. In this case there is a significant difference in transmit power (8.2 dB higher for the interferer) and even greater in the $EIRP$ (14.7 dB), suggesting it would be worth investigating further.

In the SRS/IFIC data structure, there are two powers and two power density fields in the emission table, namely, the minimum and maximum, and hence there are four possible approaches to calculate the C/I. The two most commonly used are:

a. Wanted and interferer both at maximum power
b. Wanted uses the minimum power while the interferer uses the maximum, which would result in the worst-case C/I.

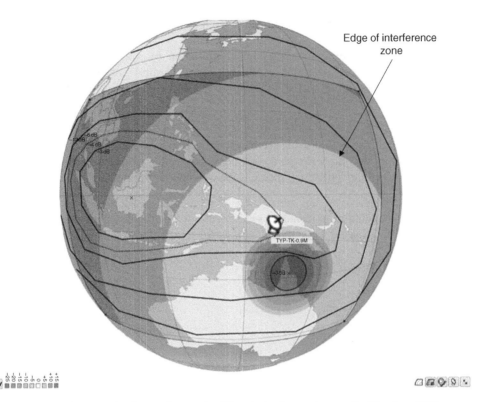

Edge of interference zone

Figure 7.30   Result of start of example *C/I* analysis. *Screenshot credit: Visualyse GSO*

Table 7.21   Wanted and interfering link budgets for *C/I* example analysis

Signal	C	I
Transmit power (dBW)	9.1	17.3
Power spectral density (dBW/Hz)	−63.9	−56.7
Transmit peak gain (dBi)	36.5	43.0
Transmit EIRP (dBW)	45.6	60.3
Transmit relative gain (dB)	−4.0	−14.5
Free space path loss (dB)	204.4	204.5
Receive peak gain (dBi)	38.2	38.2
Receive relative gain (dB)	0.0	−16.5
Bandwidth adjustment (dB)	n/a	−1.0
Received signal (dBW)	−124.6	−137.9

The actual power used on a link can (and is likely to) be less than the maximum in the filing for a number of reasons, including:

- The group for that emission record could have a set of associated Earth stations, and the link might not be using the one with the lowest gain (which would require the maximum power to close the link)
- There could be a range of modulations or coding schemes operating within the same designation of emission with differing $T(C/(N+I))$ and hence requiring different transmit powers

- There are likely to be coordination constraints limiting the maximum power that can be transmitted without causing interference
- Some services could use power control so most of the time they operate at less than maximum power
- The satellite will be constrained by the total power available.

In this case there is scope to reduce system A's transmit power to be the same power density as system B while maintaining sufficient margin to provide a service. This results in a transmit power $P_{tx} = 10.9$ dBW reducing the margin to $M(C/I) = -4$ dB.

A number of methods could be used to turn the margin positive, including:

- Considering the use of alternate polarisation between systems A and B (e.g. if system B is using LH to use LV or vice versa)
- Arguing that the side-lobe performance of the Earth stations should be modelled using the gain pattern in Rec. ITU-R S.580 rather than S.465 as this would give 3 dB of mitigation (as discussed in Section 7.5.5)
- Pointing system A's steerable beam further away from system B's coverage area.

In this case it could be sufficient to use (say) 3 dB of polarisation loss and 3 dB of antenna gain improvement, resulting in a reduced interference contour as in Figure 7.31.

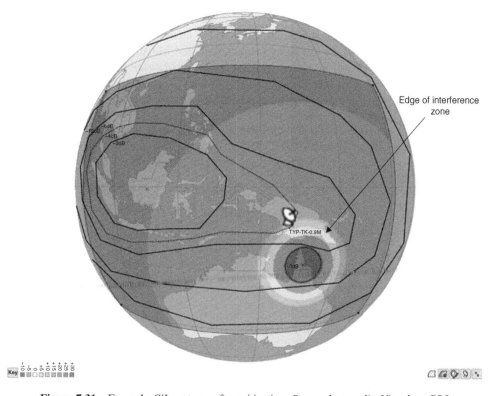

Figure 7.31 Example *C/I* contours after mitigation. *Screenshot credit: Visualyse GSO*

Other approaches to coordination could be considered, and it would be necessary to also take into account constraints to/from other satellite systems and possibly aggregation effects.

7.5.4   Coordination and Regulatory Constraints

The result of a successful detailed coordination is an agreement between the two operators involved (or, more precisely, between the two administrations) that defines constraints on one or both of them to facilitate sharing. A number of methods could be used to define these constraints including:

- Maximum EIRP limits
- Off-axis EIRP limits
- Frequency segmentation
- Geographic segmentation
- Polarisation segmentation
- Code segmentation
- Minimum antenna side-lobe performance
- Minimum dish sizes.

One of the most common formats is the off-axis EIRP constraint, which can be applied to either the satellite or Earth station transmit cases.

Example 7.26
The detailed coordination in Example 7.25 could lead to an agreement that Operator A's satellite off-axis EIRP towards the test point should not exceed 24.6 dBW/MHz based upon the calculation in Table 7.22, and in addition use the opposite polarisation to Operator B.

This approach would give Operator A the flexibility to use higher EIRPs in beams that have boresight further away from the test point and hence lower gain in that direction.

There are also constraints on the off-axis EIRP that ES are permitted to transmit in the Radio Regulations. For example, in Article 22.27 (ITU, 2012a) there are the following limits that an ES '*shall not exceed*' for off-axis angle φ in any direction, relating to the bands 12.75–13.25, 13.75–14 and 14–14.5 GHz:

$$3° \leq \vartheta \leq 7° \qquad 42 - 25\log_{10}(\vartheta) \quad \text{dBW}/40\text{KHz} \qquad (7.59)$$

$$7° < \vartheta \leq 9.2° \quad 21 \qquad\qquad\qquad \text{dBW}/40\text{kHz} \qquad (7.60)$$

Table 7.22   Example calculation of off-axis EIRP constraint

Transmit power (dBW)	10.9
Transmit peak gain (dBi)	43.0
Relative gain towards test point (dB)	−14.5
EIRP towards test point (dBW)	39.4
Bandwidth (MHz)	30.0
EIRP density towards test point (dBW/MHz)	24.6

$$9.2° < \vartheta \le 48° \qquad 45 - 25\log_{10}(\vartheta) \quad \text{dBW}/40\text{kHz} \qquad (7.61)$$
$$48° < \vartheta \le 180° \qquad 3 \qquad\qquad \text{dBW}/40\text{kHz} \qquad (7.62)$$

These limits were included to protect non-GSO FSS systems as part of a balance of regulations put in place at WRC 2000 that included the EPFD values discussed in Section 7.6.

There are also EIRP limits in Rec. ITU-R S.524 (ITU-R, 2006g): these, while similar, define the EIRP not in any direction off-axis but rather '*in any direction within 3° of the GSO*'. They are typically 3 dB tighter but only in the direction along the GSO arc (as would be of concern to GSO to GSO sharing): in other directions RR 22.27 is more relevant (as would be of concern for GSO to non-GSO sharing).

There are additional EIRP constraints in Article 21 to facilitate sharing uplinks with fixed service stations (as described in Section 7.4) and also $T(PFD)$ constraints on satellite downlinks, an example of which is shown in Table 7.23. The thresholds for *PFD*, which can be calculated using Equation 3.222, are tightened for low elevation angles where there could be less FS antenna discrimination.

Example 7.27
Satellite Operator A's spot beam in Example 7.25 would create a peak PFD subsatellite (i.e. with elevation angle $\delta = 90°$) of $-116.5\,\text{dBW}/\text{m}^2/\text{MHz}$ as calculated in Table 7.24. This is below the threshold given in Table 7.23, but there might be restrictions on operating at maximum power to low elevation angles to ensure a favourable finding by the ITU BR.

Table 7.23 Example PFD limits for the FSS in 10.7–11.7 GHz

Elevation angle δ	$0° \le \delta \le 5°$	$5° \le \delta \le 25°$	$25° \le \delta \le 90°$
PFD dBW/m^2/MHz	-129	$-129 + 0.75(\delta - 5)$	-114

Table 7.24 Example PFD calculation

Transmit power (dBW)	17.3
Transmit peak gain (dBi)	43.0
Transmit EIRP (dBW)	60.3
Bandwidth (MHz)	30.0
Transmit EIRP density (dBW/MHz)	45.5
Distance (km)	35,786
Spreading loss (dB/m^2)	162.1
PFD (dBW/m^2/MHz)	-116.5

7.5.5 Gain Patterns

7.5.5.1 Satellite Gain Patterns

The gain pattern of GSO satellites can be defined either via global beams, steerable spots (as in Operator A in the previous example) or a shaped gain pattern. The latter is where the antenna gain pattern has been designed to cover a specific area in which case it is defined via:

- A maximum or peak gain in dBi
- One or more **boresights** where the gain is a local maxima, defined relative to the maximum gain. At least one boresight has to have a relative gain of 0 dB
- A set of **contour levels** relative to the peak gain, for example, {−2, −4, −6, −8, −12, −20} dB
- A set of **contour lines** specified by a contour level and a set of (latitude, longitude) points. These contour lines can either be **closed**, in which case the line starts and ends at the same point, or **open**, in which case the start and end points are different.

An example of a GSO shaped beam is shown in Figure 7.29.

The data is often described as '.GXT' or GIMS format as the former is the text file format used to exchange data and the latter the ITU software tool used to create or edit shaped beams.

Alternatively, spot beams can be defined using the pattern in Rec. ITU-R S.672 (ITU-R, 1997a) with boresight either defined using (Az, El) from the satellite or (latitude, longitude) on the ground, together with a peak gain, beamwidth and side-lobe level, typically one of {−20 dB, −25 dB, −30 dB}.

7.5.5.2 Earth Station Gain Patterns

A number of Earth station gain patterns are available defined in ITU-R Recommendations or part of the Radio Regulations. Two often used in coordination and filing data are Rec. ITU-R S.465 (ITU-R, 2010b) and S.580 (ITU-R, 2003e) as shown in Figure 7.32 for a dish diameter = 2.4 m and frequency = 12 GHz. It can be seen that the S.580 gain pattern has a side-lobe 3 dB lower than that for S.465 because the equations for that segment are

$$G_{465}(\vartheta) = 32 - 25\log_{10}(\vartheta) \tag{7.63}$$

$$G_{580}(\vartheta) = 29 - 25\log_{10}(\vartheta) \tag{7.64}$$

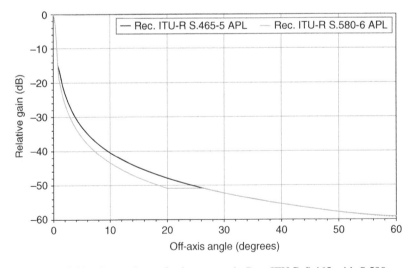

Figure 7.32 Comparison of gain patterns in Rec. ITU-R S.465 with S.580

Note that these Recommendations do not define the off-axis gain in the whole of the range [0°, 180°]. In particular, the main lobe region below about 1° or 2° off-axis is not specified. This is due to the motivation behind these Recommendations, which is for type approval: a regulator could specify that if an antenna meets this side-lobe, then it is considered acceptable for operational deployment. As noted in Section 3.7.5, there are trade-offs between aperture and beam efficiency that can change the resulting gain pattern, and having the main lobe undefined provides manufacturers with a degree of flexibility. However this lack of precision is not helpful for scenarios where it is necessary to be able to calculate the relative gain for all off-axis angles. A number of approaches can be considered during interference analysis:

- Use an alternate gain pattern, such as Rec. ITU-R S.1428 discussed in Section 7.6 relating to EPFD analysis
- Extend the gain pattern to the main lobe by simply assuming it's parabolic as in Equation 3.83
- Extend the gain pattern using more complex assumptions, such as those used by the ITU-R in their antenna pattern library (APL) as was used in Figure 7.32.

7.6 EPFD and Rec. ITU-R S.1503

7.6.1 Background

GSO satellites have proved highly successful at providing a wide range of services including broadcasting (video and sound), data and voice communications to fixed and mobile users. This success can be measured by the number of satellites on the GSO arc and the increasing difficulty in achieving coordination of new satellites.

But operating from a height of 35,786.1 km does have disadvantages, in particular:

- Increased round-trip delay and hence latency, as discussed in Section 3.8.6
- Increased free space path loss, as calculated using Equation 3.13
- The footprint of an antenna's beam is wider resulting in lower frequency reuse and hence reduced capacity for traffic per MHz.

Examples of delay, path loss and diameter for GSO and non-GSO are given in Table 7.25. The geometry of the beam size is shown in Figure 7.33 and the antenna footprint can be calculated using

$$\frac{\sin(\phi)}{R_e} = \frac{\sin\left(\frac{\pi}{2} + \epsilon\right)}{R_e + h} \tag{7.65}$$

Table 7.25 Comparison of GSO and example non-GSO characteristics

Orbit	GSO	Non-GSO
Satellite height (km)	35,786.1	1414.0
Footprint diameter of 1 deg beam (km)	625	25
Free space path loss at 11.1 GHz (dB)	204.4	176.4
Round-trip delay (mS)	238.6	9.4

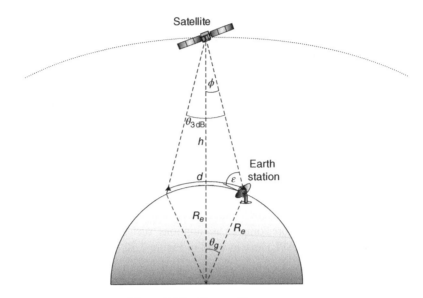

Figure 7.33   Key satellite geometry

$$\pi = \phi + \theta_g + \left(\frac{\pi}{2} + \epsilon\right) \tag{7.66}$$

$$d = 2Rg_e\theta_g \tag{7.67}$$

There are, therefore, benefits in moving to lower height orbits, though this also introduces complexities. In particular, a **constellation** of non-GSO satellites are required to cover the service area, and the user terminal must have the capacity to track the moving satellite. The task of designing the optimum non-GSO constellation can be complex, varying parameters such as the orbit height, inclination angle, number of satellites and possibly eccentricity.

A key factor in the constellation design is the minimum operating elevation angle of the user terminal: the lower the elevation angle, the greater the range its antenna must scan over, which can lead to increased terminal costs. From the elevation angle and orbital height, it is possible to calculate the geocentric angle between non-GSO satellites required to provide continual coverage using the geometry given in Figure 7.33. Assuming this separation angle is the same between satellites of the same plane and between planes the total number of satellites needed can be estimated, as shown in Table 7.26 for two example non-GSO FSS constellations, Teledesic (shown in Fig. 7.34) and SkyBridge, both proposed during the 1990s. The estimation is not exact as simplifying assumptions were required, but it shows how the constellation size is driven by these two key input parameters.

These two constellations were selected for this comparison as they approached regulatory issues in the 1990s in two very different ways, leading to the introduction of EPFD limits in the RR. It is notable that the design constraints that led to these sized constellations in the 1990s are also driving factors for the new wave of non-GSO constellations proposed at the time of writing this book.

Table 7.26 Estimation and actual constellation sizing

Constellation	Teledesic	SkyBridge
Height (km)	700	1469.3
Minimum elevation angle (deg)	40	20
Estimated constellation size	804	79
Actual constellation size	840	80

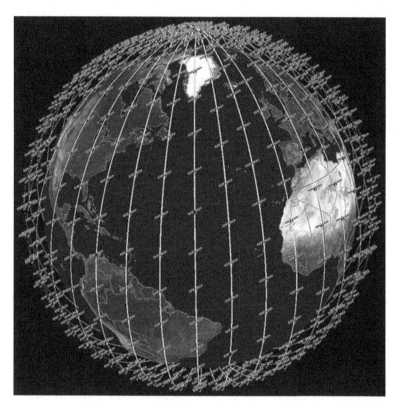

Figure 7.34 Teledesic non-GSO constellation. *Screenshot credit: Visualyse Professional. Overlay credit: NASA Visible Earth*

Two of the specific issues for large non-GSO constellations can be seen in this figure, which relate to the cost of manufacturing satellites that for most of the time are not used:

- The constellation was sized by coverage requirements at equatorial regions, but the orbital planes converge at higher latitudes, leading to many satellites having little or no traffic in polar regions
- The Earth is mostly covered by sea, and while over oceans, such as the Pacific, the satellites will again have little or no traffic.

Teledesic was the first large non-GSO FSS system to file and was one of the earliest satellite systems proposing to operate in Ka band, in particular, after WRC 97, the bands 18.8–19.3 GHz (space-to-Earth) and 28.6–29.1 GHz (Earth-to-space). It intended to use the coordination process, taking advantage of its date priority in the ITU-R queue, ahead of all but a handful of GSO satellites. However it was questioned, at WRC 97 and study group meetings prior, whether coordination was the best approach for non-GSO networks operating in bands potentially heavily used by GSO networks because:

- If the non-GSO system had priority, it would be extremely hard for GSO systems to share given the number of satellites involved and the network's global service area. The concern was that one such non-GSO system from one administration would lock out of the band all future GSO systems from all administrations
- In bands with large numbers of existing GSO systems, it would be hard for non-GSO operators to get regulatory status as they'd have to achieve a successful coordination with every one while ensuring that they did not cause what Article 22 of the RR described simply as 'unacceptable interference'.

The solution was to develop a new regulatory approach that would quantify what would be considered 'acceptable interference' by GSO systems. If the non-GSO systems met these levels, then they could operate without having to coordinate with each GSO network while providing GSO systems with confidence that their services were protected. The metric used to define this acceptable interference was the equivalent power flux density (EPFD):

$$epfd = 10\log_{10}\left[\sum_{i=1}^{N_a} 10^{P_i/10} \frac{g_{tx}(\theta_i) g_{rx}(\varphi_i)}{4\pi d_i^2 \; g_{rx,max}}\right] \tag{7.68}$$

The benefit of this metric is that it has the measurability of PFD while taking into account the high degree of directivity of FSS and BSS Earth stations, and for this reason the EPFD metric has been introduced into other scenarios, such as to protect the radio astronomy service.

There are two types of EPFD limit in Article 22 of the RR:

1. **Validation EPFD limits**, used during the non-GSO filing examination stage, which are checked via computation. The validation EPFD limits are an example of using interference analysis for regulatory approval and are hard limits in that they must be met
2. **Operational EPFD limits**, used when the non-GSO system is providing a service, which could be checked via measurement.

The EPFD levels had to take account of the variation in geometry as the satellites move. Hence rather than being a single value such as $DT/T = 6\%$, the EPFD thresholds were statistical in nature, defined with an associated percentage of time. These were derived using algorithms such as that in Rec. ITU-R S.1323 (ITU-R, 2002c), which convolved interference with rain fading and tested against unavailability requirements. The convolution worked with aggregate interference levels, which were then apportioned to single network thresholds by assuming (see Resolution 76)

$$N_{sys} = 3.5 \tag{7.69}$$

Aggregate EPFD limits are given in Resolution 140 (WRC-03).

There was some discussion in the ITU-R group set up to study non-GSO sharing, Joint Task Group (JTG) 4-9-11, about how to define a non-GSO system to avoid the case where a constellation with (say) 1000 satellites splits itself into 100 sub-constellations of 10 satellites, each generating an aggregate EPFD up to the full level permitted. One approach would be to consider the complete set of satellites and Earth stations that interoperate as a system, as in:

- Group {ES} that all are serviced by one or more satellites from group {Non-GSO} and none outside this group
- Group {Non-GSO} such that all ES serviced are in the group {ES} and none outside this group.

The EPFD limits vary depending upon the direction of interference, in particular between:

- **EPFD(↓) or EPFD(down):** interference from the non-GSO system's downlink into the GSO system's downlink. In general, this is the hardest of the three directions for the non-GSO system to meet
- **EPFD(↑) or EPFD(up):** interference from the non-GSO system's uplink into the GSO system's uplink
- **EPFD(IS):** inter-satellite (IS) interference from the non-GSO system's downlink into GSO system's uplink.

The EPFD limits also vary by service, frequency band and receiver characteristics, in particular its dish size or beamwidth and associated gain pattern.

An expert group was set up within the JTG 4-9-11 to study the algorithm required to check a non-GSO systems against the validation EPFD limits and the result was Rec. ITU-R S.1503 (ITU-R, 2013p), as described in Section 7.6.3. The key to the algorithm is the concept of the exclusion angle defined via the α parameter described in the following section.

7.6.2   Exclusion Zones and the α Angle

The EPFD(down) thresholds in Article 22.5C of the RR are defined '*for all pointing directions towards the geostationary satellite orbit*' and '*at any point on the Earth's surface visible from the geostationary-satellite*'. But how are non-GSO satellite systems operated while providing this degree of protection?

The answer is for non-GSO systems to avoid pointing their beams in directions that would result in a main beam-to-main beam alignment. A number of system designs have been proposed that meet this requirement, but one approach is shown in Figure 7.33 based upon the SkyBridge system. In this example there are three non-GSO satellites:

- Non-GSO Satellite A is directly in-line between the GSO ES and its satellite and therefore must not point any beams towards the ES. The EPFD at the GSO ES is kept low by ensuring there is a large antenna discrimination at the non-GSO satellite
- Non-GSO Satellite B is not in-line so it is able to point its beams closer to the GSO ES. However it cannot point them directly at it until the off-axis angle θ at the ES reaches a critical value, the edge of the exclusion zone

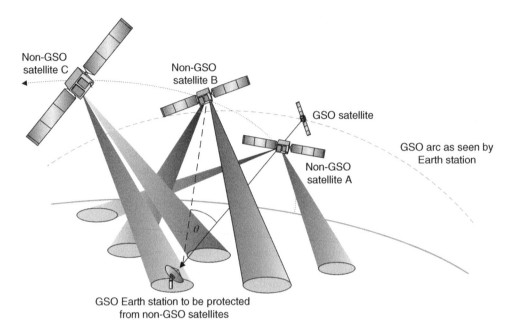

Non-GSO
satellite C

Non-GSO
satellite B

GSO satellite

GSO arc as seen by
Earth station

Non-GSO
satellite A

θ

GSO Earth station to be protected
from non-GSO satellites

Figure 7.35 In-line beam switch off tracking strategy to protect a single ES

- Non-GSO Satellite C is sufficiently far away from the line from the GSO ES to its satellite that there will be antenna discrimination at the ES. This will reduce the EPFD to within acceptable levels even when the non-GSO satellite's beam is pointed directly at the ES location.

This example considered a single ES with a single GSO satellite: to extend the geometry to the generic case, it is necessary to consider the α angle as shown in Figure 7.36. For a set of test points P_i, the angle α_i is calculated at the GSO ES (or any point on the Earth's surface visible from the GSO) between:

- The line from the GSO ES and the non-GSO satellite
- The line from the GSO ES and the test point P_i.

The α **angle** is then the minimum α_i over all test points, that is,

$$\alpha = min\{\alpha_i\} \tag{7.70}$$

There is a converse angle X, which is the minimum of the angle at the non-GSO satellite rather than at the GSO ES. When $\alpha = X = 0$, then the non-GSO satellite is directly in-line between a GSO ES and a point on the GSO arc.

One benefit of using α is that the worst-case gain at the ES can be calculated simply using α as the off-axis angle, that is, $Gain = G(\alpha)$. A downside is that there is no known analytic solution to calculate α so it must be derived using iteration and the variation of α to most inputs is non-linear.

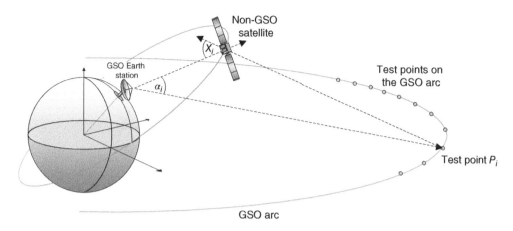

Figure 7.36 Alpha angle calculation

As seen in Figure 7.35, there is typically a critical angle where there is sufficient antenna discrimination at the ES so that the non-GSO satellite can transmit directly towards it without exceeding the EPFD limits. Whereas this figure involved just a single ES and GSO satellite, as shown in Figure 7.36, it is necessary to protect the whole GSO arc and hence the exclusion zone is typically defined via the α or X angles. The exclusion zone can be visualised as being either defined by an ES location (and hence specifying locations in its sky where a non-GSO satellite could not operate in its direction) or alternatively via the satellite (and identify locations on the ground where it could not provide a service).

Example 7.28

Figure 7.37 shows in grey the exclusion zone on the ground where $\alpha \leq 10°$ for a non-GSO satellite on the equator (i.e. latitude $= 0°$N) at a $h = 1440$ km. The latitude and longitude grid lines are spaced every $10°$ with a central black line showing where $\alpha = 0°$. The satellite would not be able to point any active beam towards locations within the exclusion zone. The circle shows the field of view visible to the satellite.

As there are potentially two α lines, the northerly one is assumed to be positive and the southerly negative. Positions east and west along these lines are defined via the Δ**longitude**, defined as the difference between the longitude of the non-GSO satellite and the point on the GSO arc that minimises α_i. Note how the α lines bend up towards the edge of the field of view.

Example 7.29

Figure 7.38 shows in grey the exclusion zone on the ground where $\alpha \leq 10°$ for a non-GSO satellite at a latitude $= 30°$N at a $h = 1440$ km. The latitude and longitude grid lines are spaced every $10°$ with a central black line showing where $\alpha = 0°$. Two ovals show the field of view visible to the satellite and the line where the elevation angle of the satellite would be $20°$. Across the Earth's surface are a set of traffic cell points: the non-GSO satellite is directing beams at those that meet the following two conditions as calculated at the traffic cell centre (an alternative approach could consider the edge of each traffic cell):

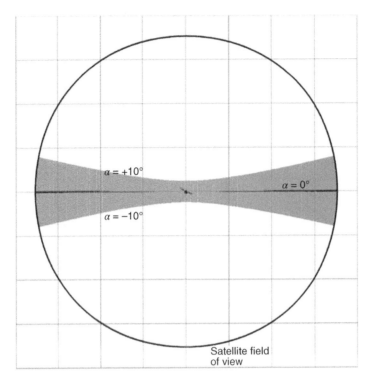

Figure 7.37   Exclusion zone based upon $\alpha = 10°$

1. The non-GSO satellite must have elevation angle $\varepsilon \geq \varepsilon_0$ where $\varepsilon_0 = 20°$ is the minimum elevation angle
2. The non-GSO satellite must have an angle to the GSO arc $\alpha \geq \alpha_0$ where $\alpha_0 = 10°$ is the size of the exclusion zone.

   These excluded zones must be served by another satellite to provide continuous coverage. This increases the requirements on the non-GSO system design to have additional capacity to handle these geometries, but the benefit is the ability to share radio spectrum with GSO systems.

Example 7.30

Figure 7.39 shows a simulation of the SkyBridge constellation, which is using beam pointing tracking strategies to provide services to North America while ensuring it meets the EPFD limits. SkyBridge proposed to use '**sticky beams**' where the pointing angles and beam shape were modified electronically so that the footprint remained approximately circular and at the same location. The satellite used to serve each cell would be selected by rules based upon the exclusion zone defined via the α_0 angle and minimum elevation angle. The alternative to sticky beams is if the beam pointing angles are fixed at the satellite.

Resource 7.4   The Visualyse Professional simulation file 'SkyBridge Example.sim' was used to create Figure 7.39.

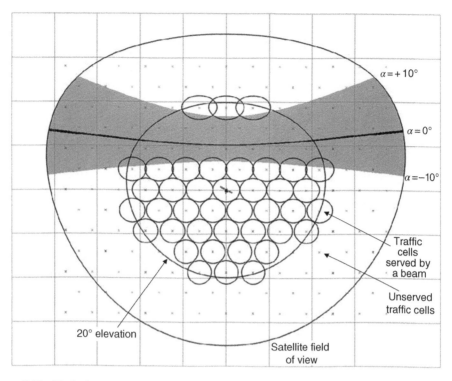

Figure 7.38 Exclusion zone and active beams for non-GSO satellite at latitude = 30°N. *Screenshot credit: Visualyse Professional*

7.6.3 EPFD Validation Methodology

The previous section describes some of the constraints on non-GSO FSS systems that must be met to ensure they do not exceed the EPFD limits, in particular avoiding pointing beams towards exclusion zones defined by GSO angles. The result is a complex network, with continually changing beam pointing angles or activities as the satellites move, with potentially larger-scale modifications as the system evolves in response to technological and market conditions. While a detailed simulation of each satellite's beam pointing and traffic can be undertaken to calculate EPFD statistics (and hence whether the thresholds are met), it would be insufficient to run just the once, as the analysis would depend upon a host of assumptions. In addition, the EPFD calculated would depend upon the specific GSO ES(s) and satellite(s) chosen, potentially requiring an extremely large number of runs to ensure that the system meets the EPFD limits for all locations and pointing angles at the GSO arc.

The requirement for the EPFD verification is to undertake a run-once analysis that would determine whether the non-GSO filing should be approved or not. Therefore an alternative methodology is required, and the one specified in Rec. ITU-R S.1503 is based upon the following two key concepts:

1. **PFD or EIRP masks**: a mask is calculated that is an envelope of all the potential EIRPs, beam pointing directions and satellite locations that would not be exceeded during a system's operation. The non-GSO operator is free to change their configuration as much as they

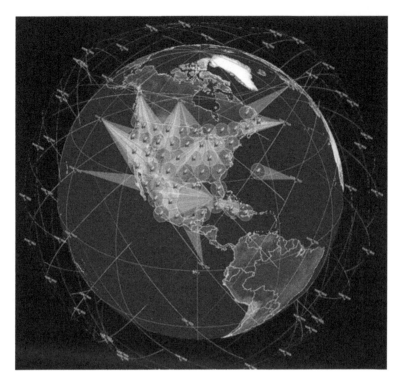

Figure 7.39   Example SkyBridge simulation. *Screenshot credit: Visualyse Professional. Overlay credit: NASA Visible Earth*

feel necessary without informing the ITU-R as long as the aggregate power does not exceed the PFD or EIRP mask. For EPFD(down), as the PFD mask is a regulatory constraint that could be measured, it is defined on the Earth's surface in units dBW/m^2 per a reference bandwidth. For the uplink and inter-satellite cases, these are EIRP masks as dBW versus off-axis angle. Each has an associated frequency range for which the PFD/EIRP mask is valid and a reference bandwidth

2. **Worst-case geometry** (WCG): given a PFD or EIRP mask and a constellation's orbit parameters, an analysis is undertaken to identify the geometrical location(s) that would result in the highest single-entry EPFD. Where there are multiple orientations with the same single-entry EPFD, then the one with the highest likelihood is selected using methods based upon Rec. ITU-R S.1257 (ITU-R, 2002b): for example, satellites seen at lower elevation angles tend to have a smaller angular velocity than those with higher elevation angles, and this should lead to larger likelihoods.

The generation of the PFD and EIRP masks is left to the non-GSO operator, and they have a free hand as to the methodology they use, though there is guidance in Rec. ITU-R S.1503. The PFD mask can be defined in a number of ways, including:

- PFD(α, Δlong, latitude)
- PFD(X, Δlong, latitude)
- PFD(Az, El, latitude).

The latitude in this case is of the non-GSO satellite, and the (Az, El) are calculated using the reference frame in Figure 3.63 (i.e. north based rather than velocity based as typically used for non-GSO as shown in Figure 3.64).

The number of parameters in the mask has evolved during the development of Rec. ITU-R S.1503 from initially just α. There is a trade-off as greater flexibility in definition of the PFD mask gives non-GSO operators more opportunity to optimise their system, but at the cost of additional complexity in the algorithm, in particular in the calculation of the WCG.

The methodology in S.1503 to check whether a non-GSO filing meets the EPFD limits in Article 22 is based on the following stages:

1. The filing and the PFD/EIRP masks it contains are examined and the relevant frequency ranges extracted
2. These frequency ranges are compared against the various EPFD limits in the RR and a set of runs generated to ensure all limits are checked (as in Example 7.31)
3. For each run the PFD/EIRP mask and constellation orbital parameters are used to derive a WCG
4. A GSO satellite and associated ES are located at the WCG in a simulation
5. For the constellation involved and the GSO satellite/ES antenna beamwidth involved, the time step and run time are calculated using an approach similar to that in Section 6.8. Note that there can be two time steps as discussed in the following
6. A dynamic analysis simulation is run until the end and EPFD statistics generated, in particular a CDF
7. The EPFD statistics generated by the run are compared against the limits in the relevant tables in Article 22 and a pass/fail flag set.

Example 7.31

A test system based upon SkyBridge was analysed using EPFD verification software. The resulting list of runs is shown in Figure 7.40 based upon the EPFD limits in Article 22 of the RR (ITU, 2012a). The gain pattern associated with each run is specified in the RR.

The result of each run is a CDF of EPFD, which can be compared against the thresholds in Article 22. The limits in Article 22 define sets of $T(EPFD, p'_\%)$ with straight lines connecting them. The percentage $p'_\%$ is given as 'percentage of time during which epfd$\downarrow$ may not be exceeded', for example, $(-160\,\text{dBW/m}^2/40\,\text{kHz}, 99.997\%)$. Alternatively, these can be defined using percentage $p_\% = $ 'percentage of time during which epfd$\downarrow$ may be exceeded' and hence this limit is $T(EPFD, p_\%) = (-160\,\text{dBW/m}^2/40\,\text{kHz}, 0.003\%)$. The resolution of the statistics is assumed to be 0.1 dB.

Example 7.32

The result of the first of the EPFD(down) runs in Example 7.31 is shown in Figure 7.41. This run would be considered a pass as for all EPFD values the percentage of time the EPFD level is exceeded is less than the associated threshold value. Or, more simply, on the chart the EPFD CDF curve is to the left of the threshold line.

A key requirement for the algorithm is that results are statistically valid, which means that all locations within an orbit have been sampled with equal likelihood. As noted in Section 6.8, orbit geometries can involve subtle effects where long run times are needed to avoid biased

Run Schedule						
Run	Frequency	Antenna	Service	Gain Pattern	Bandwidth	non-GSO
⊟-🢒 Up						
⫘ At Start	12.500	4.000	FSS	ITU-R S.672, Ls -20	40.000	SkyBridge
⫘ At Start	17.300	4.000	FSS	ITU-R S.672, Ls -20	40.000	SkyBridge
⫘ At Start	27.500	1.550	FSS	ITU-R S.672, Ls -10	40.000	SkyBridge
⫘ At Start	29.500	1.550	FSS	ITU-R S.672, Ls -10	40.000	SkyBridge
⊟-🢖 Down						
⫘ At Start	10.700	0.600	FSS	ITU-R S.1428	40.000	SkyBridge
⫘ At Start	10.700	1.200	FSS	ITU-R S.1428	40.000	SkyBridge
⫘ At Start	10.700	3.000	FSS	ITU-R S.1428	40.000	SkyBridge
⫘ At Start	10.700	10.000	FSS	ITU-R S.1428	40.000	SkyBridge
⫘ At Start	17.800	1.000	FSS	ITU-R S.1428	40.000	SkyBridge
⫘ At Start	17.800	2.000	FSS	ITU-R S.1428	40.000	SkyBridge
⫘ At Start	17.800	5.000	FSS	ITU-R S.1428	40.000	SkyBridge
⫘ At Start	17.801	1.000	FSS	ITU-R S.1428	1000.000	SkyBridge
⫘ At Start	17.801	2.000	FSS	ITU-R S.1428	1000.000	SkyBridge
⫘ At Start	17.801	5.000	FSS	ITU-R S.1428	1000.000	SkyBridge
⫘ At Start	19.700	0.700	FSS	ITU-R S.1428	40.000	SkyBridge
⫘ At Start	19.700	0.900	FSS	ITU-R S.1428	40.000	SkyBridge
⫘ At Start	19.700	2.500	FSS	ITU-R S.1428	40.000	SkyBridge
⫘ At Start	19.700	5.000	FSS	ITU-R S.1428	40.000	SkyBridge
⫘ At Start	19.701	0.700	FSS	ITU-R S.1428	1000.000	SkyBridge
⫘ At Start	19.701	0.900	FSS	ITU-R S.1428	1000.000	SkyBridge
⫘ At Start	19.701	2.500	FSS	ITU-R S.1428	1000.000	SkyBridge
⫘ At Start	19.701	5.000	FSS	ITU-R S.1428	1000.000	SkyBridge
⫘ At Start	11.700	0.300	BSS	ITU-R BO.1443	40.000	SkyBridge
⫘ At Start	11.700	0.450	BSS	ITU-R BO.1443	40.000	SkyBridge
⫘ At Start	11.700	0.600	BSS	ITU-R BO.1443	40.000	SkyBridge
⫘ At Start	11.700	0.900	BSS	ITU-R BO.1443	40.000	SkyBridge
⫘ At Start	11.700	1.200	BSS	ITU-R BO.1443	40.000	SkyBridge
⫘ At Start	11.700	1.800	BSS	ITU-R BO.1443	40.000	SkyBridge
⫘ At Start	11.700	2.400	BSS	ITU-R BO.1443	40.000	SkyBridge
⫘ At Start	11.700	3.000	BSS	ITU-R BO.1443	40.000	SkyBridge
⊟-🢔 IS						
⫘ At Start	10.700	4.000	FSS	ITU-R S.672, Ls -20	40.000	SkyBridge
⫘ At Start	17.800	4.000	FSS	ITU-R S.672, Ls -20	40.000	SkyBridge

Figure 7.40 Example list of EPFD runs. *Screenshot credit: Visualyse EPFD*

statistics. The approach taken in Rec. ITU-R S.1503 is to request that the non-GSO operator identify via their filing whether the constellation:

a. Ground track repeats, whereby satellites do not fully populate the orbit shell but rather return to the same locations (e.g. to simplify resource management). Given there will be a small amount of drift, this also requires definition of a maximum station keeping range around that repeating track, resulting in a ribbon or band of locations similar to Figure 6.28

b. Ground track drifts, whereby satellites will eventually fully populate the orbit shell. This could take a long time (as in Example 6.15) and hence artificial precession is used so the result is a uniform distribution as in Figure 6.29.

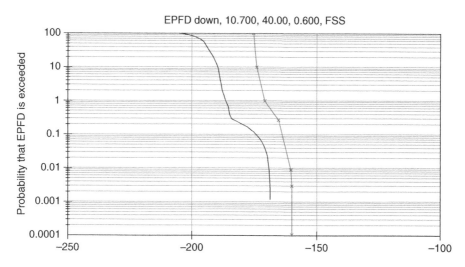

Figure 7.41 Example output of EPFD analysis run

As the run time can be quite large, the algorithm includes the concept of two time steps:

1. A fine time step for when a non-GSO satellite is within the exclusion zone and in particular near the edge or centre where the EPFD could vary rapidly with time
2. A larger time step for when all the non-GSO satellites are outside the exclusion zone and hence the EPFD is likely to vary more slowly. EPFD samples generated using the larger time step have additional weighting in the statistics compared to those from the fine time step.

7.6.4   EPFD Calculation

This section gives an overview of the core EPFD calculation methodology in Rec. ITU-R S.1503 for each of the directions of down, up and inter-satellite. In each case the run:

1. Initialises the statistics
2. Updates the simulation for the required number of time steps and at each time step:
   a. Determines the positions of the non-GSO satellites
   b. Undertakes the calculations in the following sections
3. At the end the statistics are updated and the CDF compared against the EPFD thresholds.

7.6.4.1   EPFD(down)

At each time step the EPFD(down) calculation undertakes the following (see Fig. 7.42):

1. Identifies all non-GSO satellites visible to the GSO ES
2. For each non-GSO satellite visible by the GSO ES the *EPFD* is calculated using the *PFD* mask:

Figure 7.42 Example EPFD(down) calculation. *Screenshot credit: Visualyse EPFD. Overlay credit: NASA Visible Earth*

$$EPFD_i = PFD_i + G_{rel}(\theta_i) \tag{7.71}$$

Here θ is the off-axis angle at the GSO ES and the gain pattern is either Rec. ITU-R S.1428 if the ES is in the FSS or BO.1443 if in the BSS.

3. The list of $EPFD_i$ entries is sorted by order, highest first
4. The total $EPFD$ is calculated by including the N_{co} highest entries that are outside the α_0 defined exclusion zone and above the minimum elevation angle ε_0, plus all those within the exclusion zone using

$$EPFD = \sum_i 10^{EPFD_i/10} \tag{7.72}$$

The motivation for this calculation is that in some non-GSO systems multiple satellites can provide a service to the same location without causing harmful self-interference. The N_{co} entries identify these multiple EPFD co-frequency active entries, but there will also be contributions from those non-GSO satellites that are in the main beam of the ES but not providing a service to this location (as in Fig. 7.35). Hence the following entries are included in the aggregate *EPFD* calculation:

- The largest N_{co} *EPFD$_i$* entries from those non-GSO satellites outside the exclusion zone $\alpha > \alpha_0$ and above the minimum elevation angle $\varepsilon > \varepsilon_0$ that are assumed to be actively providing a service
- All the *EPFD$_i$* entries from non-GSO satellites within the exclusion zone $\alpha \leq \alpha_0$.

7.6.4.2   EPFD(up)

The EPFD(up) calculation has an additional stage that is the deployment of the non-GSO Earth stations. In a global system there could be a very large number of such ES and it would not be feasible to model all of them. Therefore using the methodology from Section 5.7.5, a deployment is generated with a set of reference ESs separated by distance = d_{ES} specified by the non-GSO operator (e.g. the distance between traffic cell centres). These reference ESs are deployed around the GSO ES at the boresight of the GSO satellite beam to a relative gain level of 15 dB down from peak gain (see Fig. 7.43). In addition, the non-GSO operator must provide D_{ES} = the average density of ES, so that the aggregation factor A_{agg} can be derived:

$$A_{agg} = 10\log_{10}(N_{ES}) = 10\log_{10}\left(D_{ES}d_{ES}^2\right) \qquad (7.73)$$

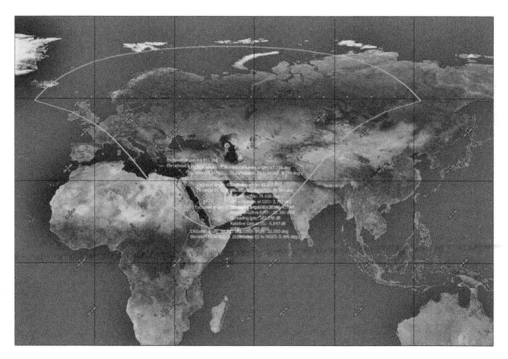

Figure 7.43   Example EPFD(up) calculation. *Screenshot credit: Visualyse EPFD. Overlay credit: NASA Visible Earth*

Other non-GSO systems could use a small number of specific ESs, so that option is provided as an alternative, in which case $A_{agg} = 0$ dB.

Then at each time step the EPFD(up) calculation undertakes the following:

1. For each non-GSO ES (i) all non-GSO satellites visible to that ES are identified
2. A list is created of those non-GSO satellites (j) that meets the criteria of minimum elevation angle $\varepsilon \geq \varepsilon_0$ and GSO avoidance angle $\alpha \geq \alpha_0$ calculated at the ES and sorted by the angle to the GSO satellite, lowest first
3. For the N_{co} entries the $EPFD_{i,j}$ is calculated using the non-GSO ES $EIRP$ mask and spread loss L_s from Equation 3.10 using

$$EPFD_{i,j} = A_{agg} + EIRP\left(\phi_{i,j}\right) - L_s + G_{rel}(\theta_i) \tag{7.74}$$

Here ϕ is the off-axis angle at the non-GSO ES and θ the off-axis angle at the GSO satellite. The GSO gain pattern is defined in Rec. ITU-R S.672 (ITU-R, 1997a).

4. The total $EPFD$ is calculated by summing over (i, j):

$$EPFD = \sum_{i,j} 10^{EPFD_{i,j}/10} \tag{7.75}$$

7.6.4.3 EPFD(IS)

This calculation simply involves the aggregation of the EPFD from each visible non-GSO satellite (see Fig. 7.44):

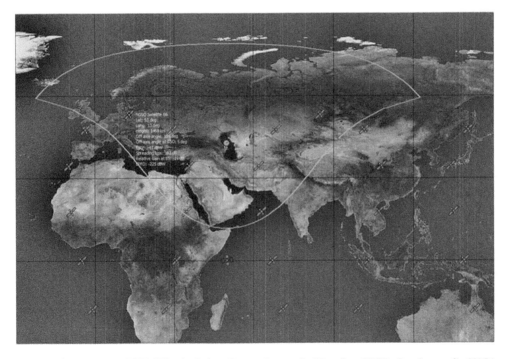

Figure 7.44   Example EPFD(IS) calculation. *Screenshot credit: Visualyse EPFD. Overlay credit: NASA Visible Earth*

1. For each non-GSO satellite (i) visible to the GSO satellite the $EPFD_i$ is calculated via the non-GSO satellite $EIRP$ mask and spreading loss L_s from Equation 3.10 using

$$EPFD_i = EIRP(\phi_i) - L_s + G_{rel}(\theta_i) \tag{7.76}$$

Here ϕ is the off-axis angle at the non-GSO satellite (assuming the boresight is the sub-satellite point) and θ the off-axis angle at the GSO satellite (assuming it is pointing at the reference GSO ES). The GSO gain pattern is defined in Rec. ITU-R S.672.

2. The total $EPFD$ is calculated by summing over (i):

$$EPFD = \sum_i 10^{EPFD_i/10} \tag{7.77}$$

7.7 The Radar Equation

All of the examples in the previous sections have used active components at both ends of a link, but the radar system is different in that one path relates to a passive object, the target, that simply reflects signals. Whereas for most interference studies it is sufficient to analyse the radar system as victim using a $T(I/N)$ (as discussed in Section 5.10.7), the operation of radar systems is driven by the **radar equation** that models reflections. This is of interest to interference analysis not just because it represents a useful methodology but also because reflections can impact a wider number of sharing scenarios than simply radars, including both terrestrial and satellite, such as:

- A remote sensing satellite's radar altimeter reflecting off the ocean into another satellite services
- A non-GSO mobile-satellite service's downlinks reflecting off metal structures into fixed links.

The key components of the radar equation are shown in Figure 7.45 where in absolute:

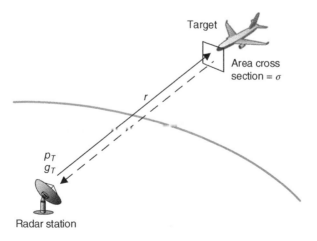

Figure 7.45 Components of the radar equation

p_T = transmit power of the radar station
g_T = gain of the radar station antenna in the direction of the target
r = distance (range) to target
σ = **radar cross section**: effective area of the target that reflects the signal.

Assuming free space propagation, the power flux density (PFD) of the radar signal at the target using Equation 3.222 is

$$pfd = \frac{p_T g_T}{4\pi r^2} \qquad (7.78)$$

The signal at the target is therefore the PFD times the radar cross-sectional area σ:

$$s' = pfd.\sigma \qquad (7.79)$$

This signal is then assumed to be reflected back to the radar station where the PFD is

$$pfd' = \frac{s'}{4\pi r^2} = \frac{p_T g_T}{4\pi r^2} \frac{\sigma}{4\pi r^2} \qquad (7.80)$$

The signal at the receiver is the PFD times the effective area of its antenna:

$$s = pfd'.a_e = \frac{p_T g_T}{4\pi r^2} \frac{\sigma}{4\pi r^2}.a_e \qquad (7.81)$$

The effective area is connected to the gain via the wavelength λ using

$$a_e = g\frac{\lambda^2}{4\pi} \qquad (7.82)$$

Hence

$$s = \frac{p_T g_T^2}{(4\pi r^2)^2} \frac{\sigma\lambda^2}{4\pi} \qquad (7.83)$$

This can be rearranged by noting that the free space path loss is (from Equation 3.11)

$$l_{fs} = \left(\frac{4\pi r}{\lambda}\right)^2 \qquad (7.84)$$

So

$$s = p_T g_T \frac{1}{(4\pi r/\lambda)^2} \frac{4\pi\sigma}{\lambda^2} \frac{1}{(4\pi r/\lambda)^2} g_T \qquad (7.85)$$

Hence in dB

$$S = P_T + G_T - L_{FS} + G_\sigma - L_{FS} + G_T \qquad (7.86)$$

where the effective gain of the target is

$$G_\sigma = 10\log_{10}\left(\frac{4\pi}{\lambda^2}\sigma\right) \tag{7.87}$$

This can be simplified to the decibel version of the radar equation:

$$S = P_T + 2G_T - 2L_{FS} + G_\sigma \tag{7.88}$$

The benefit of writing the equation in this form is that the free space path loss term can be replaced by another propagation model into the more generalised:

$$S = P_T + 2G_T - 2L_P + G_\sigma \tag{7.89}$$

It can be useful to consider alternative propagation models to free space, in particular:

- Rec. ITU-R P.528 for air to ground scenarios (as described in Section 4.5)
- Rec. ITU-R P.1812 for low-height scenarios involving terrain (as described in Section 4.3.6).

This is a simplified form of the radar equation and more information is available in specialised sources such as Skolnik (2001). In particular, there are issues relating to:

- The pulsed nature of the radar signal, including the duty cycle, pulse repetition factor and pulse width
- Additional noise at the receiver due to signals being reflected off clutter (terrain, buildings, vegetation, waves, etc.).

Example 7.33
The 13.75–14.0 GHz band has primary allocations to the FSS (Earth-to-space) and radio-location services (ITU, 2012a). Prior to WRC-03 interference into the radars was limited by a constraint on the FSS that the use of the band was restricted to ES with an antenna size of 4.5 m or larger. A study item in the cycle leading up to that WRC considered options to permit wider use of the band by the FSS, which led to the introduction of PFD constraints in 5.502. The studies were based upon parameters given in Rec. ITU-R M.1644 (ITU-R, 2003b).

The reduction in coverage of a maritime radar system with parameters given in Table 7.27 and antenna gain pattern in Figure 7.46 is shown in Figure 7.47. The target was assumed to be at a height of 1000 m and grid lines are every 10 km.

The ES was assumed to be transmitting with power and location selected such as that at the radar the $I/N = -6$ dB, and the following propagation models were used:

- Rec. ITU-R P.528 for the radar signal with $p = 50\%$
- Rec. ITU-R P.452 for the interfering signal from the FSS ES with $p = 50\%$.

Table 7.27 Example maritime radar parameters

Frequency (GHz)	13.5
Bandwidth (MHz)	10.0
Transmit power (dBW)	40.0
Antenna peak gain (dBi)	31.5
Antenna height (m)	36.0
Antenna elevation angle (deg)	4.5
Target cross-sectional area (m^2)	250.0
Receive noise temperature (K)	1445.4
$T(C/N)$ (dB) from (Skolnik, 2001)	15.75

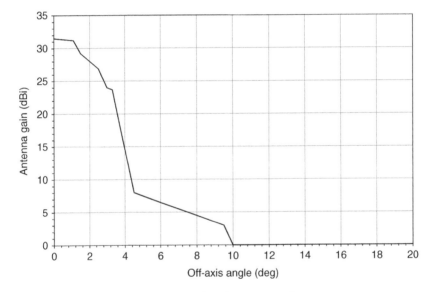

Figure 7.46 Example radar antenna gain pattern

The coverage can be seen to be reduced generally in all directions due to interference received via the radar antenna's sidelobe but with the greatest reduction in range in the direction of the interfering station. Note how the coverage is a doughnut or torus shape as the fixed elevation angle of the radar antenna results in loss of signal when the aircraft is directly over the ship.

Resource 7.5 The Visualyse Professional simulation file 'Radar Example.sim' was used to generate this example.

Rec. ITU-R M.1644 identifies that a $T(I/N) = -6\,\text{dB}$ corresponds to approximately a 6% reduction in range in the direction of the interferer, based upon the radar equation and free space path loss, because this implies a 1 dB loss in margin:

$$\Delta M = 2 \times 20 \log_{10}(1 - 0.06) \sim 1\,\text{dB} \tag{7.90}$$

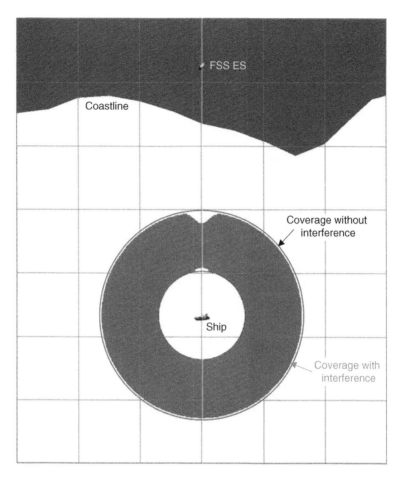

Figure 7.47 Example reduction in coverage due to interference. *Screenshot credit: Visualyse Professional*

In this case the worst-case reduction in range was from 16.5 to 14.4 km, about 12.6%. The loss of range was greater because the propagation model used was P.528 and the target was not directly in the main beam of the radar antenna. This is another example of how a $T(I/N)$ can be derived from a theoretical reduction in coverage, similar to those in Section 5.10.5. In practice the actual coverage will depend upon a number of site-specific characteristics (e.g. terrain), and so it is preferable to undertake generic analysis using an I/N threshold and use that as an input assumption in the planning process.

The I/N threshold is an aggregate but in this case it is likely that the aggregate interference will be strongly driven by the worst single-entry interferer, as both the victim service (radar) and interferer (FSS ES) use highly directional antenna, while the low FSS ES deployment density implies that there are likely to be a small number within the radar antenna's beam.

7.8   N-Systems Methodology

Regulators aim to improve spectrum efficiency and usage of the radio spectrum, but increasingly the limiting factor on the ability to deploy additional radio systems is interference. The N-Systems methodology uses simulation to predict the number of systems that can operate within a defined area by continually adding additional systems until it is not feasible to add any further due to interference constraints.

Example 7.34

Figure 7.48 shows an area of 200×200 km^2 with grid lines spaced every 10 km in which point-to-point (PtP) fixed links with characteristics as shown in Table 7.28 have been deployed one by one at random. After each has been deployed, the aggregate I/N from each into all the others is calculated. This process continued until the aggregate $T(I/N) = -6$ dB was exceeded for one link. In this case 146 links could be added before this threshold was exceeded.

In this simulation the 146th link failed, but it is unlikely that the band and area is completely full. Hence the N-Systems methodology has a loop by which alternative deployments are tried, typically at random. The algorithm ends when a predetermined number of trials have attempted to deploy the new system without success, as in the flow diagram in Figure 7.49.

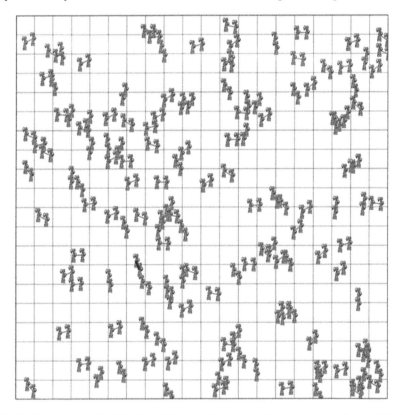

Figure 7.48   Deployment of 146th fixed link resulting in interference. *Screenshot credit: Visualyse Professional*

Table 7.28 FS system parameters

Frequency (GHz)	28.1
Bandwidth (MHz)	14
Antenna dish size (m)	0.3
Antenna gain pattern	Rec. ITU-R F.1245
Azimuth angle	Random [000°, 360°]
Hop length	5 km
Feed loss (dB)	1
Noise figure (dB)	7
Interfering propagation models	P.452
Associated percentage of time	50%

The various stages are (using the FS scenario from Example 7.34):

1. A scenario environment is defined, which in this case would be a square of size 200 by 200 km and using the ITU-R Rec. P.452 propagation model
2. The total number of systems is set to zero, that is, initially there are no PtP FS links in the simulation
3. A new system is created, which in this case would be a PtP link with random azimuth and hop length = 5 km
4. The new system is deployed within the simulation, which in this case would be at a random position within the $200 \times 200 \text{ km}^2$
5. An interference analysis is undertaken, which in this case is from each FS link's transmit stations into all other FS link's receive stations (i.e. excluding the wanted path) using P.452 and smooth Earth propagation to calculate the aggregate I/N
6. If the threshold was met for all systems in the simulation, then the process moves to step 7; otherwise step 8. In this case the threshold was an $T(I/N) = -6 \text{ dB}$
7. The count of systems is incremented by one and the process continues with step 3
8. If there are alternative locations available, then these are tried and the process continues at step 3; otherwise the process continues at step 9. In this case the FS links were being deployed at random and there could be an infinite number of potential positions to try. Therefore this check could be to try up to N_{trial} number of random locations before abandoning the attempt to add further FS links
9. The output of the N-Systems methodology is the number of links that were successfully introduced. In this case, with a simplistic $N_{trial} = 1$, the $N_{FS} = 146$.

Note that the interference analysis stage, namely, step 5, could include any methodology including static, area or Monte Carlo, and the metric could be any type, whether $X = \{I, I/N, C/I, C/(N + I)...\}$. The interference calculation would have to take account of all relevant factors including polarisation, bandwidth or mask integration.

As the process involves a random element, the result is a distribution, typically normally distributed with mean and standard deviation (μ, σ), rather than a single figure.

Aggregate interference is the appropriate metric here rather than single entry as the methodology relates to the total interference limiting the deployment of additional systems.

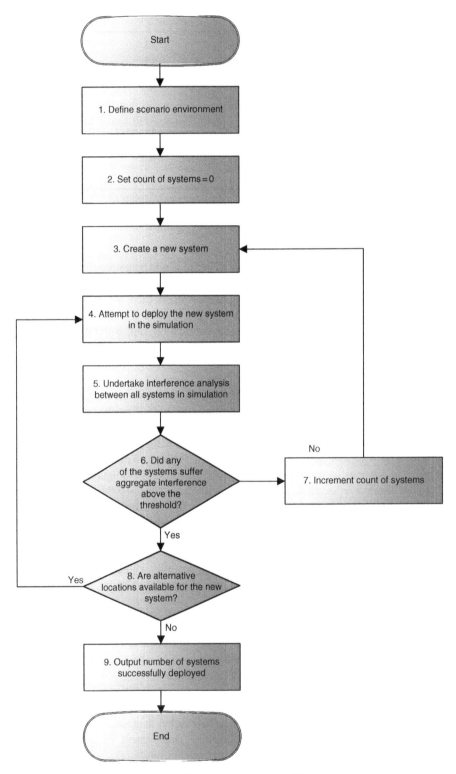

Figure 7.49 N-Systems methodology

This aggregation can be influenced by edge effects, and hence one of two geometric approaches must be considered to avoid this bias:

1. Define two zones, an inner and an outer, and only count the number of systems within the central area, sufficiently small that no part of it could suffer from edge effects
2. Use wrap around geometry, as described in Section 3.8.2.

The interference-limited density D_A of system A in number/km^2/MHz can then be calculated from the area and the system's bandwidth using

$$D_A = \frac{N_A}{Area.B_{\text{MHz}}} \tag{7.91}$$

The basic single system methodology can be used:

- To analyse the total number N_A of type A systems could be deployed within a band and geographic area, for example, to identify the maximum capacity or value of the band
- To compare predicted densities against measured occupancy to identify how close to saturation a band has become
- To analyse the difference between single-entry and aggregate entry interference, for example, when deriving single-entry thresholds from aggregate ones, as described in Section 5.9 – in particular, to determine how this ratio could vary depending on how occupied the band has become, with increasing ratios likely as the band tends towards being fully occupied.

A number of additional metrics can be derived if the N-Systems methodology is extended in one of two ways:

1. The sensitivity of N_A to changes in one of the input variables can be analysed and hence determine the impact on spectrum efficiency of that variable
2. By including a second type of system so that the number of systems of type A $= N_A$ is calculated given that there are N_B systems of type B already deployed in the test area, taking into account interference not just between systems of type A but also from type A into type B and potentially also between systems of type B.

The first approach could be used to analyse, for example, in an FS-related scenario, the impact on N_A of increasing the modulation (allowing an assessment to be made between the improved data carrying capacity against interference constraint due to larger transmit power) or the antenna dish size (more expensive but would allow more links in the band).

These factors could have spectrum policy implications, in particular for pricing. In some countries, administrations charge for site licences for systems such as PMR or FS. By using techniques such as N-Systems, the spectrum opportunity cost of a licence application could be assessed and quantified, and hence a licence price determined. Standardised pricing algorithms could also be derived based upon the results of N-Systems analysis: for example, by calculating the relationship between the antenna peak gain and spectrum utilisation.

The second approach can also be used to identify the implications for spectrum sharing between two services in terms of the restriction of one system due to deployment of the other.

Example 7.35

A study (Pahl, 2002) analysed a scenario at 18 GHz involving:

- System A = PtP fixed links
- System B = GSO receive Earth stations.

It was noted that the mean number of FS links of the type considered that could be deployed in a test area of 40 × 40 km² without any GSO ES had a mean value of 33 and standard deviation of 5.6. As the number of GSO ES in the simulation increased, the average number of FS links that could be deployed before causing harmful interference (into either the FS or GSO ES) decreased proportionally as shown in Figure 7.50.

Regression analysis was undertaken and it was determined that there was strong statistical correlation with the following relationship:

$$N_{FS} = 33 - \frac{1}{93} N_{ES} \tag{7.92}$$

Hence for every 93 ES introduced into the simulation, the average number of PtP fixed links that could be deployed decreased by one. This study could be used to argue that, for systems with the parameters involved, the cost of spectrum part of the licence price of the receive satellite ES should therefore be 1/93rd of that for the point-to-point fixed system, noting that the uplink ES price would also have to be taken into account and that there could be additional administrative costs within the total licence price.

This approach can be generalised to identify the spectrum opportunity cost of system A versus system B via the following equation:

$$N_A = N_A(0) - \alpha_{AB} N_B \tag{7.93}$$

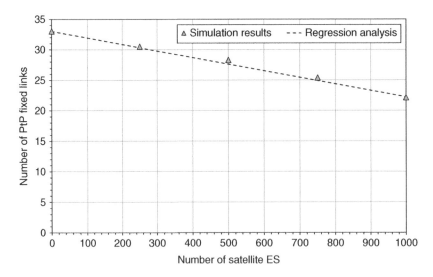

Figure 7.50 Example relationship between ES and FS spectrum opportunity cost

Here the number of systems of type A that can be introduced in an environment without any systems of type B is $N_A(0)$, and the spectrum opportunity cost of introducing each system of type B is a reduction by α_{AB} in the number of type A. The spectrum opportunity cost O_C of type B in terms of the Oc of type A is then

$$O_C(B) = \alpha_{AB}.O_C(A) \tag{7.94}$$

Example 7.36
A study for Ofcom (Aegis Systems Ltd and Transfinite Systems Ltd, 2004), which analysed spectrum occupancy at 2.4 GHz, undertook N-Systems analysis to determine the impact on the ability to deploy Wi-Fis on introducing Bluetooth (BT), microwave ovens (MO) and two types of electronic news gathering (ENG) systems. The number of Wi-Fi systems that could be deployed without any other system active in an area of 1×1 km^2 was around 25.

N-Systems analysis was used to determine the impact of introducing various interfering systems and regression analysis suggested very strong correlation with the following equation:

$$N_{WiFi} = 24.67 - \alpha_I N_I \tag{7.95}$$

The various factors of α are given in Table 7.29. It can be seen that it was significantly easier for Wi-Fi to share spectrum with BT than with ENG, and there was only a slight reduction due to microwave ovens.

In this case the interference analysis involved a detailed Monte Carlo methodology using a $T(C/(N + I))$ mapped onto derived packet error rates (BER) using the following criteria:

More than X% of locations achieving a BER < Y for more than Z% of the time

The BT systems used narrower bandwidths than that of the Wi-Fi, and so their frequency was selected at random throughout the Wi-Fi bandwidth, while systems with bandwidth wider than Wi-Fi were scaled downwards.

As well as identifying the spectrum opportunity cost, this method can also be used to assess how compatible two alternative technologies would be with another service type. For example, ENG-2 can be seen to share more readily with the Wi-Fi system than ENG-1. However it could be that ENG-1 provides a much greater service or one of higher value, so that would have to be taken into account, for example, using the value metric described in the following.

A useful derived metric is the total number of systems that can simultaneously share, defined by

$$N_{Total} = N_A + N_B = N_A(0) + (1 - \alpha_{AB})N_B \tag{7.96}$$

Table 7.29 Example values of α from 2.4 GHz N-Systems study

Interfering system	α	Units
BT	0.007	Wi-Fi/BT/km^2
MO(0.05)	0.142	Wi-Fi/MO/km^2
MO(0.1)	0.202	Wi-Fi/MO/km^2
ENG-1	21.220	Wi-Fi/ENG/km^2
ENG-2	4.890	Wi-Fi/ENG/km^2

By identifying a value V of each type of system, the total value gained by deploying a combination of systems can then be derived:

$$V_{Total} = V_A N_A(0) + (1 - \alpha_{AB}) V_B N_B \qquad (7.97)$$

These equations are considered to be valid for scenarios when the total number of systems is dominated by one or other: further study is required into more general cases, in particular the crossover point between where system A or system B dominates.

Note that the N-Systems methodology gives no information on absolute spectrum value, only relative opportunity costs. Actual values of spectrum involve externalities such as market conditions, cost of finance, value of services and cost of associated infrastructure, which are outside the scope of this book. Furthermore, in geographic areas with low levels of demand where available spectrum significantly exceeds demand, it is less important to match licence prices against spectrum denied, and cost of services or licences is predominately driven by other factors (capital, operations, etc.).

The approach in the previous examples was based upon the use of random deployments. However this assumption might not be valid as the deployments could be constrained by other factors. For example, PtP fixed links for trunk routes connect major cities, and in this case the limiting factor is not general link packing but the very specific issue of how to maximise capacity along high demand routes, for example, by using high modulations and polarisation reuse. In this case one possible approach would be to use an existing database of assignments to identify example positions and use those in sequence. This is similar to a **replanning exercise** whereby an existing set of assignments is replanned from scratch in order to identify if there are ways to assign frequencies that would require fewer channels.

Alternatively, the methodology could consider different architectures to meet a defined set of requirements. For example, ECC Report 20 (CEPT ECC, 2002) in addition to considering an N-Systems-based approach to analyse the density of FS systems also looks at the implications of the use of a star versus daisy chain arrangements of fixed links to provide mobile backhaul to a defined set of base stations.

The methodology could also have as part of the environment the grid of base stations and the system being introduced the mobile handset of various types of traffic. The N-Systems methodology could then be used to calculate the interference-limited capacity of the network.

7.9 Generic Radio Modelling Tool

The previous sections described a number of methodologies to plan services including undertaking interference analysis. Many of these interference algorithms were service specific, for example, for PMR (as in Section 7.2), broadcasting (as in Section 7.3) or fixed and satellite ES (as in Sections 7.1 and 7.4). This was consistent with an approach to spectrum management, which splits the radio spectrum into bands allocated to different services, with different software tools used to plan services and calculate interference in each band, the tool implementing the relevant specific algorithm, as in Figure 7.51.

However there are disadvantages with this approach, such as:

- The bands must be allocated in advance and it is not easy to predict demand for each service. This can lead to inefficiencies, such as excess allocations to one service and not enough to another

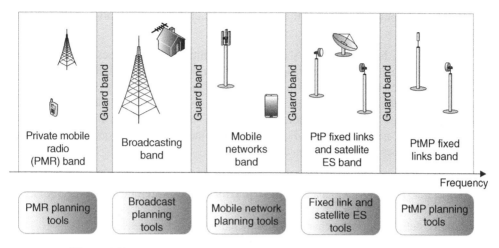

Figure 7.51   Spectrum management using service-specific bands and tools

- Some services might only be expected to operate in specific geographic locations, so allocating the whole band to that service exclusively would prevent other services operating in other regions
- There must be guard bands between services calculated using a set of assumptions that will in some circumstances be pessimistic
- Licence holders will be constrained in how they can change their system under a **licence variation** process only to characteristics that are consistent with the band allocations and the ability of the software tool.

The net result is the possibility of spectrum inefficiencies and hence the potential to improve spectrum efficiency by moving to a more flexible spectrum management regime. This was the objective of the Generic Radio Modelling Tool (GRMT) developed as a research project by the UK regulator Ofcom (Transfinite Systems, Radio Communications Research Unit, dB Spectrum, LS telcom, 2007).

The objective of the GRMT project was to develop a generic interference analysis algorithm that would permit a technology-neutral approach to spectrum licensing, introducing flexibility to allow **change of use** (CoU) between a wide range of licence classes within the same block of spectrum. For example:

- A licensee in a band allocated to one service requests CoU to another type of service
- A band is in transition with migration of the existing service to higher frequencies. New entrants of a different service type wish to start operating prior to the band being fully cleared
- A licensee wishes to operate a standard spectrum product in a band not currently allocated to that service type
- A licensee wishes to operate a new type of system that is not easily categorised into one of the existing service types
- A licensee wishes to gain a licence for a standard spectrum product in a band where there are a mix of service types – for example, because there has been at least one CoU

- Licensing is permitted in the currently empty guard bands between allocations to two different services.

The commonality between all these scenarios is flexibility of licence classes, and this leads to requirement on the underlying interference analysis algorithms (and hence software tools) to be able to handle different types of system and service, as in Figure 7.52.

In this situation the term '**technology neutral**' means 'parameter based': any technology or standard is permitted as long as its radio characteristics can be defined in the data dictionary used by the GRMT project. The data dictionary was defined based upon concepts common to all the licence classes considered, namely:

- **Licences**, comprised of one or more **Systems**
- **Systems**: transmit or receive, made up of **Deployments** and **Antennas** and using TX or RX spectrum masks
- **Antennas**: defining the gain characteristics including pointing angles
- **Deployments**: either point, line (e.g. boundary) or area.

The regulator also require a number of additional object classes, such as **Spectrum Blocks**, which defined characteristics and constraints for frequency and geographic ranges. The Spectrum Blocks also contained additional constraints such as PFD on boundary and could be used to include the requirements and databases of other administrations. In addition, both Licences and Spectrum Blocks had associated owners.

The resulting top level GRMT data dictionary is shown in Figure 7.53. These underlying structures were typically mapped onto standard licence parameters via **spectrum product templates** as part of the data migration activities and by the user interface.

In order to analyse scenarios between any type of licence class, a single interference analysis methodology was required, and this was based upon the use of a standard threshold described as the spectrum quality benchmark (SQB). The SQB was selected noting that the majority of the systems and scenarios considered in this chapter included the concept of an interference margin M_i relative to the receiver noise, resulting in an I/N threshold:

$$T\left(\frac{I_{agg}}{N}\right) = 10\log_{10}\left(10^{M_i/10} - 1\right) \tag{7.98}$$

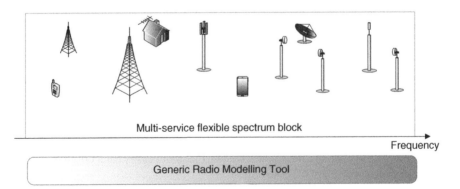

Figure 7.52   GRMT supporting flexible spectrum management

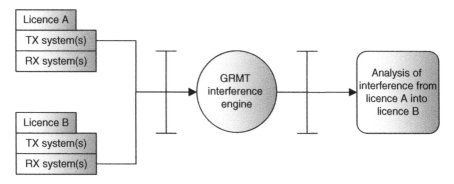

Figure 7.53   GRMT data dictionary

A fully flexible spectrum management regime would be based upon the following philosophy:

> *In a liberalised regime ideally the only constraint on spectrum use should be controls necessary to avoid harmful interference. (Ofcom, 2006b)*

If one licensee undergoes a CoU to alter its receiver noise temperature (e.g. to reduce it), then it should not cause other licences to fail the interference check if they continue to operate in a way that has already been approved. Hence the SQB metric is based upon interference at the receiver rather than I/N. The receiver could either be at a single position (such as for a fixed service station) or at any location across an area (such as a PMR handset).

The SQB was therefore defined as follows:

> *Interference at the receiver should not exceed X dBW for more than Y% of the time [at more than Z% of locations].*

The single-entry SQBs were derived from the aggregate interference via an apportionment term that took account of both number of systems and degree of bandwidth overlap as in Equation 5.155.

This approach is also a transparent, measurable and technology-neutral definition of interference, which allows each licensee to plan their service based upon expected levels of interference.

The SQB was associated with a receive system and there could be multiple SQBs for each, for example, to define short-term and long-term thresholds. Any modifications to a band such as a new licence application or CoU of an existing licence could then be checked in two directions:

1. To ensure the modified licence's transmit systems would not cause harmful interference into existing licence's receive systems
2. To ensure the modified licence's receive systems would not suffer harmful interference from existing licence's transmit systems.

Each of these two directions resulted in an interference check using the GRMT interference engine as shown in Figure 7.54.

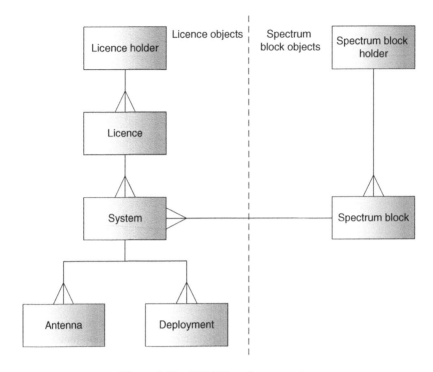

Figure 7.54 GRMT interference engine

The interference engine used a standardised algorithm that assessed each set of {TX System(s), RX System} to identify the methodology required from {Static, Area, Monte Carlo, Area Monte Carlo} using a decision tree that took account of factors such as deployment type and activity. The dependency of the methodology on the deployment type is shown in Table 7.30.

Each of these methodologies used the same standardised approach to calculate the interfering signal based upon mask integration as in Equation 5.63 plus, where appropriate, polarisation loss:

$$I = P'_{tx} + G'_{tx} - L'_p - L_{pol} + G'_{rx} - L_f + A_{MI}(tx_I, rx_W) \tag{7.99}$$

This equation could be used for both co-frequency and non-co-frequency analyses. The requirement by GRMT for flexibility and applicability over a wide range of scenarios and frequencies was one of the motivations for the development of the Rec. ITU-R P.2001 propagation model as described in Section 4.3.7.

While CoU analysis could be undertaken on a case-by-case basis, this would be resource intensive and not predictable. By using a publically available algorithm implemented in software, the result is a cost-effective, transparent and predictable spectrum management regime that would facilitate flexible licensing.

Table 7.30 GRMT methodology selection by deployment type

Methodology Interferer TX system	Victim RX system	
	Deployment type = point	Deployment type = area
All deployment types = point	Static	Area
At least one deployment type = area	Monte Carlo	Area Monte Carlo

Example 7.37

As described in Section 6.2.1, there has been significant interest in the use of the 3.6–4.2 GHz band for a wide range of services including:

- Receive satellite ES
- PtP fixed service (e.g. low-latency links)
- PtMP fixed service (e.g. to connect LTE base stations)
- Public LTE networks, most likely low-height, low-power base stations for urban small cell deployments or indoors
- Other LTE networks (e.g. for utilities or emergency services)
- Potential new services in the future.

One approach could be to have a flexible licensing regime that is able to handle all of these services via a single web-based portal, such as that shown in Figure 7.55, which provides:

- User management including security controls
- Capture of licence parameters including graphical data via maps
- Support for licence planning tasks (such as coverage of point-to-area licence types)
- Frequency assignment (where applicable)
- Workflow support
- Automated interference assessment using the GRMT methodology
- Reporting tools including licence search facilities
- Issuing of eLicences (e.g. via PDFs).

The portal could also support the payment process, with licence fees derived using methodologies such as N-Systems described in the previous section.

A GRMT-based licensing tool would permit this band to be used for LTE in dense urban areas but other services, including a mix of ES, PtP FS and PtMP FS at other locations, with vigorous interference checks to ensure all receivers are protected. Each new assignment or change to an existing assignment would be subject to an assessment that would include:

1. Checks on parameters, such as frequencies, powers, heights and locations
2. Checks that the new assignment would not cause harmful interference into existing assignments whether co-frequency or non-co-frequency
3. Checks that the new assignment would not suffer harmful interference from existing assignments whether co-frequency or non-co-frequency
4. Check any PFD on boundary constraints

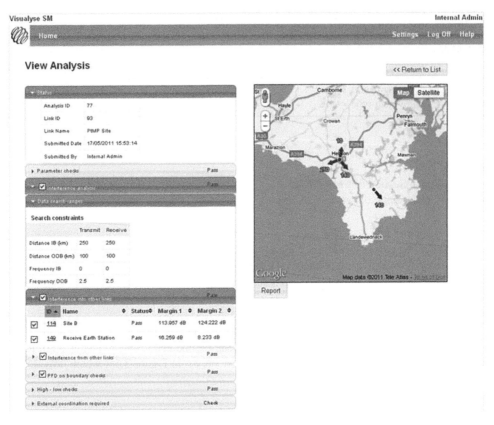

Figure 7.55   Web-based licensing tool using GRMT interference engine. *Screenshot credit: Visualyse Spectrum Manager. Image credit: Google © 2011 Tele Atlas*

5. Check any high/low constraints
6. Check whether external coordination is required (e.g. internationally or with another spectrum management organisation).

   While the web interface shown was designed for human interaction, the process would also permit a machine-to-machine (M2M) interface, for example, to automate assigning frequencies to FS or ES in shared bands, as discussed in Section 7.4. This form of auto-mated assignment is the basis of the authorisation of WSDs, as discussed in the following section.

   This approach is an example of **licensed shared access** (LSA), in which new users of a radio band are authorised to operate in accordance with rules that ensure that incumbents are protected from harmful interference. The GRMT framework would provide a flexible, multiband, parameter-based approach to LSA, allowing a wide range of services to share spectrum taking into account both co-frequency and non-co-frequency interference paths.

7.10   White Space Devices

7.10.1   Background and Services

The previous section described an approach to licensing that would permit a wide range of services to share the same band without causing or suffering harmful interference, described as licensed shared access. The benefit of this approach is that the regulator maintains control of the process, keeping records of all the transmitters and receivers within a band and their characteristics, in particular location and frequency. The drawback is that each site has to be individually licensed, and this has a cost implication. This overhead can be minimised by providing eLicensing portals (as in Fig. 7.55) and can be acceptable when the value per site is sufficiently high to cover the price of spectrum and administrative costs.

However when there are large numbers of devices, the management cost can become a burden and it can become preferable to have a fully automated process whereby spectrum authorisations are made without human involvement, for example, via sensing or M2M communication, such as would be used by WSDs.

WSDs are often associated with sharing with television and hence can also be known as **television band devices** (TVBDs), though the technology can be used to operate in bands used by a wide range of other services. The key concept in sharing with TV services is shown in Figure 7.56.

Terrestrial television services are characterised by high-power transmitters often with high-height antennas. This provides coverage over a wide area but also sterilises a larger area, as

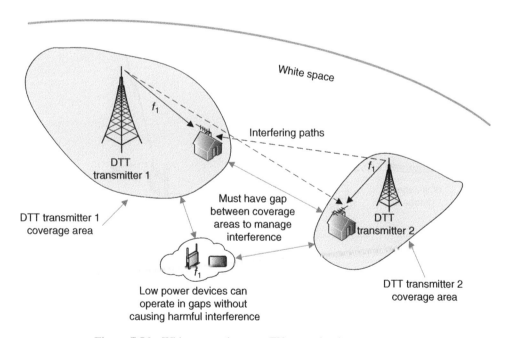

Figure 7.56   White spaces between TV transmitter's coverage areas

there must be gaps between coverage of different transmitters to avoid interference.[1] If the areas that are used for TV services are shaded (as in Fig. 7.56), then the spaces in between will be shown as white space, hence the name. Potentially these areas could then be used by other services, in particular if they transmit with low power and operate at low heights so that the range over which they could cause interference into the primary service is limited.

A methodology such as that in Section 7.3 can be used to predict the coverage of each broadcasting transmitter, and once planned the network typically changes slowly. The locations of these white spaces, both geographically and in frequency, are predictable and also sufficiently stable to be usable by devices on a day-by-day basis. The location of these white spaces can be determined by one of two methods:

1. **Spectrum sensing**, whereby the WSD listens for any TV signal and avoids those channels that are currently being used in its vicinity
2. **Geolocation database** (or geodatabase), whereby the WSD identifies its location and then contacts a database server that provides it with a list of frequencies appropriate for its position.

Spectrum sensing has to overcome a number of difficulties, such as when there are hidden nodes, as shown in Figure 3.22, and hence this section will focus on the geolocation database approach, as in Figure 7.57.

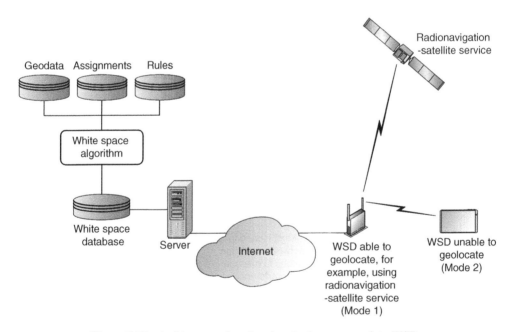

Figure 7.57   Architecture of geolocation database approach to WSDs

---

[1] Single-frequency networks (SFNs) would have multiple transmitters within each coverage area, but there would still have to be gaps between coverages of co-frequency networks.

The key components and processes are as follows:

- A WSD that is able to geolocate (in the FCC rules this is described as a Mode 1 device) identifies its position, for example, using a radionavigation-satellite service (RNSS) such as GPS or Galileo.
- This WSD accesses the Internet to contact a predefined server giving its location as a (latitude, longitude) pair together with its device type (e.g. fixed, personal/portable).
- This server queries a white space database for that location and type and returns a list of suitable channels
- The white space database is updated regularly using the white space algorithm (described in the following sections) and databases for:
  ○ Geodata, containing data on borders, terrain heights, land use codes, etc.
  ○ Assignments, such as the locations and channels used by TV transmitters or other licensed services
  ○ Rules, such as those described in the following sections
- If a WSD is unable to geolocate (described in the FCC rules as a Mode 2 device), it is still able to access white space frequencies via a Mode 1 device that provides it with a list of suitable channels.

The VHF and UHF bands used for TV services have the advantage that they provide good coverage, as propagation loss is lower than at higher frequencies (e.g. the 2.4 GHz band used by Wi-Fi) and there is also lower building penetration loss. This is particularly important for rural areas that might have lesser coverage of the latest mobile services than urban areas. A WSD device could provide coverage over (say) a farm including inside remote buildings in a way that Wi-Fi could not. They are, therefore, a good candidate to provide similar services to Wi-Fi on a licence-exempt basis.

However these beneficial propagation characteristics mean that these frequency bands can be attractive to a wide range of other services. Examples of other services that operate in the relevant bands in the United States include (FCC, 2010):

- TV receive sites (TVRS) including translator and booster stations: using high-gain fixed antennas to receive video signals outside the standard coverage area for retransmission on the same or another frequency
- Multichannel video programming distributor (MVPD) receive sites, for example, using high-gain fixed antennas to receive video signals outside the standard coverage area for retransmission over a cable system
- Fixed broadcast auxiliary service (BAS), for example, transmitting programmes from the studio to the TV broadcast transmitter
- PMR and public mobile services
- Offshore radiotelephone service, for example, to provide telephone service to offshore oil or gas rigs
- Low-power auxiliary services, including wireless microphones and control/synchronisation systems auxiliary to broadcasting services
- Radio astronomy services, including the Very Large Array (VLA)
- Unlicensed devices such as medical telemetry and remote control equipment.

Other countries can have a different set of services to protect – for example, in the United Kingdom the principal services are digital terrestrial television (DTT) and programme making and special events (PMSE). It is also necessary to consider how to protect receivers in adjacent countries from harmful interference and also adjacent bands (e.g. mobile services).

In order to protect these services, an algorithm is required to calculate which channels are available at which location and, in particular, what is the maximum transmit power or EIRP that can be permitted on each to avoid harmful interference. A number of methodologies could be considered, including:

- FCC methodology, based on contour plus distance, as described in Section 7.10.2
- Ofcom methodology, based on reduction in DTT coverage, as described in Section 7.10.3.

A comparison for selected scenarios is given in Section 7.10.4. Additional information is available in ERC Report 185 (CEPT, 2011a) and ERC Report 186 (CEPT, 2011b).

7.10.2   FCC Methodology

This section describes the contour plus distance-based approach specified in the Federal Communications Commission (FCC) Rules Part 15.700 onwards, and the key concepts are shown in Figure 7.58. Two types of TVBD are considered:

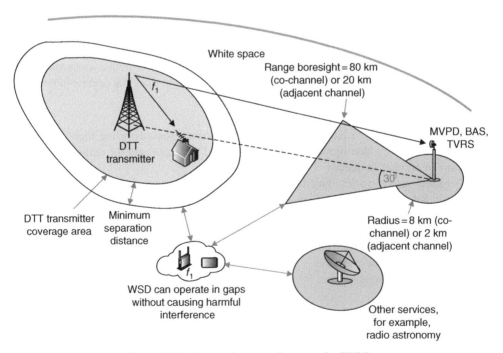

Figure 7.58   Types of protected contour for TVBDs

- Fixed: operate from a static (and known) location with antenna height not more than 30 m above ground at a location where the ground height above average terrain (HAAT) is no more than 76 m. Fixed TVBDs have access to a wider range of channels than personal/portable devices for fixed-to-fixed communication and must operate as Mode 1
- Personal/portable: may operate as Mode 1 (if it has geolocation capabilities) or Mode 2 (if it doesn't) and may only operate on frequency bands 512–608 MHz (TV channels 21–36) or 614–698 MHz (TV channels 38–51).

Channel 37 between 608 and 614 MHz (note that the US TV channel bandwidth is 6 MHz) is excluded in order to protect the radio astronomy service. There is also the need to protect services in adjacent countries: this is done by including their receivers in the database along with those in the United States. The accuracy of the geolocation should be within 50 m.

The key concept is that of the protected contour, which is one of:

1. Polygon defining the edge of the TV coverage area
2. Keyhole shape around the MVPD, BAS or TVRS receiver
3. Rectangle or circle around another service, such as radio microphones or radio astronomy site.

The TV coverage area polygon is generated using the prediction approach in FCC Rules Part 73 with edge defined by the field strengths in Table 7.31. The propagation model used in this case is that given in Report R-6602 (FCC, 1966) that has similarities with Rec. ITU-R P.370 (used to develop P.1546) in that it is based upon field strength curves with adjustment, for example, for terrain roughness and transmitter height. It gives the field strength $F(q, p)$ **exceeded** for $q\%$ of locations and $p\%$ of the time. It specifies values for 10 and 50% of the time and suggests that values for 90% can be derived by assuming the time variation follows the normal distribution around the mean value. The polygons are generated using the radial method and typically result in a single closed loop.

Given the protected contour, a separation distance is required, which depends upon whether the TVBD would be operating in the same or adjacent channel as the TV service, as specified in Table 7.32.

For receive stations used by a MVPD, BAS or TVRS system, the protected contour is a keyhole shape with:

Table 7.31 Protected contour for TV services

Type of station	Channel	Protected contour	
		Field strength (dBµV)	Propagation curve
Analogue	Low VHF (2–6)	47	F(50, 50)
	High VHF (7–13)	56	F(50, 50)
	UHF (14–69)	64	F(50, 50)
Digital	Low VHF (2–6)	28	F(50, 90)
	High VHF (7–13)	36	F(50, 90)
	UHF (14–51)	41	F(50, 90)

Table 7.32   Required separation distance from TVBD to DTT protected contour

Height (*h*) of TVBD	Required separation from TV protected contour (km)	
	Co-channel	Adjacent channel
$h < 3$ m	6.0	0.1
3 m $\leq h < 10$ m	8.0	0.1
10 m $\leq h \leq 30$ m	14.4	0.74

- Distance from the receiver in the direction of the relevant TV transmitter of 80 km (co-channel) or 20 km (adjacent channel)
- Angle of keyhole: 30° either side of the line from the receiver to the relevant TV transmitter
- Distance around the receiver in all other directions: 8 km (co-channel) or 2 km (adjacent channel).

The separation distance and range for other services depends upon their characteristics. For example, for wireless microphones the separation distance is 1 km for fixed devices and 400 m for personal/portable TVBDs, while for the radio astronomy service the separation distance is 2.4 km.

The power limit for fixed devices is 1 W, that is, 0 dBW, with an antenna gain of 6 dBi permitted, giving a maximum EIRP of 6 dBW within the channel bandwidth of 6 MHz. Higher-gain antennas are permitted but the power must be reduced to ensure the EIRP limit is not exceeded. The EIRP limit for personal/portable devices in most cases is 100 mW, that is, 20 dBm/6 MHz, though there is a tighter EIRP limit of 16 dBm/6 MHz, for example, for Mode 2 TVBDs operating closer or within the protected contours in Table 7.32. Adjacent channel interference is managed via a requirement for the power spectral density to be reduced by 55 dB compared to the maximum mean value in-channel, both measured in terms of dBW/100 kHz.

7.10.3   Ofcom Methodology

Another approach to identifying where and how WSDs can operate has been proposed in the United Kingdom based upon a threshold specified as a reduction in availability of the coverage of a DTT network. While the Ofcom method is more involved than that proposed by the FCC, making a few assumptions it can be shown that there are similarities in the resulting operational constraints.

Some differences between the two approaches are simple, such as the use of 8 MHz channels for DTT in the United Kingdom. Another difference relates to the planning algorithm used to predict TV coverage between:

- US: polygons derived by radial area analysis
- UK: grid area analysis, with the signal strength mean and standard deviation available at each 100×100 m pixel.

In addition, slightly different terminology is used in the Ofcom methodology, taken from ETSI EN 301 598 (ETSI, 2014b):

- Type A: fixed outdoors
- Type B: portable/mobile, either indoors or outdoors.

The algorithm is designed to protect DTT services within the United Kingdom, DTT services in adjacent countries, PMSE within the United Kingdom and services in adjacent bands within the United Kingdom (e.g. mobile) as shown in Figure 7.59. For each constraint there is a minimum permitted power, and the resulting WSD maximum power is then

$$P_{WSD} = min\{P_{WSD,DTT}, P_{WSD,PMSE}, P_{WSD,Int}, P_{UA}, 36 \text{ dBm}\} \qquad (7.100)$$

where

$P_{WSD,DTT}$ = maximum power permitted by DTT protection algorithm
$P_{WSD,PMSE}$ = maximum power permitted by PMSE protection algorithm
$P_{WSD,Int}$ = maximum power permitted to protection international borders
P_{UA} = additional Ofcom provided constraint for unscheduled adjustments.

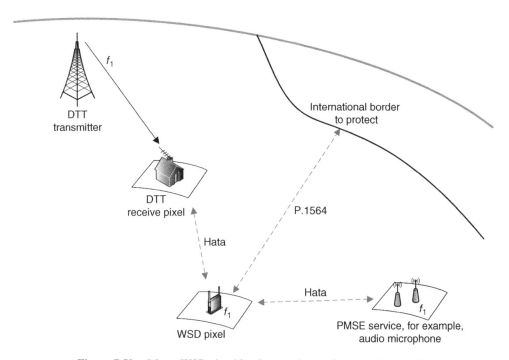

Figure 7.59 Ofcom WSD algorithm key services and propagation models

The approach to calculate the $P_{WSD,DTT}$ is based upon a threshold reduction in coverage probability q of $\Delta q = 7\%$ as in

$$q_2 = q_1 - \Delta q_T \tag{7.101}$$

In the United Kingdom, the minimum coverage probability is $q_1 = 70\%$ before interference, so the value including interference would be $q_2 = 63\%$. With a location variation with a standard deviation of 5.5 dB, this represents a decrease in margin of

$$\Delta M = 2.8 - 1.8 = 1 \text{ dB} \tag{7.102}$$

Note that for the edge of coverage case this would represent a threshold I/N of

$$T\left(\frac{I}{N}\right) = -6 \text{ dB} \tag{7.103}$$

For pixels at other locations the $T(I/N)$ would be higher, potentially significantly higher (see Example 7.38). The permissible WSD interference z at the pixel can then be calculated from the following inputs (in absolute, hence lower case):

- Required DTT signal level $p_{S,min}$
- Predicted wanted signal with normal distribution $p_S = N(m_s, \sigma_s^2)$
- Predicted DTT unwanted signals with normal distribution $p_{U,k} = N(m_{U,k}, \sigma_{U,k}^2)$
- Protection ratio for DTT-to-DTT interference, $pr_{DTT}(c/i, \Delta f)$
- Protection ratio for WSD-to-DTT interference, $pr_{WSD}(c/i, \Delta f)$
- Decrease in coverage within pixel $\Delta q_T = 7\%$.

The two coverage probabilities P_r can be specified as follows:

$$q_1 = P_r\left\{p_S \geq p_{S,min} + \sum_{k=1}^{K} pr_{DTT} \cdot p_{U,k}\right\} \tag{7.104}$$

$$q_2 = P_r\left\{p_S \geq p_{S,min} + \sum_{k=1}^{K} pr_{DTT} \cdot p_{U,k} + z \cdot pr_{WSD}\right\} \tag{7.105}$$

$$\Delta q_T = q_1 - q_2 \tag{7.106}$$

Note that using the standard notation for this book, the first two equations are

$$q_1 = P_r\left\{\frac{c}{n + \sum_{k=1}^{K} i_{DTT,k}} \geq t\left(\frac{c}{n+i}\right)\right\} \tag{7.107}$$

$$q_2 = P_r \left\{ \frac{c}{n + \sum\limits_{k=1}^{K} i_{DTT,k} + i_{WSD}} \geq t\left(\frac{c}{n+i}\right) \right\} \tag{7.108}$$

The adjacent channel case of Equation 7.108 could be written as

$$q_2 = P_r \left\{ \frac{c}{n + \sum\limits_{k=1}^{K} i_{DTT,k} + i_{WSD,co} \cdot a_{MI}} \geq t\left(\frac{c}{n+i}\right) \right\} \tag{7.109}$$

where a_{MI} is the mask integration adjustment or from Equation 5.82

$$a_{MI} = \frac{1}{acir} \tag{7.110}$$

As both the DTT wanted and interfering signals involve normal distributions, this requires iteration and a Monte Carlo simulation to derive for this pixel the threshold value $t(i)$ of i_{WSD} that meets Equation 7.106.

The interfering signal link budget equation can then be used to derive the maximum permissible power of the WSD to protect this DTT pixel in decibels:

$$T(I) = P_{WSD,DTT} - L_{Hata} + G_{DTT} \tag{7.111}$$

Or

$$P_{WSD,DTT} = T(I) + L_{Hata} - G_{DTT} \tag{7.112}$$

where L_{Hata} is the propagation loss calculated using the Hata/COST 231 propagation model and G_{DTT} is the DTT receive gain derived using the antenna pattern in Rec. ITU-R BT.419 as shown in Figure 7.13 assumed to be pointed at the relevant DTT transmitter. Additional terms could also be included to take account (say) of building penetration loss or body loss.

This value would have to be calculated for each WSD pixel and each channel, and in each case take into account all relevant DTT pixels and DTT transmitters on all channels, and hence involve multiple loops. In addition, it is necessary to expand the calculation to take account of the fact that:

- The WSD location has a given uncertainty
- The WSD could be serving slave devices over a service area.

Hence a single WSD device could require the checking of multiple locations rather than a single WSD pixel, as shown in Figure 7.60. In addition, special consideration would be required when the WSD and DTT receiver are in the same or adjacent pixels where the total path loss and receive relative gain would vary significantly across the pixels. In this case a

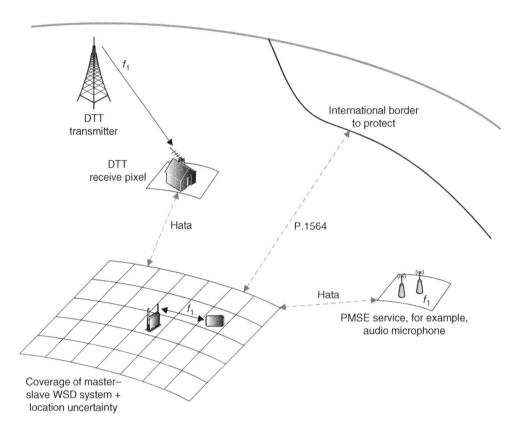

Figure 7.60 Range of WSD pixels to check for each DTT/PMSE location pixel

Table 7.33 GE06 trigger field strengths for international coordination

Band	Band IV	Band V	Band VI
Channels	21–34	35–51	51–69
Frequencies (MHz)	470–582	582–718	718–862
Field strength (dBµV/m)	21	23	25

deployment-type Monte Carlo analysis can be used to work out a total coupling gain $G = L_p - G_{DTT}$ exceeded for no more than a given percentage of deployments.

The loops have to include both the co-channel case and the adjacent channels, which has an impact on how far to search in pixels, namely, within 20 km (co-channel) or 2 km (adjacent channels). The adjacent channels can be modelled via adjusting the $PR(C/I)$ or including the ACIR in Equation 7.112.

The maximum WSD power to protect international borders $P_{WSD,Int}$ can be derived using the Rec. ITU-R P.1546 propagation model and the GE06 thresholds to trigger international coordination shown in Table 7.33. These represent an example of boundary analysis with the calculation of field strength given in Equation 7.21.

The PMSE services to protect include wireless microphones, talkback systems, in-ear monitors, program audio links and some video/data links. The principal method is via a threshold:

$$T(I) = -104\,\text{dBm}/200\,\text{kHz} \tag{7.113}$$

The calculation of maximum power for the WSD to protect the PMSE is then

$$P_{WSD,PMSE} = T(I) + L_{Hata} - G_{PMSE} \tag{7.114}$$

There is a further special case of channel 38 used for wireless microphones where their location is not known and so a reference geometry is used. Note this implies that the $T(I/N) = +10$ dB, significantly above the receiver noise level, assuming a noise figure of 7 dB.

More information on this algorithm can be found in Ofcom (2015).

7.10.4   Comparison of Approaches

The FCC approach specifies a minimum distance from the edge of the protected contour of a TV transmitter so that co-frequency, co-coverage operation of the TVBD is not permitted. The Ofcom methodology, in contrast, defines a maximum reduction in pixel coverage likelihood, which could, in theory, permit co-frequency, co-coverage operation. However it is also necessary to consider the reverse direction and whether the WSD device could operate within the DTT coverage area.

Example 7.38
A WSD is located within 1 km of a DTT transmitter. The link budgets of the DTT receiver with and without interference from the WSD are given in Table 7.34. The interfering signal strength was selected by iteration to reduce the location availability by 7%.

Table 7.34   Example WSD interference level calculation co-coverage co-frequency

DTT link budget	No interference	With interference	Notes
Frequency (MHz)	610	610	From Example 7.11
Bandwidth (MHz)	7.9	7.9	From Example 7.11
TX power P_{tx} (dBW)	40	40	Assumption
Distance (km)	1	1	Assumption
Free space path loss L_{fs} (dB)	88.3	88.3	Equation 3.13
RX antenna gain G_{rx} (dBi)	9.15	9.15	From Table 7.13
RX noise figure (dB)	7	7	From Example 7.11
RX noise N (dBW)	−128.0	−128.0	Equation 3.60
RX interference I (dBW)	n/a	−69.0	By iteration
RX $(N+I)$ (dBW)	−128.0	−69.0	Calculated from N, I
I/N (dB)	n/a	59.0	$= I - N$
Receive signal C (dBW)	−39.1	−39.1	$= P_{tx} - L_{fs} + G_{rx}$
$C/(N+I)$ (dB)	88.9	29.9	$= C - (N+I)$
Threshold $T(C/(N+I))$ (dB)	21.6	21.6	From Example 7.11
Wanted signal variation (dB)	5.5	5.5	Standard
Coverage likelihood (%)	100	93	Derived

It is noticeable how the I/N is significantly greater than the $-6\,$dB figure given in Equation 7.103. The maximum WSD power can be derived assuming a reference geometry involving a separation distance of $20\,$m using the calculations in Table 7.35 using Equation 7.112.

For this power the wanted signal at a distance of just $5\,$m can be calculated, together with the interfering signal from the DTT transmitter $1\,$km away, as shown in Table 7.36. Hence the $C/(N+I)$ was calculated to be $-18\,$dB.

It can therefore be seen that even though the Ofcom methodology would permit the WSD to operate co-frequency / co-coverage with the DTT system, it is unlikely to do so as it would either suffer harmful interference or have a minimal coverage.

Given that both methodologies result (either directly or indirectly) in co-channel sharing between WSD and DTT requiring a degree of geographic separation, the question is then how do the distances compare.

Example 7.39

A WSD wishes to operate co-frequency with a DTT service as in the geometry in Figure 7.61. What is the necessary separation distance d to achieve a reduction in location availability of 7% assuming the WSD is operating at its maximum power of $6\,$dBW?

For a DTT receiver in the pixel at the edge of coverage the $T(I/N) = -6\,$dB as in Equation 7.103. Using Equation 7.114 the required propagation loss can be derived from

Table 7.35   Example WSD maximum power from interference limit

Interfering signal (dBW)	−69.0
RX gain (dBi)	9.15
TX gain (dBi)	0.0
Distance (m)	20.0
Free space path loss (dB)	54.2
Maximum power (dBW)	−24.0

Table 7.36   Example WSD wanted and interfering signal link budgets

Signal	WSD wanted	DTT interfering
Frequency (MHz)	610.0	610.0
Bandwidth (MHz)	7.9	7.9
TX power (dBW)	−24.0	40.0
Distance (km)	0.005	1.0
Free space path loss (dB)	42.2	88.3
RX antenna gain (dBi)	0.0	0.0
RX noise figure (dB)	7.0	7.0
RX noise (dBW)	−128.0	−128.0
RX interference (dBW)	−48.3	−69.0
RX $(N+I)$ (dBW)	−48.3	−69.0
Receive signal (dBW)	−66.2	−48.3

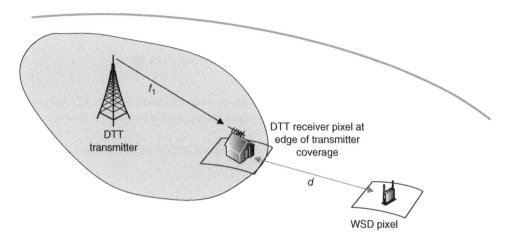

Figure 7.61   Restricted geometry for DTT example calculation

Table 7.37   Example required WSD separation distances to DTT

Environment	Open	Suburban	Urban
Distance (km)	21.9	6.52	3.77

$$L_{Hata} = P_{WSD} + G_{DTT} - N - T\left(\frac{I}{N}\right) \qquad (7.115)$$

Given the DTT receiver is at the edge of coverage, it must be pointing inwards and hence away from the WSD. Therefore the antenna gain will be far off-axis, which from Rec. ITU-R BT.419 is 16 dB down from the peak of 9.15 dBi. Hence the required path loss is

$$L_{Hata} = 6 + (9.15 - 16) - (-128) - (-6) = 131.1 \text{ dB}$$

The distance this corresponds to depends upon the heights and environment involved, but for transmit and receiver heights of 10 m the various distances are as given in Table 7.37.

These distances are comparable to the value of 14.4 km defined for the equivalent FCC case in Table 7.32. The use of an environment parameter means the distance can vary rather than being a single value.

Example 7.40
A WSD wishes to operate co-frequency with a PMSE service operating at a height of 1.5 m. What is the necessary separation distance d assuming the WSD is a fixed device at a height of 10 m and operating at its maximum power of 6 dBW?

The Ofcom's PMSE threshold is -104 dBm/200 kHz $= -134$ dBW/200 kHz and the required path loss can be calculated using

$$L_{Hata} = P_{WSD} + G_{PMSE} + A_{BW} - T(I) \qquad (7.116)$$

Assuming the receive antenna is isotropic then

$$L_{Hata} = 6 + 0 + 10\log_{10}\left(\frac{200}{8000}\right) - (-134) = 124 \text{ dB}$$

Using the Hata/COST 231propagation this relates to a separation distance of 1.26 km (suburban environment) or 0.725 km (urban environments), very similar to the FCC distance of 1 km given in Section 7.10.2.

It can be seen that though the methodologies have significant differences, the resulting separation distances can be very similar.

7.11   Final Thoughts

The objective of this book was to inform the reader in the techniques and methodologies that can be used to analyse interference. It has covered topics such as antennas, propagation models, the interference calculation, mitigation methods and methodologies. The final chapter has covered a number of specific algorithms and services, both as examples of techniques covered earlier and also to introduce additional ideas.

However interference analysis is a very wide subject field, one that is continually growing, and it was not feasible for this book to cover every aspect in detail. It is worth, therefore, being aware of general principles that can be applied in multiple scenarios and sources of further information, such as ITU-R Recommendations.

Often new methodologies are created by borrowing ideas from those already existing, adapting where necessary. It can also be useful to have whiteboard discussions with colleagues to create new ideas, often using a figure as a visualisation tool, such as the space–frequency figures described in this book, or identifying the key questions to answer.

I hope that this book will be a resource that can be built upon, using interference analysis as a vital tool in the effective management of the radio spectrum.

References

3GPP, 2010. *3GPP TR 36.814: Technical specification group radio access network; evolved universal terrestrial radio access (E-UTRA); further advancements for E-UTRA physical layer aspects*, Sophia-Antipolis: 3GPP.

ACMA, 1998. *Microwave fixed services frequency coordination*, Canberra: ACMA.

ACMA, 2000. *Frequency assignment requirements for the land mobile service*, Canberra: ACMA.

Aegis Systems Ltd, 2011. *The co-existence of LTE and DTT services at UHF: A field trial*, London: Ofcom.

Aegis Systems Ltd and Transfinite Systems Ltd, 2004. *Evaluating spectrum percentage occupancy in licence-exempt allocations*, London: Ofcom.

Aegis Systems Ltd, Transfinite Systems Ltd and Indepen Ltd, 2006. *Technology-neutral spectrum usage rights*, London: Ofcom.

Auer, G., et al., 2012. *How much energy is needed to run a wireless network*, s.l.: EU FP7 Project Earth.

Barclay, L., 2003. *Propagation of radiowaves*, 2nd ed., London: The Institute of Electrical Engineers.

Bate, R. R., Mueller, D. D., and White, J. E., 1971. *Fundamentals of astrodynamics*, 1st ed., New York: Dover.

Bousquet, M., and Maral, G., 1986. *Satellite communication systems*, 1st ed. Chichester: John Wiley & Sons, Ltd.

CEPT, 1997. *The Chester 1997 multilateral coordination agreement relating to technical criteria, coordination principles and procedures for the introduction of terrestrial digital video broadcasting (DVB-T)*, Chester: CEPT.

CEPT, 2009. *CEPT ECC Report 131: Derivation of a block edge mask (BEM) for terminal stations in the 2.6 GHz frequency band (2 500–2 690 MHz)*, Dublin: CEPT.

CEPT, 2011a. *ECC Report 185: Further definition of technical and operational requirements for the operation of white space devices in the band 470–790 MHz*, Copenhagen: CEPT.

CEPT, 2011b. *ECC Report 186: Technical and operational requirements for the operation of white space devices under geo-location approach*, Copenhagen: CEPT.

CEPT, 2011c. *ERC Recommendation 74-01: Unwanted emissions in the spurious domain*, Cardiff: CEPT.

CEPT ECC, 2002. *ECC Report 20: Methodology to determine the density of fixed service links*, Sesimbra: CEPT ECC.

CEPT ECC, 2010. *Recommendation T/R 13-02: Preferred channel arrangements for fixed service systems in the frequency range 22.0–29.5 GHz*, s.l.: s.n.

CEPT ERC, 1999. *CEPT ERC Report 101: A comparison of the minimum coupling loss method, enhanced minimum coupling loss method, and the Monte Carlo simulation*, Menton: CEPT ERC.

CEPT ERC, 2002. *Report 68: Monte-Carlo radio simulation methodology for the use in sharing and compatibility studies between different radio services or systems*, s.l.: s.n.

Interference Analysis: Modelling Radio Systems for Spectrum Management, First Edition. John Pahl.
© 2016 John Wiley & Sons, Ltd. Published 2016 by John Wiley & Sons, Ltd.
Companion website: www.wiley.com/go/pahl1015

Clarke, A. C., 1945. Extra-terrestrial relays – can rocket stations give worldwide radio coverage. *Wireless World*, Issue October, pp. 305–308.

Cockell, C. S., 2006. *Project Boreas: A station for the Martian geographic north pole*, London: British Interplanetary Society.

Commission of the European Communities, 2007. Commission Decision of 21 February 2007 on allowing the use of the radio spectrum for equipment using ultra-wideband technology in a harmonised manner in the community. *Official Journal of the European Union*, **55**, pp. 33–36.

COST Action 231, n.d. *Final report*, s.l.: s.n.

Craig, K. H., 2004. *Theoretical assessment of the impact of the correlation of signal enhancements on area-to-point interference*, s.l.: Radiocommunications Agency.

Crane, R. K., 1996. *Electromagnetic wave propagation through rain*, s.l.: Wiley.

DTG Testing, 2014. *Lab measurements of WSD-DTT protection ratios*, London: Ofcom.

EBU, 2005. *EBU technical report 24: SFN frequency planning and network implementation with regard to D-DAB and DVB-T*, Geneva: EBU.

EBU, 2014. *Tech 3348: Frequency and network planning aspects of DVB-T2*, Geneva: EBU.

ECC and ETSI, 2011. *The European regulatory environment for radio equipment and spectrum*. [Online] Available at: http://www.cept.org/files/1051/ECC_ETSI_2011.pdf (Accessed 31 October 2014).

Egli, J. J., 1957. Radio propagation above 40 MC over irregular terrain. *Proceedings of the IRE (IEEE)*, **45**(10), pp. 1383–1391.

ERC, 1999. *ERC report 101: A comparison of the minimum coupling loss method, the enhanced mininimum coupling loss method and the monte Carlo simulation*, Menton: CEPT ERC.

ETSI, 2005. *ETSI TR 101 854; V1.3.1: Technical report; fixed radio systems; point-to-point equipment; derivation of receiver interference parameters useful for planning fixed service point-to-point systems operating different equipment classes and/or capacities*, Sophia Antipolis: ETSI.

ETSI, 2009. *EN 300 744; V1.6.1; european standard (telecommunications series) digital video broadcasting (DVB); framing structure, channel coding and modulation for digital terrestrial television*, Sophia Antipolis: ETSI.

ETSI, 2010. *EN 302 217-4-2 V1.5.1: Fixed radio systems; characteristics and requirements for point-to-point equipment and antennas; Part 4-2: Antennas; harmonized EN covering the essential requirements of article 3.2 of the R&TTE directive*, Sophia Antipolis: ETSI.

ETSI, 2014a. *EN 301 893; V1.7.2: Broadband radio access networks (BRAN); 5 GHz high performance RLAN; harmonized EN covering the essential requirements of article 3.2 of the R&TTE directive*, Sophia Antipolis: ETSI.

ETSI, 2014b. *ETSI EN 301 598: White space devices (WSD); wireless access systems operating in the 470 MHz to 790 MHz TV broadcast band; harmonized EN covering the essential requirements of article 3.2 of the R&TTE directive*, Sophia Antipolis: ETSI.

ETSI, 2014c. *EN 302 217-2-2 V2.2.1: Fixed radio systems; characteristics and requirements for point-to-point equipment and antennas; Part 2-2: Digital systems operating in frequency bands where frequency co-ordination is applied*, Sophia Antipolis: ETSI.

ETSI/3GPP, 2010. *ETSI TS 36.101; version 8.10.0; evolved universal terrestrial radio access (E-UTRA); user equipment (UE) radio transmission and reception*, Sophia Antipolis: ETSI/3GPP.

ETSI/3GPP, 2011. *ETSI TS 36.141; version 10.1.0; evolved universal terrestrial radio access (E-UTRA); base station (BS) conformance testing*, Sophia Antipolis: ETSI/3GPP.

ETSI/3GPP, 2014. *ETSI 36.104: LTE; v11.10.0; evolved universal terrestrial radio access (E-UTRA); base station (BS) radio transmission and reception*, Sophia Antipolis: ETSI/3GPP.

FCC, 1966. *Report R-6602: Development of VHF and UHF propagation curves for TV and FM broadcasting*, Washington, DC: FCC.

FCC, 2004. *FCC 04-168: In the matter of improving public safety communications in the 800 MHz band consolidating the 800 and 900 MHz industrial/land transportation and business pool channels*, s.l.: s.n.

FCC, 2010. *FCC 10-174: Second memorandum opinion and order*, Washington, DC: FCC.

Federal Communications Commission, 2002. *First report and order in the matter of the revision of Part 15 of the Commission's rules regarding ultra-wideband transmission systems*, s.l.: s.n.

Flood, I., 2013. *Graph theoretic methods for radio equipment selection*, Cardiff: Cardiff University.

Flood, I., and Bacon, D., 2006. Towards more spectrally efficient frequency assignment for microwave fixed link. *International Journal of Mobile Network Design and Innovation*, **1**(2), pp. 147–152.

Goodman, M., n.d. *The Radio Act of 1927 as a product of progressivism*. [Online] Available at: http://www.scripps.ohiou.edu/mediahistory/mhmjour2-2.htm (Accessed 26 October 2014).

Haslett, C., 2008. *Essentials of radio wave propagation*, 1st ed., Cambridge: Cambridge University Press.

Hata, M., 1980. Empirical formula for propagation loss in land mobile radio services. *IEEE Transactions on Vehicular Technology*, **VT-29**(3), pp. 317–325.

HCM Administrations, 2013. *Agreement on the co-ordination of frequencies between 29.7 MHz and 43.5 GHz for the fixed service and the land mobile service*, s.l.: s.n.

Ho, C., Golshan, N., and Kliore, A., 2002. *Radio wave propagation handbook for communication on and around mars*, Los Angeles: JPL.

Holma, H., and Toskala, A., 2010. *WCDMA for UMTS – HSPA evolution and LTE*, 5th ed., Chichester: John Wiley & Sons, Ltd.

IEEE, 2009a. *802.11n-2009: Amendment 5: Enhancements for higher throughput*, s.l.: s.n.

IEEE, 2009b. *IEEE 802.16m evaluation methodology document*, New York: IEEE.

International Commission on Non-Ionizing Radiation Protection (ICNIRP), 1998. Guidelines for limiting exposure to time-varying electric, magnetic and electromagnetic fields (up to 300 GHz). *Health Physics*, **74**, pp. 494–522.

International Launch Service, 2000. *Sirius 1: Launch on the proton launch vehicle*, s.l.: s.n.

ITU, 2004. *Radio Regulations*, 2004 ed., Geneva: ITU.

ITU, 2008. *Radio Regulations*, 2008 ed., Geneva: ITU.

ITU, 2011. *Collection of the basic texts of the International Telecommunications Union adopted by the Plenipotentiary Conference*, 2011 ed., Geneva: ITU.

ITU, 2012a. *Radio Regulations*, 2012 ed., Geneva: ITU.

ITU, 2012b. *Resolutions of the Radiocommunications Assembly*, Geneva: ITU.

ITU, 2014. *Final Acts of the Plenipotentiary Conference (Busan, 2014)*, Geneva: ITU.

ITU, 2015. *Provisional Final Acts WRC-15*, Geneva: ITU.

ITU, n.d. *Overview of ITU's history*. [Online] Available at: http://www.itu.int/en/history/Documents/ITU-HISTORY-Overview.pdf (Accessed 27 Octobter 2012).

ITU-R, 1986. *Report ITU-R M.739-1: Interference due to intermodulation products in the land mobile service between 25 and 1 000 MHz*, Geneva: ITU-R.

ITU-R, 1992a. *Recommendation ITU-R BT.419-3: Directivity and polarization discrimination of antennas in the reception of television broadcasting*, Geneva: ITU-R.

ITU-R, 1992b. *Recommendation ITU-R S.738: Procedure for determining if coordination is required between geostationary-satellite networks sharing the same frequency bands*, Geneva: ITU-R.

ITU-R, 1992c. *Recommendation ITU-R S.740:Technical coordination methods for fixed-satellite networks*, Geneva: ITU-R.

ITU-R, 1993. *Recommendation ITU-R SF.1006: Determination of the interference potential between earth stations of the fixed-satellite service and stations in the fixed service*, Geneva: ITU-R.

ITU-R, 1994a. *Recommendation ITU-R F.1101: Characteristics of digital fixed wireless systems below about 17 GHz*, Geneva: ITU-R.

ITU-R, 1994b. *Recommendation ITU-R P.525-2: Calculation of free space attenuation*, Geneva: ITU-R.

ITU-R, 1994c. *Recommendation ITU-R S.741-2: Carrier-to-interference calculations between networks in the fixed-satellite service*, Geneva: ITU-R.

ITU-R, 1995. *Recommendation ITU-R S.728-1: Maximum permissible level of off-axis e.i.r.p. density from very small aperture terminals (VSATs)*, Geneva: ITU-R.

ITU-R, 1997a. *Recommendation ITU-R S.672-4: Satellite antenna radiation pattern for use as a design objective in the fixed-satellite service employing geostationary satellites*, Geneva: ITU-R.

ITU-R, 1997b. *Recommendation ITU-R S.736: Estimation of polarization discrimination in calculations of interference between geostationary-satellite networks in the fixed-satellite service*, Geneva: ITU-R.

ITU-R, 1999a. *Recommendation ITU-R P.341-5: The concept of transmission loss for radio links*, Geneva: ITU-R.

ITU-R, 1999h *Recommendation ITU-R P.1058: Digital topographic databases for propagation studies*, Geneva: ITU-R.

ITU-R, 1999c. *Recommendation ITU-R SF.1395: Minimum propagation attenuation due to atmospheric gases for use in frequency sharing studies between the fixed-satellite service and the fixed service*, Geneva: ITU-R.

ITU-R, 2000a. *Recommendation ITU-R M.1454: E.i.r.p. density limit and operational restrictions for RLANS or other wireless access transmitters in order to ensure the protection of feeder links of non-geostationary systems in the mobile-satellite service in the frequency band 5 150–5 250 MHz*, Geneva: ITU-R.

ITU-R, 2000b. *Recommendation ITU-R SM.1448: Determination of the coordination area around an Earth station in the frequency bands between 100 MHz and 105 GHz,* Geneva: ITU-R.

ITU-R, 2000c. *Recommendation ITU-R V.431-7: Nomenclature of the frequency and wavelength bands used in telecommunications,* Geneva: ITU-R.

ITU-R, 2001a. *Recommendation ITU-R F.748-4: Radio-frequency arrangements for systems of the fixed service operating in the 25, 26 and 28 GHz bands,* Geneva: ITU-R.

ITU-R, 2001b. *Recommendation ITU-R P.1510: Annual mean surface temperature,* Geneva: ITU-R.

ITU-R, 2001c. *Recommendation ITU-R S.1428-1: Reference FSS earth-station radiation patterns for use in interference assessment involving non-GSO satellites in frequency bands between 10.7 GHz and 30 GHz,* Geneva: ITU-R.

ITU-R, 2001d. *Recommendation ITU-R S.1528: Satellite antenna radiation patterns for non-geostationary orbit satellite antennas operating in the fixed-satellite service below 30 GHz,* Geneva: ITU-R.

ITU-R, 2001e. *Recommendation ITU-R S.1529: Analytical method for determining the statistics of interference between non-geostationary-satellite orbit FSS systems and other non-geostationary-satellite orbit FSS systems or geostationary-satellite orbit FSS networks,* Geneva: ITU-R.

ITU-R, 2002a. *ITU-R Report SM.2028-1: Monte Carlo simulation methodology for the use in sharing and compatibility studies between different radio services or systems,* Geneva: ITU-R.

ITU-R, 2002b. *Recommendation ITU-R S.1257-3: Analytical method to calculate short-term visibility and interference statistics for non-geostationary satellite orbit satellites as seen from a point on the Earth's surface,* Geneva: ITU-R.

ITU-R, 2002c. *Recommendation ITU-R S.1323-2: Maximum permissible levels of interference in a satellite network (GSO/FSS; non-GSO/FSS; non-GSO/MSS feeder links) in the fixed-satellite service caused by other codirectional FSS networks below 30 GHz,* Geneva: ITU-R.

ITU-R, 2003a. *Recommendation ITU-R M.1184-2: Technical characteristics of mobile satellite systems in the frequency bands below 3 GHz for use in developing criteria for sharing between the mobile-satellite service (MSS) and other services,* Geneva: ITU-R.

ITU-R, 2003b. *Recommendation ITU-R M.1644: Technical and operational characteristics, and criteria for protecting the mission of radars in the radiolocation and radionavigation service operating in the frequency band 13.75–14 GHz,* Geneva: ITU-R.

ITU-R, 2003c. *Recommendation ITU-R M.1654: A methodology to assess interference from broadcasting-satellite service (sound) into terrestrial IMT-2000 systems intending to use the band 2 630–2 655 MHz,* Geneva: ITU-R.

ITU-R, 2003d. *Recommendation ITU-R RA.769-2: Protection criteria used for radio astronomical measurements,* Geneva: ITU-R.

ITU-R, 2003e. *Recommendation ITU-R S.580-6: Radiation diagrams for use as design objectives for antennas of earth stations operating with geostationary satellites,* Geneva: ITU-R.

ITU-R, 2003f. *Recommendation ITU-R S.1325-3: Simulation methodologies for determining statistics of short-term interference between co-frequency, codirectional non-geostationary-satellite orbit FSS systems in circular orbits and other non-geostationary FSS networks,* Geneva: ITU-R.

ITU-R, 2005a. *Recommendation ITU-R F.1108-4: Determination of the criteria to protect fixed service receivers from the emissions of space stations operating in non-geostationary orbits in shared frequency bands,* Geneva: ITU-R.

ITU-R, 2005b. *Recommendation ITU-R M.1141-2: Sharing in the 1–3 GHz frequency range between non-geostationary space stations operating in the mobile-satellite service,* Geneva: ITU-R.

ITU-R, 2005c. *Recommendation ITU-R M.1143-3: System specific methodology for coordination of non-geostationary space stations (space-to-Earth) operating in the mobile-satellite service with the fixed service,* Geneva: ITU-R.

ITU-R, 2005d. *Recommendation ITU-R P.838-3: Specific attenuation model for rain for use in prediction methods,* Geneva: ITU-R.

ITU-R, 2006a. *Final Acts of the Regional Radiocommunication Conference for planning of the digital terrestrial broadcasting service in parts of Regions 1 and 3, in the frequency bands 174–230 MHz and 470–862 MHz.* Geneva: ITU-R.

ITU-R, 2006b. *Recommendation ITU-R F.699-7: Reference radiation patterns for fixed wireless system antennas for use in coordination studies and interference assessment in the frequency range from 100 MHz to about 70 GHz,* Geneva: ITU-R.

ITU-R, 2006c. *Recommendation ITU-R F.1760: Methodology for the calculation of aggregate equivalent isotropically radiated power (a.e.i.r.p.) distribution from point-to-multipoint high-density applications in the fixed service operating in bands above 30 GHz,* Geneva: ITU-R.

ITU-R, 2006d. *Recommendation ITU-R F.1766: Methodology to determine the probability of a radio astronomy observatory receiving interference based on calculated exclusion zones to protect against interference from p-mp high-density applications in the fixed service,* Geneva: ITU-R.

ITU-R, 2006e. *Recommendation ITU-R M.1460-1: Technical and operational characteristics and protection criteria of radiodetermination radars in the 2 900–3 100 MHz band,* Geneva: ITU-R.

ITU-R, 2006f. *Recommendation ITU-R M.1739: Protection criteria for wireless access systems, including radio local area networks, operating in the mobile service in accordance with Resolution 229 (WRC-03) in the bands 5 150–5 250 MHz, 5 250–5 350 MHz and 5 470–5 725 MHz,* Geneva: ITU-R.

ITU-R, 2006g. *Recommendation ITU-R S.524-9: Maximum permissible levels of off-axis e.i.r.p. density from earth stations in geostationary-satellite orbit networks operating in the FSS transmitting in the 6 GHz, 13 GHz, 14 GHz and 30 GHz frequency bands,* Geneva: ITU-R.

ITU-R, 2006h. *Recommendation ITU-R S.1432-1: Apportionment of the allowable error performance degradations to fixed-satellite service (FSS) hypothetical reference digital paths arising from time invariant interference for systems operating below 30 GHz,* Geneva: ITU-R.

ITU-R, 2006i. *Recommendation ITU-R SM.328-11: Spectra and bandwidth of emissions,* Geneva: ITU-R.

ITU-R, 2007a. *Recommendation ITU-R F.1094-2: Maximum allowable error performance and availability degradations to digital fixed wireless systems arising from radio interference from emissions and radiations from other sources,* Geneva: ITU-R.

ITU-R, 2007b. *Recommendation ITU-R M. 1583-1: Interference calculations between non-geostationary mobile-satellite service or radionavigation-satellite service systems and radio astronomy telescope sites,* Geneva: ITU-R.

ITU-R, 2007c. *Recommendation ITU-R P.368-9: Ground-wave propagation curves for frequencies between 10 kHz and 30 MHz,* Geneva: ITU-R.

ITU-R, 2007d. *Recommendation ITU-R P.1147-4: Prediction of sky-wave field strength at frequencies between about 150 and 1 700 kHz,* Geneva: ITU-R.

ITU-R, 2007e. *Recommendation ITU-R P.1791: Propagation prediction methods for assessment of the impact of ultra-wideband devices,* Geneva: ITU-R.

ITU-R, 2007f. *Recommendation ITU-R SM.1134-1: Intermodulation interference calculations in the land-mobile service,* Geneva: ITU-R.

ITU-R, 2008b. *Recommendation ITU-R SM.337-6: Frequency and distance separations,* Geneva: ITU-R.

ITU-R, 2010a. *Recommendation ITU-R M.1319-3: The basis of a methodology to assess the impact of interference from a TDMA/FDMA MSS space-to-Earth transmissions on the performance of line-of-sight FS receiver,* Geneva: ITU-R.

ITU-R, 2010b. *Recommendation ITU-R S.465-6: Reference radiation pattern for earth station antennas in the fixed-satellite service for use in coordination and interference assessment in the frequency range from 2 to 31 GHz,* Geneva: ITU-R.

ITU-R, 2011a. *Recommendation ITU-R BT.1895: Protection criteria for terrestrial broadcasting systems,* Geneva: ITU-R.

ITU-R, 2011b. *Recommendation ITU-R M.1652-1: Dynamic frequency selection in wireless access systems including radio local area networks for the purpose of protecting the radiodetermination service in the 5 GHz band,* Geneva: ITU-R.

ITU-R, 2011c. *Report ITU-R M.2235: Aeronautical mobile (route) service sharing studies in the frequency band 960–1 164 MHz,* Geneva: ITU-R.

ITU-R, 2012a. *Recommendation ITU-R 329-12: Unwanted emissions in the spurious domain,* Geneva: ITU-R.

ITU-R, 2012b. *Recommendation ITU-R F.746-10: Radio-frequency arrangements for fixed service systems,* Geneva: ITU-R.

ITU-R, 2012c. *Recommendation ITU-R F.1245-2: Mathematical model of average and related radiation patterns for line-of-sight PtP fixed wireless system antennas for use in certain coordination studies and interference assessment in the frequency range 1 to about 70 GHz,* Geneva: ITU-R.

ITU-R, 2012d. *Recommendation ITU-R P.453-10: The radio refractive index: Its formula and refractivity data,* Geneva: ITU-R.

ITU-R, 2012e. *Recommendation ITU-R P.528-3: Propagation curves for aeronautical mobile and radionavigation services using the VHF, UHF and SHF bands,* Geneva: ITU-R.

ITU-R, 2012f. *Recommendation ITU-R P.837-6: Characteristics of precipitation for propagation modelling,* Geneva: ITU-R.

ITU-R, 2012g. *Recommendation ITU-R P.1238-7: Propagation data and prediction methods for the planning of indoor radiocommunication systems and radio local area networks in the frequency range 900 MHz to 100 GHz*, Geneva: ITU-R.

ITU-R, 2012h. *Recommendation ITU-R RS.2017: Performance and interference criteria for satellite passive remote sensing*, Geneva: ITU-R.

ITU-R, 2012i. *Report ITU-R BT.2265: Guidelines for the assessment of interference into the broadcasting service*, Geneva: ITU-R.

ITU-R, 2013a. *Guidelines for the working methods of the RA, the SGs and related groups*, s.l.: s.n.

ITU-R, 2013b. *Recommendation ITU-R BT.2036: Characteristics of a reference receiving system for frequency planning of digital terrestrial television systems*, Geneva: ITU-R.

ITU-R, 2013c. *Recommendation ITU-R P.372-11: Radio noise*, Geneva: ITU-R.

ITU-R, 2013d. *Recommendation ITU-R P.452-15: Prediction procedure for the evaluation of interference between stations on the surface of the Earth at frequencies above about 0.1 GHz*, Geneva: ITU-R.

ITU-R, 2013e. *Recommendation ITU-R P.526-13: Propagation by diffraction*, Geneva: ITU-R.

ITU-R, 2013f. *Recommendation ITU-R P.530-15: Propagation data and prediction methods required for the design of terrestrial line-of-sight systems*, Geneva: ITU-R.

ITU-R, 2013g. *Recommendation ITU-R P.618-11: Propagation data and prediction methods required for the design of Earth-space telecommunication systems*, Geneva: ITU-R.

ITU-R, 2013h. *Recommendation ITU-R P.676-10: Attenuation by atmospheric gases*, Geneva: ITU-R.

ITU-R, 2013i. *Recommendation ITU-R P.835-5: Reference standard atmospheres*, Geneva: ITU-R.

ITU-R, 2013j. *Recommendation ITU-R P.836-5: Water vapour: surface density and total columnar content*, Geneva: ITU-R.

ITU-R, 2013k. *Recommendation ITU-R P.1411-7: Propagation data and prediction methods for the planning of short-range outdoor radiocommunication systems and radio local area networks in the frequency range 300 MHz to 100 GHz*, Geneva: ITU-R.

ITU-R, 2013l. *Recommendation ITU-R P.1546-5: Method for point-to-area predictions for terrestrial services in the frequency range 30 MHz to 3 000 MHz*, Geneva: ITU-R.

ITU-R, 2013m. *Recommendation ITU-R P.1812-3: A path-specific propagation prediction method for point-to-area terrestrial services in the VHF and UHF bands*, Geneva: ITU-R.

ITU-R, 2013n. *Recommendation ITU-R P.2001-1: A general purpose wide-range terrestrial propagation model in the frequency range 30 MHz to 50 GHz*, Geneva: ITU-R.

ITU-R, 2013o. *Recommendation ITU-R P.2040: Effects of building materials and structures on radiowave propagation above about 100 MHz*, Geneva: ITU-R.

ITU-R, 2013p. *Recommendation ITU-R S.1503-2: Functional description to be used in developing software tools for determining conformity of non-geostationary-satellite orbit fixed-satellite system networks with limits contained in Article 22 of the Radio Regulations*, Geneva: ITU-R.

ITU-R, 2013q. *Report ITU-R M.2290-0: Future spectrum requirements estimate for terrestrial IMT*, Geneva: ITU-R.

ITU-R, 2013r. *Report ITU-R M.2292: Characteristics of terrestrial IMT-advanced systems for frequency sharing/interference analyses*, Geneva: ITU-R.

ITU-R, 2013s. *Recommendation ITU-R P.833-8: Attenuation in vegetation*, Geneva: ITU-R.

ITU-R, 2014a. *Document JTG 4-5-6-7/715: Report on the sixth and final meeting of Joint Task Group 4-5-6-7*, s.l.: s.n.

ITU-R, 2014b. *Draft New Report ITU-R [FSS-IMT C-BAND DOWNLINK]: Sharing studies between International Mobile Telecommunication-advanced systems and geostationary satellite networks in the FSS in the 3 400–4 200 MHz and 4 500–4 800 MHz frequency bands*, s.l.: s.n.

ITU-R, 2014c. *Recommendation ITU-R BO.1443-3: Reference BSS earth station antenna patterns for use in interference assessment involving non-GSO satellites in frequency bands covered by RR Appendix 30*, Geneva: ITU-R.

ITU-R, 2014d. *Recommendation ITU-R BT.1368-11: Planning criteria, including protection ratios, for digital terrestrial television services in the VHF/UHF bands*, Geneva: ITU-R.

ITU-R JTG 4-5-6-7, 2014. *Document 715 Annex 05: Draft new Report ITU-R BT.[MBB_DTTB_470_694] – sharing and compatibility studies between digital terrestrial television broadcasting and terrestrial mobile broadband applications, including IMT, in the frequency band 470–694/698 MHz*, s.l.: s.n.

Joint Frequency Planning Project, 2012. *Technical parameters and planning algorithms*, London: Ofcom.

Kraus, J. D., and Marhefka, R. J., 2003. *Antennas for all applications*, 3rd ed., New York: McGraw Hill.

Lee, W. C. Y., 1993. *Mobile communications design fundamentals*. s.l.: Wiley Interscience.

NASA, n.d. *Shuttle radar topography mission*. [Online] Available at: http://www2.jpl.nasa.gov/srtm/index.html (Accessed 10 December 2014).

NASA JPL, n.d. *ASTER global digital elevation Map*. [Online] Available at: http://asterweb.jpl.nasa.gov/gdem.asp (Accessed 10 December 2014).

Nelsen, R., 1999. *An introduction to copulas*, s.l.: Springer.

Ofcom, 2006a. *3M/164-E: Information paper: Modelling multiple interfering signals*, Geneva: ITU-R.

Ofcom, 2006b. *Spectrum usage rights: Technology and usage neutral access to the radio spectrum*, London: Ofcom.

Ofcom, 2007. *Spectrum co-existence document: Broadband fixed wireless access (BFWA) and spectrum access – sub national (SA-SN) – in 28 GHz*, London: Ofcom.

Ofcom, 2008a. *Document 3K/10-E: Diffraction model comparison using cleaned 3K1 correspondence group database*, Geneva: ITU-R.

Ofcom, 2008b. *OfW 164: Business radio technical frequency assignment criteria*, London: Ofcom.

Ofcom, 2012a. *3G coverage obligation verification methodology*, London: Ofcom.

Ofcom, 2012b. *4G coverage obligation notice of compliance verification methodology: LTE*, London: Ofcom.

Ofcom, 2013. *OfW 446 technical frequency assignment criteria for fixed point-to-point radio services with digital modulation*, London: Ofcom.

Ofcom, 2014. *Business radio assignment sharing criteria review*, London: Ofcom.

Ofcom, 2015. *Implementing white spaces: Annexes 1 to 12*, London: Ofcom.

Ofcom and ANFR, 2004. *MoU concluded between the administrations of France and the UK on coordination in the 47–68 MHz frequency band*, London and Paris: Ofcom and ANFR.

PA Knowledge Limited, 2009. *Predicting areas of spectrum shortage*, London: Ofcom.

Pahl, J., 2002. *Analysis of the spectrum efficiency of sharing between terrestrial and satellite services*, London: IEE Conference on Radio Spectrum.

Pahl, J., 2006. Communications and navigation networks for pole station. In: C. S. Cockell, ed. *Project Boreas: A station for the Martian geographic north pole*, London: British Interplanetary Society, pp. 123–146.

Petroff, E., et al., 2015. Identifying the source of perytons at the Parkes radio telescope. arXiv:1504.02165v1, 9 April.

Rappaport, T. S., 1996. *Wireless communications principle & practices*, 1st ed., s.l.: Prentice Hall.

Scientific Generics Limited, 2005. *Cost and power consumption implications of digital switchover*, London: Ofcom.

Skolnik, M. I., 2001. *Introduction to radar systems*, 3rd ed., Singapore: McGraw-Hill Higher Education.

Telecommunications Industry Association, 1994. *Technical service bulletin 10F: Interference criteria for microwave systems*, Washington, DC: Telecommunications Industry Association.

Transfinite Systems, Radio Communications Research Unit, dB Spectrum, LS telcom, 2007. *Final report: GRMT phase 2*, London: Ofcom.

Vallado, D. A., Crawford, P., Hujsak, R., and Kelso, T. S., 2006. *Revisiting spacetrack report #3: Rev 2*, Reston, VA: American Institute of Aeronautics and Astronautics.

Wikipedia, 2014a. *Boltzmann constant*. [Online] Available at: http://en.wikipedia.org/wiki/Boltzmann_constant (Accessed 23 October 2014).

Wikipedia, 2014b. *Speed of light*. [Online] Available at: http://en.wikipedia.org/wiki/Speed_of_light (Accessed 23 October 2014).

Wikipedia, 2014c. *Tragedy of the commons*. [Online] Available at: http://en.wikipedia.org/wiki/Tragedy_of_the_commons (Accessed 26 October 2014).

Wikipedia, 2014d. *World geodetic system*. [Online] Available at: http://en.wikipedia.org/wiki/World_Geodetic_System (Accessed 23 October 2014).

Acronyms, Abbreviations and Symbols

Acronyms and Abbreviations

3GPP	3rd Generation Partnership Project
AA	area analysis
ABW	Allocated bandwidth
ACMA	Australian Communications and Media Authority
ACIR	adjacent channel interference ratio
ACLR	adjacent channel leakage ratio
ACS	adjacent channel selectivity
AEIRP	aggregate equivalent isotropically radiated power
AF	activity factor
AM	amplitude modulation
AMS	aeronautical mobile service
AMSS	aeronautical mobile-satellite service
ANFR	L'Agence Nationale des Fréquences
APC	automatic power control
APT	Asia-Pacific Telecommunity
ASMG	Arab Spectrum Management Group
ASTER	Advanced Spaceborne Thermal Emission and Reflection Radiometer
ATPC	adaptive transmit power control
ATU	African Telecommunications Union
BAS	broadcast auxiliary service
BEM	block edge mask
BER	bit error rate
BPSK	binary phase shift keying

Interference Analysis: Modelling Radio Systems for Spectrum Management, First Edition. John Pahl.
© 2016 John Wiley & Sons, Ltd. Published 2016 by John Wiley & Sons, Ltd.
Companion website: www.wiley.com/go/pahl1015

BR	business radio/radio bureau
BS	base station/broadcasting service
BSS	broadcasting-satellite service
CA	collision avoidance
CAC	channel availability check
CAPEX	capital expenditure
CCDP	co-channel dual-polar
CCV	Coordination Committee for Vocabulary
CD	collision detection
CDF	cumulative distribution function
CDMA	code division multiple access
CEPT	European Conference of Postal and Telecommunications Administrations
CITEL	Inter-American Telecommunication Commission
COST	Cooperation in Science and Technology
CoU	change of use
CPM	Conference Preparatory Meeting
CS	central station
CSMA	collision sensing multiple access
CW	continuous wave
DFS	dynamic frequency selection
DG	drafting group
DL	downlink
DMS	degrees, minutes, seconds
DNR	draft new recommendation
DPSK	differential phase shift keying
DTT	digital terrestrial television
DVB	Digital Video Broadcasting
EBU	European Broadcasting Union
ECC	Electronic Communications Committee
ECI	Earth-centred inertial
ECP	European Common Position
EESS	Earth exploration-satellite service
EHF	extremely high frequency
EIRP	equivalent isotropically radiated power
ENG	electronic news gathering
EPFD	equivalent power flux density
ERC	European Radiocommunications Committee
ERP	effective radiated power
ES	Earth station
ESIM	Earth station in motion
ESOMP	Earth station on mobile platform
ESV	Earth station on a vessel
ETSI	European Telecommunications Standards Institute
E-UTRA	Evolved Universal Terrestrial Radio Access
FAT	frequency allocation table
FCC	Federal Communications Commission
FCFS	first-come, first-served
FDMA	frequency division multiple access
FDP	fractional degradation in performance
FDR	frequency-dependent rejection

FEC	forward error correction
FFT	fast Fourier transform
FRA	Federal Radio Agency
FM	frequency modulation
FS	fixed service
FSK	frequency shift keying
FSS	fixed-satellite service
FWA	fixed wireless access
GIMS	Graphical Interference Management System
GIS	geographic information system
GPS	Global Positioning System
GRMT	Generic Radio Modelling Tool
GSM	Global System for Mobile Communications
GSO	geostationary satellite orbit
HAAT	height above average terrain
HCM	Harmonised Common Methodology
HEO	highly elliptical orbit
HF	high frequency
ICAO	International Civil Aviation Organisation
IDWM	ITU Digitized World Map
IEEE	Institute of Electrical and Electronics Engineers
IF	intermediate frequency
IFIC	International Frequency Information Circular
IMO	International Maritime Organisation
IMT	International Mobile Telecommunications
IS	intersatellite
ITU	International Telecommunication Union
ITU-R	International Telecommunication Union–Radiocommunication Sector
JEG	Joint Expert Group
JPL	Jet Propulsion Laboratory
JRG	Joint Rapporteur Group
JTG	Joint Task Group
LEO	low Earth orbit
LF	low frequency
LH	linear horizontal
LHC	left-hand circular
LM	land mobile
LMS	land mobile service
LNM	log-normal methodology
LSA	licensed shared access
LT	long-term
LTE	Long Term Evolution
LV	linear vertical
M2M	machine to machine
MASTS	Mobile Assignment Technical System
Mbps	megabit per second
MCL	minimum coupling loss
MF	medium frequency
MIFR	Master International Frequency Register
MIMO	multiple-input/multiple output

MMS	maritime mobile service
MO	microwave oven
MOLA	Mars Orbiter Laser Altimeter
MoU	memorandum of understanding
MPtMP	multipoint-to-multipoint
MS	mobile service/mobile station
MSS	mobile-satellite service
MVPD	multichannel video programming distributor
NASA	National Aeronautics and Space Administration
NFD	net filter discrimination
NM	nautical mile
NTIA	National Telecommunications and Information Administration
OBW	occupied bandwidth
Ofcom	Office of Communications
OFDM	orthogonal frequency division multiplexing
OFDMA	orthogonal frequency division multiple access
OFR	off-frequency rejection
OR	off-route
OS	Ordnance Survey
OSGB	Ordnance Survey Great Britain
OTR	on-tune rejection
PC	personal computer
PCA	predicted coverage area
PDF	probability density function or Portable Document Format
PDNR	preliminary draft new recommendation
PER	packet error rate
PFD	power flux density
PMR	private mobile radio
PMSE	programme making and special events
PSA	protected service area
PSK	phase shift keying
PtMP	point-to-multipoint
PtP	point-to-point
QAM	quadrature amplitude modulation
QoS	quality of service
QPSK	quadrature phase shift keying
R	route
RA	Radio Assembly
RAG	Radiocommunication Advisory Group
RAL	Rutherford Appleton Laboratory
RALI	Radiocommunications Assignment and Licensing Instruction
RAS	radio astronomy service
RCC	Regional Commonwealth In the Field of Communications
Rec.	Recommendation
RED	Radio Equipment Directive
RF	radio frequency
RHC	right-hand circular
RLAN	radio local area network
RPE	radiation pattern envelope
RR	Radio Regulations

RRB	Radio Regulations Board
RRC	Regional Radiocommunication Conference
RSA	requested service area
RSC	Radio Spectrum Committee
RSL	receiver sensitivity level
RSPG	Radio Spectrum Policy Group
RX	receive
SAB	services auxiliary to broadcasting
SAP	services auxiliary to programme making
SC	Special Committee on Regulatory/Procedural Matters
SFN	single-frequency network
SG	study group/subgroup
SHF	super high frequency
SM	spectrum manager
SQB	spectrum quality benchmark
SRR	short-range radar
SRS	Space Radiocommunication Stations
SRTM	Shuttle Radar Topography Mission
SSB	single sideband
ST	short-term
TDMA	time division multiple access
TFAC	Technical Frequency Assignment Criteria
TRP	total radiated power
TS	terminal station
TV	television
TVBD	television band devices
TVRS	TV receive sites
TX	transmit
UHF	ultra high frequency
UKPM	UK Planning Model
UL	uplink
ULF	ultra low frequency
UMTS	Universal Mobile Telecommunications System
UN	United Nations
UWB	ultra wideband
VHF	very high frequency
VLA	very large array
VLBI	very long baseline interferometry
VLF	very low frequency
VSAT	very small aperture terminal
VSB	vestigial sideband modulation
WARC	World Administrative Radio Conference
WCDMA	wideband code division multiple access
WCG	worst-case geometry
WGS	World Geodetic System
WP	working party
WRC	World Radiocommunication Conference
WRPM	wide-range propagation model
WSD	white space device
XPD	cross-polar discrimination

Principal Symbols

a	orbit semimajor axis
a_e	effective Earth radius
A_{BW}	bandwidth adjustment factor
A_c	amplitude of signal or intermodulation coupling loss
A_e	effective area of antenna
A_h	clutter loss
A_I	intermodulation conversion loss
A_{MI}	mask integration adjustment factor
A_t	average traffic in Erlangs
A_{tc}	average traffic per channel
A_U	traffic per user in Erlangs
Az	azimuth
B	bandwidth
B_{3dB}	bandwidth within 3 dB of peak power density
B_I	interferer bandwidth
B_M	measurement bandwidth
B_V	victim bandwidth
B_W	wanted bandwidth
c	wanted signal in absolute or speed of light
C	wanted signal in dB
C_0	wanted signal in dB in 1 Hz
C_a	channel capacity
c/i	ratio of wanted to interfering signals in absolute
C/I	ratio of wanted to interfering signals in dB
C_n	number of channels
c/n	ratio of wanted signal to noise in absolute
C/N	ratio of wanted signal to noise in dB
C_0/N_0	ratio of wanted signal to noise in dB in 1 Hz
$c/(n+i)$	ratio of the wanted signal to noise plus interference in absolute
$C/(N+I)$	ratio of the wanted signal to noise plus interference in dB
$C_0/(N_0+I_0)$	ratio of the wanted signal to noise plus interference in dB in 1 Hz
D	dish size in metres
DT	change in temperature (K)
DT/T	ratio of change in temperature to temperature as a percentage
e	orbit eccentricity
E	orbit eccentric anomaly
e_b	energy per bit in absolute
E_b	energy per bit in dB
e_c	efficiency of carrier
El	elevation
e_v	field strength in volts/m
$E_{\mu v}$	field strength in dBµV/m
f_c	central frequency of carrier
F_c	coverage ratio
f_{MHz}	frequency in MHz
f_n	noise factor in absolute
F_N	noise figure in dB
F_P	pollution ratio

G_p	processing gain in dB
g_{rx}	receive gain in absolute
G_{rx}	receive gain in dB
g_{tx}	transmit gain in absolute
G_{tx}	transmit gain in dB
H	average call duration
h	height
h_{rx}	receive antenna height
h_{tx}	transmit antenna height
i	interfering signal in absolute or index or orbit inclination
I	interfering signal in dB
I_0	interfering signal in dB in 1 Hz
i/n	ratio of interfering signal to noise in absolute
I/N	ratio of interfering signal to noise in dB
k	Boltzmann's constant
l_f	feed loss in absolute
L_f	feed loss in dB
L_{fs}	free space path loss in dB
L_p	propagation loss in dB
L_{pol}	polarization loss in dB
L_s	spreading loss in dB
M	orbit mean anomaly
M_{fade}	fade margin in dB
M_i	interference margin in dB
m_{rx}	receive spectrum mask in absolute
M_{rx}	receive spectrum mask in dB
m_{tx}	transmit spectrum mask in absolute
M_s	system margin in dB
M_{tx}	transmit spectrum mask in dB
$M(X)$	margin for metric X in dB
n	noise in absolute or orbit mean motion
N	noise in dB or number of cases
n_0	noise in 1 Hz in absolute
N_0	noise in 1 Hz in dB
N_c	number of channels
N_{hit}	required number of samples within receiver main beam
N_{PCA}	number of pixels in the predicted coverage area
N_{PSA}	number of pixels in the protected service area
N_r	noise rise in dB
N_{RSA}	number of pixels in the requested service area
N_{sys}	number of systems
p	percentage of time, for example, threshold or propagation model
P	orbit period
PR	protection ratio in dB
p_{rx}	receive power in absolute
P_{rx}	receive power in dB
$p_t(X)$	threshold percentage of time for metric X
p_{tx}	transmit power in absolute
P_{tx}	transmit power in dB
q	percentage of locations, for example, propagation model

r_a	apogee radius of orbit
R_c	chip rate
R_{cs}	channel sharing ratio
R_d	data rate
R_e	radius of the Earth
r_p	perigee radius of orbit
s	signal strength in absolute
S	signal strength in dB
t	time
T	temperature in Kelvin
T_a	antenna temperature in Kelvin
T_b	bit duration
T_e	effective temperature in Kelvin
T_f	feed temperature in Kelvin
T_o	reference temperature = 290°K
T_r	receiver temperature in Kelvin
$T(X)$	threshold for metric X
U	unwanted signal in dB
U_n	number of users
v	orbit velocity
W	wanted signal in dB
X	metric, typically one of $\{C, I, C/I, C/N, C/(N+I), I/N, DT/T, PFD, EPFD\}$
α	minimum angle between line and GSO arc
Δf	difference in frequency
ε	elevation angle
γ	amplification gain
η	antenna efficiency
φ	off-axis angle
λ	wavelength, typically in metres, or average call requests
μ	mean
ν	orbit true anomaly
θ_{3dB}	half power beamwidth of antenna
$\theta_{3dB,a}$	half power beamwidth of elliptical beam along semimajor axis
$\theta_{3dB,b}$	half power beamwidth of elliptical beam along semiminor axis
σ	standard deviation
ω	orbit argument of perigee
Ω	longitude of ascending node

The table below shows the letters in the Greek alphabet:

A	α	alpha	H	η	eta	N	ν	nu	T	τ	tau
B	β	beta	Θ	θ	theta	Ξ	ξ	xi	Υ	υ	upsilon
Γ	γ	gamma	I	ι	iota	O	o	omicron	Φ	ϕ	phi
Δ	δ	delta	K	κ	kappa	Π	π	pi	X	χ	chi
E	ε	epsilon	Λ	λ	lambda	P	ρ	rho	Ψ	ψ	psi
Z	ζ	zeta	M	μ	mu	Σ	σ	sigma	Ω	ω	omega

Index

3GPP, 36, 87, 99, 229, 244

ACMA *see* Australian Communications and Media Authority
adaptive systems, 258–263
   adaptive coding and modulation, 63, 262
   antennas, 95–96
   automatic power control, 259–262
   dynamic frequency selection, 258–259, 449
adjacent band *see* non-co-frequency
adjacent channel *see* non-co-frequency
adjustments
   bandwidth, 221–223, 227, 239, 290, 354, 448, 458
   mask integration, 232–239, 244, 353, 498, 509
aeronautical, 35, 108, 307, 353, 363, 379, 483
   propagation models, 205
aggregate interference, 64, 137, 207, 210, 281, 284, 300, 308, 339, 437, 440–441, 464, 487
   apportionment, 283–292, 300, 412, 470, 497
   modelling techniques, 103, 105, 266–269, 275–276, 399–401, 489
allocation, 10, 12, 16, 19–21, 23, 24, 34, 246, 283, 331
   primary, 19–20, 283
   secondary, 19–20, 283
   super-primary, 19

allotment, 12, 16
ALOHA, 68
amateur–satellite service, 17
amateur service, 17
ANFR, 41
α angle, 471–476, 480–482
antennas, 44–46, 82–101
   adaptive, 95–96
   aperture efficiency, 91
   azimuth and elevation slices, 99–100
   azimuth patterns, 96–98
   beam efficiency, 91
   beamwidth, 85–86, 91
   dipole, 48, 49, 52, 96–97, 432
   directivity, 84
   dish diameter, 88–89, 139
   downtilt, 94–95
   efficiency, 84, 88–89, 91
   elevation patterns, 98–99
   elliptical patterns, 92–95
   far field, 83
   front to back ratio, 98
   gain definition, 51, 84
   gain pattern, 83–100, 256, 316–317, 414–415, 435, 465–467
   half power beamwidth, 85–86, 91
   isotropic, 48–52, 88

Interference Analysis: Modelling Radio Systems for Spectrum Management, First Edition. John Pahl.
© 2016 John Wiley & Sons, Ltd. Published 2016 by John Wiley & Sons, Ltd.
Companion website: www.wiley.com/go/pahl1015

Printed and bound by CPI Group (UK) Ltd, Croydon, CR0 4YY

16/04/2025

14658385-0001